Stable Design Patterns for Software and Systems

Stable Design Patterns for Software and Systems

Mohamed E. Fayad

CRC Press
Taylor & Francis Group
Boca Raton London New York

CRC Press is an imprint of the
Taylor & Francis Group, an **informa** business

AN AUERBACH BOOK

CRC Press
Taylor & Francis Group
6000 Broken Sound Parkway NW, Suite 300
Boca Raton, FL 33487-2742

© 2018 by Taylor & Francis Group, LLC
CRC Press is an imprint of Taylor & Francis Group, an Informa business

No claim to original U.S. Government works

Printed on acid-free paper

International Standard Book Number-13: 978-1-4987-0330-7 (Hardback)

Visit the Taylor & Francis Web site at
http://www.taylorandfrancis.com

and the CRC Press Web site at
http://www.crcpress.com

To all the Software Engineering and Software Design Communities, especially the ones who apply my work.

Contents

SECTION I Introduction

SECTION II   SDPs' Detailed Documentation Template

SECTION III   SDPs' Mid-Size Documentation Template

SECTION IV   SDPs' Short-Size Documentation Template

Preface

The software pattern that administers design patterns is unquestionably one of the fastest growing communities in the field of software engineering. Possibly, one of the main reasons for such huge and enormous growth is a strong faith that using made to measure solutions for solving frequently occurring problems encountered throughout the design phase will greatly reduce the total cost and the time of developing software products. Nevertheless, existing design patterns are yet to fulfill the confidence that developers always wanted to possess. The perceived instability is considered one of the major problems in today's design patterns. Therefore, achieving limitless and enduring stability, improving the factors of reusability, and increasing the wide applicability of design patterns are extremely important and they are major challenges that need immediate solutions in the near future. This book presents the novel concept of *Stable Design Patterns for Software and Systems* as a new and fresh approach for creating stable, reusable, and widely applicable design patterns. This book also deals with the concept of *Stable Design Patterns based on software stability* as a contemporary approach for building stable and highly reusable and widely applicable design patterns. This book also overcomes the factor of immaturity in existing software patterns in general and design patterns in particular, and is presented in many different existing software patterns pitfalls that can be found in Chapters 1 and 2.

This book intends to provide a true understanding of the solution space and a deep focus on ultimate solution. This book also shows that this new formation approach of discovering/creating stable design patterns (SDPs) accords so nicely with Alexander's current understanding of architectural patterns. This agreement is not accidental, but very fundamental. The SDPs are a kind of—knowledge patterns that underline human problem solving methods—and appeal to the pattern community to look for patterns from building a foundation of architectures on demand from a broader system perspective. Each chapter of this book concludes with an open research issue, review questions, exercises, and projects.

This book examines software SDPs with respect to *four central themes*:

1. How do we develop a solution for the problem through software stability concepts?

    This book offers a direct application of using software stability concepts for modeling solutions for the problem or the concept requirements. It shows a software developer how to design and create an appropriate solution for the problem, in terms of its enduring business themes (EBTs) and business objects (BOs). In addition, it also shows how to gel and combine these components together to design a stable model. A few of the stable patterns presented in this book will help readers acquire the necessary knowledge for applying the software stability concepts in actual practice.

    What are the unique roles of SDPs in modeling the accurate solution of the problem at hand and in providing stable and undisputed design for such problems?

    This book enumerates a complete and domain-less list of SDPs that are useful for designing and modeling solutions for frequently recurring problems. Each pattern will illustrate methods to address the issues that accompany these problems. In addition, this book will also illustrate and display different implementation issues that will smoothly transfer the SDP into the detailed solution space (the implementation phase).

2. How do we achieve software stability over time and design SDPs that are effective to use?

    Hitherto, the use of the EBTs and the BOs has assisted software developers to construct models that are both *stable over time*, and *stable across various paradigm shifts within a given domain or application context*. Since the main idea behind the SDPs is to design the problem under consideration in terms of its EBTs and the BOs, the resulting stable patterns are reusable in modeling the same problem under any given context.

3. What is the most efficient way to document the SDPs in order to ensure efficient reusability? This book is an extension to the contemporary templates that are used in documenting design patterns. These templates are provided in the appendices section of this book, which include a comprehensive description for each of the presented design pattern. This description utilizes several modeling artifacts to guarantee the lucidity and easy understandability of the pattern. It also provides us the dynamic and static models and behaviors of the patterns, the use-case diagram, the use-case description, the behavior diagrams, the collaboration and responsibility cards (CRC). This detailed list of descriptions will help increase the reusability of the book patterns.

OUTSTANDING FEATURES

This book gives a pragmatic and a novel approach toward understanding the problem domain and in proposing stable solutions called SDPs for engineering the right and stable software systems, components, and frameworks. Besides the value of reusing the presented patterns, this book also assists developers to acquire the core knowledge needed to analyze and extract design patterns for their domain of interest.

Moreover, readers will learn how to:

1. Ensure software stability over a period and to build SDPs that can be effectively reused.
2. Use SDPs to seek an accurate or precise solution to the underlying problem.
3. Achieve the necessary level of abstraction that makes the resultant design patterns effectively reusable, yet easy to comprehend and understand.
4. Use the necessary information that design patterns provide in order to ensure and guarantee a smooth transition from the design phase to the implementation phase.
5. Transform stable pattern model to any other models, such as formal model or implementation model.

In the future, this book is expected to provide a very significant contribution to the computing field for several important reasons. It will be the first and the only *complete reference manual* on the topic of stable design patterns. It will also be the first book on handling the true understanding of the ultimate solution space, and it will teach the reader how to create an ultimate solution to any recurring problem, and methods to build a myriad of cost-effective and highly maintainable systems using SDPs.

1. It will be the first and the only *complete reference manual* written on the important topic of SDPs.
2. It will document and highlight the *major aspects, techniques, and processes* related to providing an ultimate solution to any recurring problem or set of requirements.
3. It will also illustrate many delicate problems with existing design patterns.
4. It will also highlight innumerable *workable solutions to the most controversial and debatable questions* that the software design patterns face today.
5. It will enumerate a diversity of domain-less SDPs that can be easily comprehended and reused to model similar problems and solutions in any given context.
6. It will demonstrate different methods to link the design pattern to the unlimited applications phase. It will also emphasize the main design issues that are necessary for a smooth transition between analysis, design, and implementation phases.
7. It will supply a new template for improving the communication of design patterns among all developers. This template aims to capture the static and dynamic behavior of the pattern, while maintaining the simplicity of reading and understanding the pattern. Subsequently, the reusability of the patterns will be enhanced further.

8. It will also help readers attain the required knowledge to generate and document their own SDPs.

9. It will also be *the most important book for many undergraduate and graduate* courses on topics of object-oriented analysis and design, requirements engineering, software design techniques, software reuse, software patterns and architectures, and software modeling approaches.

10. It will also discuss and evaluate *lessons learned, experiences gained, and future trends* in SDPs.

TARGET READERSHIP

This book intends to cater to the needs of a big community of computer and software professionals who are involved in the management, research, and development of software systems, components, and enterprise and application frameworks. Software researchers, system architects, software designers, and software engineers (both systems level and application level) will also greatly benefit from this book. We also anticipate that the new concepts presented in this book will definitely influence the development of old and new software systems, components, enterprise frameworks, and application frameworks for the next one or two decades.

Potential buyers of this book will be found among established software firms and from those people who are currently analyzing, designing, and developing different types of software for different domain. In addition, the UML and communities, who are active in designing software architecture, components, software engineering, enterprise frameworks, application frameworks, and the analysis pattern and design pattern, may also evince a keen interest in this book.

In a bookshop and library, the librarian would index and store this book in the Computer Science/Software Engineering/Object-Orientation/Component Software/Enterprise Frameworks/Application Frameworks category. This book is very suitable for anyone dealing with knowledge in their domain, software engineering academic and software development, sciences, engineering, business, and other communities.

INDUSTRY AND TECHNOLOGY TRENDS

1. Software design patterns play a vital role to create an enduring solution to any recurring problem. It will also help in reducing the overall cost and in abridging and compressing the time of software project life cycles. However, building unlimitedly reusable and SDPs is still a major and sensitive challenge. This book proposes the novel concept of Stable Design patterns based on software stability [1, 2, 3, 4, 5], as a modern approach for building stable and highly reusable and widely applicable design patterns. This book is expected to overcome the factor of immaturity that exists among software patterns in general and design patterns in particular.

2. a. Stabilizes the process lines of design patterns' development and creation activities.

   b. Introduces a direct application of using software stability concepts to model core knowledge of the ultimate solution space, which fits all existing scenarios of any recurring problem.

   c. Presents a comprehensive, complete, and domain-less lists of SDPs that are useful to understand the ultimate solution for any recurring problems and their modeling. Each of these patterns will illustrate avenues and modes to capture the core knowledge and to address the issues that are related to the ultimate solution space.

   d. In addition, this book will also illustrate different design issues that will smoothly transfer the SDP into the detailed solution space (the unlimited applications phase).

   e. We always write for the benefit of masses. This book will be of great help (frankly speaking is a must) to a large community of computing and modeling academics, and students, software technologists and methodologists, software patterns' communities, component developers, and software reuse communities and software professionals (analysts, designers, architects, programmers, testers, maintainers, developer), who are involved in the management, research, and development of methodologies and software patterns. Industry agents, who work on any technology project and who want to improve the project's reliability and cost-effectiveness, will also benefit hugely by reading this book.

BOOK SUPPLEMENT

The author provides a book supplement of 250+ pages containing solutions to the exercises and projects in each chapter. In addition to problem statements for team projects, exams, quizzes, and modeling tips, heuristics are also given. The supplement also provides over 20–30 numbers of special design patterns. Static, dynamic, and behavior models for each of the patterns are also given. Similarly, the author will also provide a number of scenarios to illustrate how to do them in the supplement section. This book supplement also has several private links for PowerPoint presentation of all the sections of this book.

Acknowledgments

This book would not have been completed without the help of many great people; I *thank them all*. Special thanks to my friend and one of my best student Dr. Haitham Hamza, Cairo University, Egypt for his excellent thesis work on Stable Design Patterns. Special thanks to my friend Srikanth G. K. Hegde. I would also like to thank all of my student assistants, Vishnu Sai Reddy Gangireddy, Mansi Joshi, Siddharth Jindal, Polar Halim, Charles Flood III for their work on the figures and diagrams and long discussions on some of the topics of this book.

This was a great and fun project because of your tremendous help and extensive patience. Special thanks to my San Jose State University students for forming teams and work on some of the exercises and projects in this book. Special thanks to my wife Raefa, my lovely daughters Rodina and Rawan, and my son Ahmad for their great patience and understanding. Special thanks to Srikanth's wife, Kumuda Srikanth, for help with reviewing some of the chapters. Special thanks to all my friends all over the world for their encouragement and long discussions about the topics and the issues in this book. Thanks to all my students and coauthors of many articles related to this topic with me, in particular Ahmed Mahdy (Texas A&M University at Corpus Christi, Corpus Christi, TX), Shasha Wu (Spring Arbor University, Michigan), and Shivanshu Singh, to my friends, Davide Brugali and Ahmed Yousif for their encouragement during this project, to the *Communications of the ACM* staff—my friends Diana Crawford, the executive editor, Thomas E. Lambert, the managing editor, and Andrew Rosenbloom, the senior editor.

I would like to acknowledge and thank all of those who have had a part in the production of this book. First, and foremost, we owe our families a huge debt of gratitude for being so patient while we put their world in a whirl by injecting this writing activity into their already busy lives. We also thank the various reviewers and editors who have helped in so many ways to get this book together. We thank our associates who offered their advice and wisdom in defining the content of this book. We also owe special thanks to those who have worked on the various projects covered in the case studies and examples.

Finally, we would like to acknowledge and thank the work of some of the people who helped us in this effort: John Wyzalek, acquisition editor, Stephanie Place-Retzlaff, editorial assistant, and Robert Sims, project editor, for their excellent and quality support and work done to produce this book and a special acknowledgment and thanks to Ragesh K, project manager at Nova Techset who did a tremendous job for proofreading and copyediting of all the chapters in detail and the elegant and focused ways of taking care of day-to-day handling of this book, and special thanks to all the people in marketing, design, and support staff at Taylor & Francis.

Author

Mohamed E. Fayad is a professor of computer engineering at San Jose State University since 2002. From 1999 to 2002, Dr. Fayad was J. D. Edwards professor of software engineering in the Department of Computer Science and Engineering at the University of Nebraska-Lincoln. Between 1995 and 1999, he was an associate professor of computer science and a faculty of computer engineering at the University of Nevada. He has more than 15 years of industrial experience that include 10 years as a software architect in companies, such as McDonnell Douglas and Philips Research Laboratory. His reputation has grown by his achievements in the industry—he has been an IEEE distinguished speaker, an associate editor, editorial advisor, a columnist for *The Communications of the ACM* (his column is Thinking Objectively), a columnist for *Al-Ahram Egyptians Newspaper* (2 million subscribers), an editor-in-chief for IEEE Computer Society Press—Computer Science and Engineering Practice Press (1995–1997), a general chair of IEEE/Arab Computer Society (ACS) International Conference on Computer Systems and Applications (AICCSA 2001), Beirut, Lebanon, June 26–29, 2001, and the founder and president of ACS from April 2004 to April 2007.

Dr. Fayad is a well-known and recognized authority in the domain of theory and the applications of software engineering. Fayad's publications are in the very core, archival journals and conferences in the field of software engineering. Dr. Fayad was a guest editor on 11 theme issues: CACM's OO Experiences, October 1995, IEEE Computer's Managing OO Software Development Projects, September 1996, CACM's Software Patterns, October 1996, CACM's OO Application Frameworks, October 1997, ACM Computing Surveys—OO Application Frameworks, March 2000, IEEE Software—Software Engineering in-the-small, September/October 2000, and *International Journal on Software Practice and Experiences*, July 2001, *IEEE Transaction on Robotics and Automation*—Object-Oriented Methods for Distributed Control Architecture, October 2002, *Annals of Software Engineering Journal*—OO Web-Based Software Engineering, October 2002, *Journal of Systems and Software*, Elsevier, *Software Architectures and Mobility*, July 2010, and *Pattern Languages: Addressing the Challenges*, the *Journal of Software, Practice and Experience*, March–April 2012.

Dr. Fayad has published more than 300 high-quality articles, which include profound and well-cited reports (more than 50 in number) in reputed journals, and over 100 advanced articles in refereed conferences, more than 25 well-received and cited journal columns, 16 blogged columns, 11 well-cited theme issues in prestigious journals and flagship magazines, 24 different workshops in very respected conferences, over 125 tutorials, seminars, and short presentations in more than 25 States in the United States since 1978 and 30 different countries, such as, Hong Kong (April 1996), Canada (12 times), Bahrain (2 times), Saudi Arabia (4 times), Egypt (25 times), Lebanon (2), UAE (2 times), Qatar (2 times), Portugal (October 1996, July 1999), Finland (2 times), UK (3 times), Holland (3 times), Germany (4 times), Mexico (October 1998), Argentina (3 times), Chile (2000), Peru (2002), Spain (2002), Brazil (2004), China (4), Morocco (March 2017), and Poland (April 2017). He is the founder of seven new online journals, *NASA Red Team Review of QRAS* and NSF-USA Research Delegations' Workshops to Argentina and Chile, Expert witness for major cases between Alcatel and Cisco 2000–2002 and patent infringement matter—*iRobot Corporation v. Hoover, Inc.*, and more than 10 authoritative books, of which 3 of them are translated into different languages such as Chinese. Dr. Fayad is also filling for 8 new, valuable, and innovative patents and has developed over 800 stable software patterns and brought a breakthrough in the software engineering field. Dr. Fayad earned an MS and a PhD in computer science from the University of Minnesota at Minneapolis. His research topic was OO Software Engineering: Problems and Perspectives. He is the lead author of several classic Wiley books: *Transition to OO Software Development* (1998), *Building Application Frameworks* (1999), *Implementing Application Frameworks* (1999), *Domain-Specific Application Frameworks* (1999), and several books by CRC Press, Taylor & Francis Group: *Software Patterns, Knowledge Maps, and Domain Analysis* (2014) and *Stable Analysis Pattern for Software and Systems* (2017), *Stable Design Pattern for Software and Systems* (2017). Several new books in progress for Taylor & Francis—*Unified Business Rules Standard* (UBRS) (expected 2018), *Software Architecture On Demand*, and *Unified Software Engineering Reuse* (USER), *Unified Software Engineering* (USE) (expected to be published during 2018), and several books in production and progress with German Omniscriptum Publishing Group (expected to be published during 2017 and 2018).

Section I

Introduction

This part consists of 4 chapters and 20 sidebars.

Chapter 1 is titled "The Impact of Stability on Design Patterns' Implementation." Stable design patterns (SDPs) are reusable constructs. They are stable and adaptable by definition. Unfortunately, to achieve usability, the elegant characteristics displayed in design patterns, such as stability, self-adaptability, and generality are severely diminished in the implemented models. Thus, a discrepancy is likely to be revealed among design patterns and these models. This chapter suggests the use of Software Stability [1–3] as a solution for resolving the observed inconsistency between the design patterns and their implementation.

Chapter 2 is titled "Pitfalls Categories Overview: Pitfalls in Traditional Software Patterns: The Factor of Immaturity." This chapter focuses on 15 pitfalls of traditional patterns, such as the Gang of Four [4], and Coad [5], and Fowler analysis patterns [6].

Chapter 3 is titled "Engineering Stable Atomic Knowledge (SAK) Patterns."

This chapter introduces the concept of SAK patterns as a solution for enabling the reuse of domain-independent knowledge across domains. An SAK pattern captures the core knowledge of a recurring problem, and hence it can be used to model this problem wherever it appears, regardless of the application it appears. We have also elucidated and described a high-level engineering process for developing SAK pattern, such as stable analysis [7] and design patterns.

Chapter 4 is titled "Stable Analysis and Design Patterns: Unified Software Engine (USE)." This chapter presents a new approach to use analysis and design patterns to develop USE and software systems [8]. The approach uses the approach of designing stable analysis patterns to capture the core concepts of the problem; hence, design patterns are used to transform analysis patterns into design modules. These design modules are integrated to form the final design of the system.

Each of the chapters concluded with a summary and an open research issue. This chapter also provides a number of review questions, exercises, and projects.

Chapter 1 has three sidebars (SB1.1–SB1.3):

Sidebar 1.1 is titled "Software Stability Model" and it explores SSM, which promotes a way of building software, such that the software is not tied to any specific domain or application context [1–3].

Sidebar 1.2 is titled "Knowledge Maps" and it discusses briefly different ways to do a domain analysis related to any problem domain, in a way this is not tied to a specific case or a specific context; rather, the domain analysis is done in a holistic and conceptual sense [9].

Sidebar 1.3 is titled "Stable Analysis Patterns" and it presents a brief overview of SAPs which play a major and decisive role in reducing the overall cost and in condensing the time duration of software project life cycles [7].

Chapter 2 has 15 sidebars (SB2.1–SB2.15):

Sidebar 2.1 is titled "Skill and Experience: The Magical Wands!"

Sidebar 2.2 is titled "Same Problem, but Multiple Patterns!: The Common Problem of Duplication."

Sidebar 2.3 is titled "Choosing the Right Pattern-Real Challenges."

Sidebar 2.4 is titled "Drawing a Fine Line between an Analysis Pattern and Design Pattern."

Sidebar 2.5 is titled "Keeping it Very Simple!"

Sidebar 2.6 is titled "Focus! Focus! Focus! Golden Words for Developing Meaningful Patterns."

Sidebar 2.7 is titled "A Brief Summary of Pattern Compositions: Understanding the Insider's Secret."

Sidebar 2.8 is titled "Lack of Patterns Connectivity across Development Phases."

Sidebar 2.9 is titled "Untraceable Patterns—The Pitfall of the Pattern Systems."

Sidebar 2.10 is titled "No Guidelines for Extracting Patterns."

Sidebar 2.11 is titled "Chasing an Elusive Vocabulary: Importance of Vocabulary in Pattern Development."

Sidebar 2.12 is titled "Outdated and Jaded Patterns or Fresh and Everlasting Patterns: Choose your Pick!"

Sidebar 2.13 is titled "Modeling Problems are Nagging Obstacles to Creating Meaningful Patterns."

Sidebar 2.14 is titled "Reinventing the Wheel: An Undesirable and Dangerous Development."

Sidebar 2.15 is titled "Analyzing Pitfalls to Seek a Permanent Solution."

Chapter 3 will have one sidebar (SB3.1):

Sidebar 3.1 is titled "Extracting Domain-Specific and Domain-Independent Patterns."

Chapter 4 has one sidebar (SB4.1):

Sidebar 4.1 is titled "Unified Software Engines."

REFERENCES

1. M.E. Fayad and A. Altman. Introduction to software stability. *Communications of the ACM*, 44(9), 2001, 95–98.
2. M.E. Fayad. Accomplishing software stability. *Communications of the ACM*, 45(1), 2002, 111–115.
3. M.E. Fayad. How to deal with software stability. *Communications of the ACM*, 45(4), 2002, 109–112.
4. E. Gamma, R. Helm, R. Johnson, and J. Vlissides. *Design Patterns: Elements of Reusable Object-Oriented Software*, Addison-Wesley, Boston, MA, 1994.
5. P. Coad, D. North, and M. Mayfield. *Object Models—Strategies, Patterns, & Applications*, Yourdon Press, Prentice-Hall, Inc., NJ, 1995.
6. M. Fowler. *Analysis Patterns—Reusable Object Models*, Addison-Wesley Professional, Boston, MA, 1997.
7. M.E. Fayad. *Stable Analysis Patterns for Software and Systems*, Auerbach Publications, Boca Raton, FL, 2017.
8. M.E. Fayad and S. Singh. Unified software engine (USE), Master's thesis, San Jose State University (SJSU), 2013.
9. M.E. Fayad, H.A. Sanchez, S.G.K. Hegde, A. Basia, and A. Vakil. *Software Patterns, Knowledge Maps, and Domain Analysis*, Auerbach Publications, Boca Raton, FL, 2014.

1 Impact of Stability on Design Patterns' Implementation

I find that stability is good for my creativity.

Ellen Forney [1]

1.1 OVERVIEW OF STABLE DESIGN PATTERNS

The community that manages design patterns is unquestionably one of the quickest growing communities in the domain of software engineering. Possibly, one of the main reasons for such huge and enormous growth is a strong hope that using made to order solutions for solving regularly repeating problems encountered throughout the design phase will to a great extent reduce the total cost and the time of developing software products. However, existing design patterns are yet to fulfill the cheerfulness that developers always dreamt of. The professed lack of stability is considered to be one of the major problems in today's design patterns. Therefore, achieving limitless and enduring stability, improving the factors of reusability, and increasing the wide applicability of design patterns are extremely important and they are major challenges that need to be solved immediately in the near future too. *Stable design patterns* (*SDPs*) is an original concept that comes with a new and fresh approach for creating stable, reusable, and widely applicable design patterns. *SDPs based on software stability* [2–6] is a modern approach for building stable and highly reusable and widely applicable design patterns.

The rapid growth of modern technology, coupled with the constricted software development time and production cost limitations, has forced tremendous pressure and an intense desire for software enterprises to create new and innovative designs, which respond to speedily altering business and operating environments. Enterprises must invest in building stable architectures that are flexible and that can be easily adapted. We refer to these emerging trends of architectures as *Architectures on Demand* as they can be *adaptable, customizable, extensible, personalizable, self-configurable*, and *self-manageable* accordingly to meet the future requirements and changes in the operating environments. In nutshell, adaptability refers to the level software system architecture can accommodate changes in its environment, constrained by the hardware and software. Customizability refers to the ability of the architecture to be managed and customized by an agent, its users, and benefiting applications. Extensibility means that the architecture is designed to include mechanisms for expanding/enhancing the system with new capabilities without having to assemble major changes to the architecture and the underlying infrastructure. A good architecture provides the design principles to ensure this—a roadmap for that portion of the road yet to be built. Self-configurable and self-manageable architectures refer to architectures that can manage and "self-heal" their properties dynamically at runtime at the components, connectors, and the underlying infrastructure.

1.2 INTRODUCTION

SDPs are reusable constructs. They are stable and adaptable by definition. Unfortunately, in order to achieve usability, the elegant characteristics displayed in design patterns, such as stability, adaptability, and generality are diminished in the implemented models. Thus, a discrepancy is revealed between the design patterns and these models. This chapter suggests the use of software stability [2–6] as a solution for resolving the inconsistency between the design patterns and their implementation.

1.3   CONTEXT

The basic message of this chapter is that the implementation of design patterns leaves the programmer with no trace of kind of patterns that were applied and the type of design decisions that were made. Thus, when changes have to be made, the entire design has to be almost entirely reconstructed.

We will explain this with an analogy to driving rules. The pattern instances correspond to actual driving and the problem context can be thought of as different driving situations. The rules are always constant, although the actual driving may change under different circumstances. To drive safely and efficiently, people must learn the rules first and then apply them according to the different circumstances. Every time they drive, they must keep the rules in mind in order to adapt to different situations. That is the key to stable and safe driving. Expecting these models to be stable under changes is similar to applying past driving sequence actions to different times and locations.

1.4   PROBLEM

"A pattern is a plan, rather than a specific implementation" [7]. They are "descriptions of communicating objects and classes that are customized to solve a general design problem in a particular context" [7]. Consequently, design patterns must be stable, abstract, and common. This makes them unsuitable for detailed implementation on specific problems. So, they are not directly used to guide programming. They guide the modeling process and detailed design process. By instantiation, the original abstract objects of design patterns are replaced with concrete objects representing instances of a specific domain; plus, other objects. Those instances are the actual guide for programming. Figure 1.1 shows the mechanism of the adaptation process of a design pattern.

In current approaches of implementing design patterns, design patterns and their instances are separate models; abstract and concrete domains. After instantiation, the instances are less abstract, less stable, and difficult to extend, compared to their design patterns. This is because a pattern gains its quality factors from being a recurrent design issue and solution in multiple contexts/domains. Design patterns are not traceable from the resulting implementation model. Thus, the solution loses an important quality of patterns: *stability in the face of changes*.

The instantiation process can be used to solve this discrepancy. But, the description of this process is not traceable from the resulting implementation model too. This means a specific piece of code (implementation model) implementing a given solution, lacks the ability to trace back to its blueprint design pattern. This disallows the extension of the model for future changes. Hence, the theme of this chapter is that the pattern itself is lost in the implementation. Is there a way to convey the pattern characteristics to its instances to achieve both usability and stability in the resulting models?

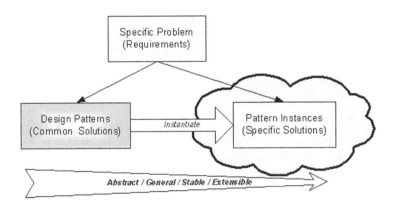

FIGURE 1.1   Current approach of implementing design patterns.

1.5 SOLUTION

This chapter suggests the use of stable models and EBTs [2–6] to help restore what would otherwise be lost in the implementation. The model loses its generality and abstraction after instantiation, causing it to be weak in adaptability and extensibility. When developers want to extend the software at a later stage, they cannot directly reuse this model, because the implementation of pattern does not allow tracing back to the abstract design pattern. They have to reconstruct the whole model from the original design patterns.

The software stability approach [2–6] has the potential to build such models. "A Software Stability Model (SSM) can be triply partitioned into levels: Enduring Business Themes (EBTs), Business Objects (BOs), and Industrial Objects (IOs). EBTs represent intangible objects that remain stable internally and externally. BOs are objects that are internally adaptable, but externally stable, and IOs are the external interface of the system. In addition to the conceptual differences between EBTs and BOs, a BO can be distinguished from an EBT by tangibility. While EBTs are completely intangible concepts, BOs are partially tangible. These artifacts develop a hierarchal order for the system objects, from totally stable at the EBTs level to unstable at the IOs level, through adaptable though stable at the BOs level. The stable objects of the system are those do not change with time" [8]. From an abstractness aspect, the EBTs are completely abstract, the BOs are mostly abstract, and the IOs are not abstract. Hence, the EBTs and BOs are common among applications with similar core, while the IOs are those object differentiate an application from another. Figure 1.2 shows the SSM structure.

The EBTs and BOs are abstract-like design patterns, but they do not disappear during implementation. They describe a common solution to the problem. The IOs are the same as the pattern instance. They are concrete, problem specific, and unstable. When we include the original design patterns in the final instance models and provide their collaborations, the model bears a striking resemblance to the SSM. By associating these two parts in the SSM instead of separating them as in the current design patterns approach, we add a stable core to the resulting model. This keeps the model stable over changes.

Future developers can trace the ideas of the original model designers and extend the model safely, as the core remains stable. Combining the abstraction and generalization from the original design patterns (i.e., EBTs and BOs) and the specification of the instances (i.e., IOs), the SSM achieves stability and usability concurrently. It shoots two birds with one bullet.

1.6 EXAMPLE

In Reference 7, Figures 6.23 and 6.33, Design pattern of transaction and Design pattern instances on order present a form of analogy that indicates both figures are alike—the original pattern and

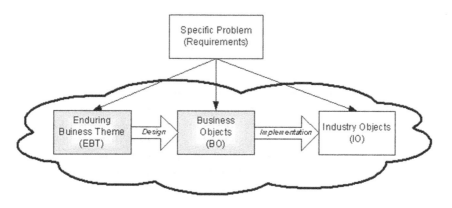

FIGURE 1.2 Stability model structure.

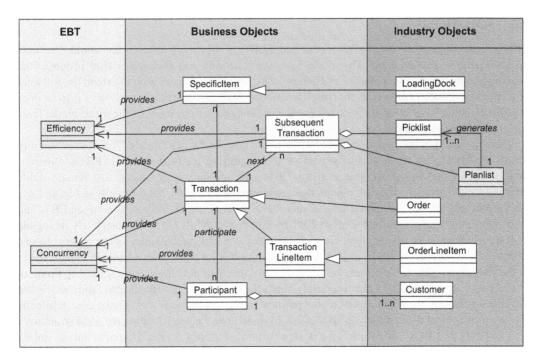

FIGURE 1.3 Stability model for transaction design patterns on order instance.

its instances. Both are two different things: the analogy between the heart and a pump and between the human and a car.

Figure 1.3 describes the same problem in Reference 7 using the software stability approach. Using this new model, we can easily change the IOs without worrying about destroying the whole structure of the model. Suppose, the circumstances change and we need to introduce "Planlist" to satisfy a new requirement. By using SSM, we can easily instantiate "Planlist" from "SubsequentTransaction" as an IO to extend the model without modifying the whole structure. The EBTs and BOs remain constant during this process. This efficiently keeps the core structure and design ideas of this model unchanged over time.

1.7 SUMMARY

"Patterns explicitly capture expert knowledge and design tradeoffs, and make this expertise more widely available" [12–14]. However, the implementation of design patterns has difficulty in constructing stable software products because much of the design abstractions are lost in the implementation, with no traceability back to the design patterns to accommodate new variations. An SSM has the ability to extend its usage of pattern implementation without modifying its whole structure.

The software stability approach provides a practical method of explicitly describing the two-way mapping relationship between design patterns and their implementations/instances. The EBTs and BOs represent the core of the model and are constantly under change. This allows the model to remain stable. The IOs can be modified easily and safely according to the specific problem and their original design patterns (i.e., EBTs and BOs). The abstract parts (EBTs and BOs) and concrete parts (IOs) are separated clearly, but connected closely in the SSM. Therefore, designers have a means of tracing back and re-executing the instantiation process to extend the model. Accordingly, the stability of the model becomes feasible.

1.8 REVIEW QUESTIONS

1. What are some of the positive aspects of Architectures on Demand?
2. What are SDPs?
3. Define some of the contexts where they can be used.
4. Name some of the problems that surround creation of design patterns.
5. What is software stability approach?
6. What are EBTs?
7. What are BOs?
8. What are IOs?
9. Define their characteristics.
10. Demonstrate some basic examples of design patterns.
11. Highlight some of the problems that are associated with these design patterns.
12. What are some of the significant benefits of using SDPs?
13. Are there any disadvantages of using SDPs?

1.9 EXERCISES

1. Separate out EBT, BOs, and IOs for a direct sales business model with a sequence diagram.
2. Provide a documentation development for design pattern for a freight carrier model.

1.10 PROJECTS

1. Create an SDP scenario with a normal banking scenario.
2. Design an SDP with a subscription business model.
3. Formulate an SDP for auction business model.

REFERENCES

1. E. Forney. *Marbles: Mania, Depression, Michelangelo, & Me: A Graphic Memoir*, City of Westminster, London, UK, Gotham/Penguin Books, 2012.
2. M.E. Fayad and A. Altman. Introduction to software stability, *Communications of the ACM*, 44(9), 2001, 95–98.
3. M.E. Fayad. Accomplishing software stability, *Communications of the ACM*, 45(1), 2002, 111–115.
4. M.E. Fayad. How to deal with software stability, *Communications of the ACM*, 45(3), 2002, 102–106.
5. M.E. Fayad and S. Wu. Merging multiple conventional models in one stable model, *Communications of the ACM*, 45(9), 2002, 105–106.
6. A. Mahdy and M.E. Fayad. A software stability model pattern, in *Proceedings of 9th Conference on Pattern Languages of Programs 2002 (PLoP02)*, Monticello, IL, September 2002.
7. P. Coad, D. North, and M. Mayfield, *Object Models—Strategies, Patterns, & Applications*, Yourdon Press, Prentice-Hall, Inc., New Jersey, 1995.
8. E. Gamma, R. Helm, R. Johnson, and J. Vlissides, *Design Patterns: Elements of Reusable Object-Oriented Software*, Addison-Wesley, New York, 1995.
9. H. Hamza and M.E. Fayad. A pattern language for building stable analysis patterns, in *Proceedings of 9th Conference on Pattern Languages of Programs 2002 (PLoP02)*, Monticello, IL, September 2002.
10. M.E. Fayad, H.A. Sanchez, S.G.K. Hegde, A. Basia, and A. Vakil. *Software Patterns, Knowledge Maps, and Domain Analysis*, Auerbach Publications, Boca Raton, FL, Taylor & Francis Catalog #: K16540, December 2014, ISBN-13: 978-1466571433.
11. M.E. Fayad. *Stable Analysis Patterns for Software and Systems*, Auerbach Publications, Boca Raton, FL, Taylor & Francis Catalog #: K24627, May 2017, ISBN-13: 978-1-4987-0274-4.
12. D.L. Levine and D.C. Schmidt. *Introduction to Patterns and Frameworks*, Department of Computer, Science Washington University, St. Louis, available at: http://www.cs.wustl.edu/~schmidt/PDF/patterns-intro4.pdf.

13. A. Mahdy, M.E. Fayad, H. Hamza, and P. Tugnawat. Stable and reusable model-based architectures, in *12th Workshop on Model-Based Software Reuse, 16th ECOOP 2002*, Malaga, Spain, 2002.
14. D.C. Schmidt, M.E. Fayad, and R. Johnson. Software patterns, *The Special Issues in Communications of the ACM*, 39(10), 1996, 37–39.

SIDEBAR 1.1   SOFTWARE STABILITY MODEL

Software stability model (SSM) [1–4] is a radically new and a disruptive innovation in the field of software engineering. It promotes a holistic approach of how any software development effort should be undertaken, different from the approaches that we see today. An SSM can be used to build systems in any domain. The SSM promotes a way of building software such that the software is not tied to any specific domain or application context. The SSM fundamentally targets the problem of low reuse, high costs, phasing out software solutions, components with great impedance mismatch and more. Research and development done for the SSM will enable practitioners and researchers of software engineering to leverage new ways of developing software that solve these problems. The methods employed in designing software using SSM methods aid in advancing our understanding of how to do requirement analysis that is long lived and hits right at what is being aimed at. SSM will not only advance the field of software engineering but also benefit Knowledge Engineering in general. SSM concepts hold true for many engineering disciplines, not just software.

Self-Adaptable, Self-Scalable, Stability, Self-Manageable, Easy Extensible, Self-Configurable with Unlimited Reuse

Software that is architected and designed on the lines of SSM is done in a way that makes it and self-adaptable in nature, that is, separation of concerns between functionality sets is extremely well done. As SSM steers the designer toward a solution-strategy that employs a generic stable core, SSM enables the solution to be well suited and adaptable to a variety of typical changes in the application context by ensuring that adaptation is limited to just the application context-specific outer layer of the software longevity, high returns on investments, self-configurability, self-customizability, self-manageability, easy extensibility with unlimited reusability of the artifacts developed and much more.

Advantages of SSM

1. With SSM there is no need for revenging the wheels comparing to the traditional software methodologies such as spiral model [5] and reusable model [6]; that means there is no maintenance.
2. SSM creates products and components that are machine and platform independent.
3. SSM creates unlimited reuse of unified software architectures on demand (USA on Demand).
4. SSM generates assets that stable over time and have high return on investments.
5. SSM represents the core knowledge [7] of unlimited applicability that is self-adaptable, self-scalable, stability, self-manageable, easy extensible, self-configurable with unlimited reuse.
6. SSM is used as a method of domain analysis, analysis, and understands the software requirements that we called the problem and generates an unlimited solution to the problem.

Product Technologies

SSM is the base of Knowledge Maps [7, refer to SB 1.2], Stable Analysis Patterns (SAPs) [8, refer to SB 1.3], Stable Design Patterns (SDPs) the subject of this book, Unified Software Architectures on Demand (USA on Demand) [9], Unified Software Engines (USEs) [10].

References

1. M.E. Fayad and A. Altman. Introduction to software stability, *Communications of the ACM*, 44(9), 2001, 95–98.
2. M.E. Fayad. Accomplishing software stability, *Communications of the ACM*, 45(1), 2002, 111–115.
3. M.E. Fayad. How to deal with software stability, *Communications of the ACM*, 45(3), 2002, 109–112.
4. M.E. Fayad and S. Wu. Merging multiple conventional models in one stable model, *Communications of the ACM*, 45(9), 2002, 102–106.
5. B. Boehm. A spiral model of software development and enhancement, *ACM SIGSOFT Software Engineering Notes. ACM*, 11(4), 1986, 14–24.
6. K.C. Kang (Pohang University of Science and Technology), S.G. Cohen, R.R. Holibaugh, J.M. Perry, A.S. Peterson. CMU/SEI Report Number: CMU/SEI-92-SR-004. Software Engineering Institute, January 1992.
7. M.E. Fayad, H.A. Sanchez, S.G.K. Hegde, A. Basia, and A. Vakil. *Software Patterns, Knowledge Maps, and Domain Analysis*, Auerbach Publications, Boca Raton, FL, Taylor & Francis Catalog #: K16540, December 2014, ISBN-13: 978-1466571433.
8. M.E. Fayad. *Stable Analysis Patterns for Software and Systems*, Auerbach Publications, Boca Raton, FL, Taylor & Francis Catalog #: K24627, May 2017, ISBN-13: 978-1-4987-0274-4.
9. M.E. Fayad and P. Halim. Unified software architectures on-demand (USA on-Demand). Master thesis, San Jose State University, 2014.
10. M.E. Fayad and P. Halim. Unified software Arch. Master thesis, San Jose State University, 2017.

SIDEBAR 1.2   KNOWLEDGE MAPS

Knowledge Maps (KMs) [1] enable us to do a domain analysis related to any problem domain in a way this is not tied to a specific case or a specific context, rather the domain analysis is done in a holistic and conceptual sense. The core knowledge related to any domain is recorded, thus enabling comprehensive requirement analysis. The benefits do not stop there. KM's properties like intersection of different KMs, using remote KMs in association with the KM of any one domain under consideration lets us analyze the knowledge across various such KMs and thus multiple domains, as a whole and help us in coming up with requirements and design of any system that spans across more than one domain of application.

Any software that is architected and designed from the knowledge, captured by KMs, is done in a way that makes it highly adaptable in nature, that is, separation of concerns between functionality sets is extremely well done. Each part of the functionality can be taken out or some other can be plugged into the architecture, scaling it in whatever way it is required. A KM lets us come up with a stable core software system rather quickly which consists of the core knowledge for one or more applications. By starting with a generic stable core that can be reused in multiple scenarios, the domain analysis is more readily suited and adapted to a change in the application context by changing just the application context-specific parts (which are relatively a small percentage of the total solution). All of this is possible since it is built on the foundation of SSM (refer to SB 1.1). This brings along a number of benefits, such as longevity, high returns on investments, self-configurability, self-customizability, self-scalability, self-manageability with unlimited reuse of the artifacts developed and much more.

Reference

1. M.E. Fayad, H.A. Sanchez, S.G.K. Hegde, A. Basia, and A. Vakil. *Software Patterns, Knowledge Maps, and Domain Analysis*, Auerbach Publications, Boca Raton, FL, Taylor & Francis Catalog #: K16540, December 2014, ISBN-13: 978-1466571433.

SIDEBAR 1.3   STABLE ANALYSIS PATTERNS

Software analysis patterns play a major and decisive role in reducing the overall cost and in condensing the time duration of software project life cycles. However, building reusable and SAPs is still a major challenge. *SAPs* are a new and fresh approach for building stable and reusable analysis patterns based on the concept of software stability [1].

Software stability concepts, introduced and pioneered by me, have demonstrated great promise and immense hope in the area of software reuse and life cycle improvement. In practice, SSMs apply the concepts of "Enduring Business Themes" (EBTs) or "Goals" and "Business Objects" (BOs) or "Capabilities to achieve the Goals" (refer to SB 1.1). These revolutionary concepts have shown to produce and yield models that are both stable over time and stable across various paradigm shifts, within a given domain or application context. By applying the enduring concepts of stability model to the notion of analysis patterns, we are proposing the concept of SAPs. Here, we attempt to analyze the problem under consideration, in terms of its EBTs and the BOs, with the ultimate goal of reaching increased stability and broader reuse. By analyzing the problem in terms of its EBTs and BOs, the resulting pattern models structure the core knowledge of the problem. The ultimate goal, therefore, is *stability*. As a result, these stable patterns could be easily understood and reused to model the same problem in any context.

Major Advantages of SAPs

1. Highlight the *major aspects, techniques, and processes* related to software problem understanding.
2. Illustrate many delicate problems with existing analysis patterns.
3. Provide *workable solutions to the most controversial and debatable questions* facing software analysis patterns today.
4. Provide a diversity of domain-less SAPs that can be easily comprehended and reused to model similar problems in any given context.
5. Show how to link the analysis pattern to the design phase. It will also provide the main design issues necessary for a smooth transition between analysis and design.
6. Provide a new template for improving the communication of analysis patterns among all developers. This template aims to capture the static and dynamic behavior of the pattern, while maintaining the simplicity of reading and understanding the pattern. Subsequently, the reusability of the patterns will be enhanced further.

Reference

1. M.E. Fayad. *Stable Analysis Patterns for Software and Systems*, Auerbach Publications, Boca Raton, FL, Taylor & Francis Catalog #: K24627, May 2017, ISBN-13: 978-1-4987-0274-4.

2 Pitfalls Categories Overview
Pitfalls in Traditional Software Patterns—The Factor of Immaturity

Our element is unending immaturity.

Witold Gombrowicz [1]

One of the major pitfalls in developing meaningful and convenient patterns is the perceived factor of immaturity; thus, most of the patterns developed are yet to fulfill the expectations for their use in facilitating the development of software systems. As a result, it becomes a major concern to investigate why software patterns have not yet developed the level of maturity that is so much needed for establishing the stability of a given software system. Even with best of the software architecture, a piece of software system is bound to face some problems in its lifetime. This column attempts to highlight and deliberate a number of common problems found in today's traditional software patterns like the Gang of Four (GOF), Siemens Group, and the others.

2.1 INTRODUCTION

Of late, software patterns have emerged as a promising technique for facilitating the development of highly sophisticated software systems. Even with the best of the pattern developmental methodologies, patterns developed are yet to fulfill the expectations software developers always envisioned. In spite of some pronounced glitches in the produced patterns, patterns, as concepts, still hold an important key in developing state-of-the-art software systems in the near future.

Developing meaningful patterns is a thing of art and a system of perfect skills; improving the overall quality of patterns is never easy and quick. More often, developers take an inordinately long time to design perfect and meaningful patterns. To develop meaningful and robust patterns, a developer may need to design them in phased manner. The most important and critical of all these phases is the *diagnostic phase*, using which one can understand and comprehend the main problems that come in the way of development of today's patterns. Once a pattern developer identifies and notes all the bottlenecks, it becomes very easy to explore the causes of pattern immaturity and their subsequent usability.

In the forthcoming columns, we have attempted to investigate more than 16 problems and bottlenecks that play a critical role in diminishing the overall effectiveness of today's traditional software patterns. Today, most of the pattern developers tend to use the word "experience," which is often an improperly used keyword, among the pattern community. Currently, usage of this important keyword is resulting in producing and preserving very low-quality patterns. In addition to this pitfall, simultaneous existence of a number of different patterns addressing almost the same type of problems will hinder the dream of developing a common vocabulary for patterns.

Another critical problem is the paucity of required guidelines and hints for choosing the appropriate patterns from a large inventory of alternatives. This particular bottleneck may result in a challenging situation for both the expert and novice developers. Many a time a pattern developer is often confused differentiating between analysis and design patterns; mostly, inherited from the usual confusion between analysis and design in a general sense. Such a feeling of confusion is actually very dangerous, as it may result in creation of defective systems.

As the factor of time starts getting very critical and crucial in developing a software system, other factors of *simplicity* and *clarity* tend to become highly essential qualities. It is a well-known fact that developers tend to get discouraged and skeptical, while using complex, confusing and unclear type of patterns.

A typically good and meaningful pattern is clearly focused in the sense that pattern addresses a specific problem or challenge with well-defined perimeter boundaries. It is also true that developing patterns that aim to solve a big collection of problems and challenges will usually result in large-scale problems with very limited applicability.

Thus, the most obvious solution would be to focus more on developing effective patterns that contribute positively to the future development of software systems. In the forthcoming columns, we will highlight on specific problems that hinder today's pattern developmental efforts.

2.2   TRADITIONAL PATTERNS' PITFALLS

SIDEBAR 2.1   SKILL AND EXPERIENCE: THE MAGICAL WANDS!

Skill is the unified force of experience, intellect, and passion in their operation.

John Ruskin [1]

Traditional software patterns include all the existing patterns like the GOF [2], Siemens Group [3], and the others [4,5]. Experience is more of a subjective topic rather than an objective property. As of now, neither there are well-defined metrics, by which experience can be quantitatively measured, nor does it always make sense to define specific metrics. However, when this experience plays a crucial role within a specific domain, it becomes necessary to identify a set of standard qualifications that govern this experience. For example, in medicine domain, surgeons are selected for a particular surgical procedure based on their past surgical record. A pilot's experience can also be quantified as being issued a license based on the total number of hours previously flown.

In the patterns community, experience is highly critical, and more so for writing patterns, since it relates itself directly to the important task of documenting the developer's past experience. However, it is very hard to quantify experience with clear and well-set metrics because software development experience involves more than just measuring the number of the projects previously developed. Developers can produce workable software systems without necessarily employing the best of practices.

> It is very hard to quantify experience with clear and well set metrics, because software development experience involves more than just simply measuring the number of the projects previously developed.

The inaccurate use of the word "experience" can result in many potential and unforeseen problems that would seriously undermine the benefits of reusing patterns in software development.

Some of these major problems are

> *It hinders pattern improvement*: The word "experience" tends to make pattern writers immune from undue criticism. When challenged on the validity and veracity of the pattern, the writer can claim that he/she has successfully used the pattern on developing many projects. However, simply using a pattern may not ensure that the project was correctly designed and developed. More often, developers tend

to overestimate their own abilities, while designing and creating a pattern, which ultimately leads to inferior pattern structures. Meaningful pattern improvement is possible, only when the pattern writer accepts the undeniable fact of need for gaining years of experience of understanding the advanced concepts of patterns and their usage in developing reusable, stable, and authentic patterns.

It produces low-quality patterns: Without a clear definition of experience, it is common to read patterns from novice and inexperienced developers who lack the required knowledge and skill to develop high-quality solutions. Despite the fact that various patterns are readily available, it is still the sole responsibility of the developer to choose the right pattern for the new system. Usage of low-quality patterns will only further complicate the choice of the right pattern.

It discourages validation and verification techniques: A pattern that has been used in creating successful projects is not necessarily a good pattern. On the flipside, a pattern that has been used in failed projects is not necessary a bad pattern. Clearly, a pattern's limited usage is an inaccurate measure for validation or certification. The only way to claim that a pattern is working properly and correctly is by multiple experiences gained by using the pattern. To date, no serious attempts have been made to develop a theoretical foundation for validating and verifying software patterns.

It is not obvious how to measure software development experience appropriately: Note that physicians must successfully complete undergraduate school, medical school, internship, residency, and specialist training, all of which are supervised by more experienced physicians that are working under extensive sets of operating protocols, in order to begin to operate independently on their own. Pilots must have logged in a certain number of hours in flying their aircraft and must have multiple levels of training to operate different types of commercial or military aircraft. They must also operate under explicit governmental and organizational rules and regulations. Both pilots and doctors must continue to upgrade their training and experience in order to keep and retain their licenses. Therefore, while there is no objective method to assess the quality of their experience, there is some definite assurance that they will satisfy some mutually agreeable standards.

Nothing similar exists for software developers: No minimum standards of experience or skills are needed while designing patterns. The SEI's PSP, TSP, and the IEEE Computer Society's certification do not strike me as being adequate or skilled. Most university programs never require supervised and administered training. As far as we know, patterns do not automatically undergo any sort of peer review by people who are acknowledged to have enough experience to judge the quality of patterns and their relevance. Conversely, just a few high-quality peer reviewers and experienced technical experts can objectively analyze patterns that are presented for criticism and opinions.

If you want to specify a minimum standard for pattern writers, how about the following points:

- Formal and established training in software system development (BSCS, certificate programs, or similar)
- At least a minimum software development experience of 7 years including knowing how to model and architect in multiple real-time projects
- At least, 2 years of using patterns in real-time projects
- Peer review of previously published patterns

Discipline-specific patterns will require more verified and mature experience in the discipline. Note that experience with failed software projects could be valuable (if not more) than those experiences with successful projects, as long as the failure experience is critically analyzed.

References

1. John Ruskin (February 8, 1819–January 20, 1900), last modified on June 19, 2017, https://en.wikipedia.org/wiki/John_Ruskin
2. E. Gamma et al. *Design Patterns: Elements of Reusable Object-Oriented Software*, Addison-Wesley Professional Computing Series, Addison-Wesley, New York, 1995.
3. F. Buschmann et al. *Pattern-Oriented Software Architecture, A System of Patterns*, John Wiley & Sons Ltd, Chichester, 1996.
4. M. Fowler. *Analysis Patterns: Reusable Object Models*, Addison-Wesley, Boston, Massachusetts, 1997.
5. E. Freeman, E. Freeman, B. Bates, and S. Kathy. *Head First Design Patterns: A Brain-Friendly Guide*, O'Reilly Media, Sebastopol, California, p. 694, Final Release Date: October 2004.

SIDEBAR 2.2   SAME PROBLEM, BUT MULTIPLE PATTERNS: THE COMMON PROBLEM OF DUPLICATION

In trying to make something new, half the undertaking lies in discovering whether it can be done. Once it has been established that it can be done, duplication is inevitable.

Helen Gahagan [1]

With the tremendous growth of the patterns community, it becomes natural to find similar patterns that attempt to address almost the same problem. The nature and format of the design, as the solution space, makes it perfectly acceptable for developers to find different patterns, which provide dissimilar approaches to solve the same problem. However, it is not quite normal in analysis, as the problem space, to find multiple analysis models, which analyze the similar problem, which result in patterns that are remarkably different in their structure and architecture.

For instance, now consider the "Account" pattern. There is more than one pattern that models this "Account" problem; yet they are all quite different (Figures SBF2.1 and SBF2.2). This definitely suggests that something is wrong with these models, as patterns. Perhaps, they have many redundant objects that have nothing to do with the underlying problem; therefore, the absence of these objects from other models, still allow them to work properly. Assuming that all of these different models have worked properly in real-life projects, the use of analysis patterns becomes much more complex. The only way that a developer can choose the right analysis pattern is to critically analyze the problem and check which pattern makes the most sense for the given application. Consequently, the advantage of reusing analysis patterns simply vanishes.

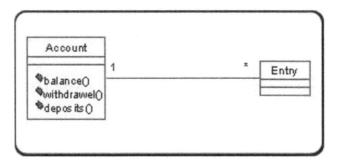

FIGURE SBF2.1   One model for the Account patterns.

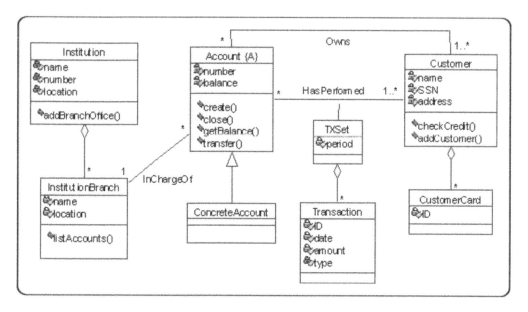

FIGURE SBF2.2   Another pattern for the Account problem [2].

The unsavory situation of having many different patterns (either analysis or design) for the same problem will always generate a number of other complications when adopting patterns in practice. One of these several complications is the difficulty of developing the pattern's vocabulary, either in the analysis or design spheres. With the tremendous increase in the number of patterns that essentially do the same thing, the pattern's names can no longer serve as the common vocabulary, as they did a few years ago. We will discuss this problem in a lengthy detail at a later stage (pitfall 10). In addition, the problem of ways to figure out, which pattern is the right pattern (in the case of analysis) or the appropriate pattern (in the case of design) is still a major problem to solve (pitfall 4).

References

1. Helen Gahagan Douglas (November 25, 1900–June 28, 1980), last modified on June 19, 2017, https://en.wikipedia.org/wiki/Helen_Gahagan_Douglas
2. E. Fernandez and Y. Liu. The account analysis pattern, in *Proceeding of the 7th European Conference on Pattern Languages of Programs*, EuroPloPP02, Kloster Irsee, Bavaria, Germany, 2002.

SIDEBAR 2.3   CHOOSING THE RIGHT PATTERN: REAL CHALLENGES

A key ingredient in innovation is the ability to challenge authority and break rules.

Vivek Wadhwa [1]

The existence of a number of patterns that address the same problem will naturally raise the perplexing and mysterious question: Which pattern shall we choose? Today, the decision to choose a particular pattern has become a crucial issue, while applying patterns.

The problem of pattern selection falls in two main categories. The first is the selection among different patterns that sound more relevant to the problem existing in existence. (For instance, separating the models of the system from their views is possible through several different patterns. One might use the Model–View–Control pattern, while others may choose to utilize the Pipe and Filter design pattern. Each pattern has its own philosophy and idea, and using each of them will require different considerations in the design of the system, although both are still valid solutions for the problem.) The second is selecting among the same patterns, which claim to address the problem at hand. (e.g., choosing between the Account problem in Figure SBF2.1 of pitfall 2 and the one in Figure SBF2.2 in pitfall 2; they both have the same name but are very different in structure.)

The pattern community is well aware of the first type of the selection problem (selection among different patterns), and thus have proposed some advanced techniques that can facilitate the selection process. For example, Gamma et al. (the authors popularly known as the GOF) proposed six different approaches to choose a design pattern [2–4]. Another example is the selection procedure, as proposed by Buschmann in Reference 5. However, these approaches have many and severe limitations. The main limitation with these approaches is that designers developed them for local use, for selection of only the patterns presented, within this book and not for any group of patterns. In fact, these designers tied these approaches with patterns classifications and documentation style. Thus, these approaches are not applicable in a general way to select among different patterns. Another sever limitation is their dependence on searching mechanisms, where the developer needs to scan and probe different parts of the template before making the final choice. With the tremendous growth of a number of patterns today, this type of searching will impose and force a great burden on developers and complicate the use of patterns in development. Apart from time-consuming, searching patterns could be very difficult, since the pattern, templates are not standard, and some of them have additional fields that are not present in others. In addition, pattern descriptions depend heavily on the pattern writer style. Therefore, the problem description might be too vague for the pattern reader, and force him to skip the pattern as an option, simply because the reader did not understand the nature of problem the pattern was addressing.

> There should be processs that are more effective or relevant tools that help developers to select the right patterns for the given job. These processes and techniques should not be restricted to specific kinds or groups of patterns, as is the case today.

On the other hand, designers have not attempted to address the second type of selection problem. That is, the problem of selecting among the same pattern that addresses the problem at hand, as shown in the Accounts example above. In design patterns, several factors can assist in choosing the appropriate pattern for such situations. One of these factors relate to the implementation issues. The implementation language used in the development of a project could make the implementation of one pattern look less complex, compared to the other. Thus, a decision on which pattern to choose becomes more obvious and apparent.

Another aspect that might help in choosing the appropriate design pattern relates to the solution that it proposes, and how it resolves the forces of the problem it solves. It is very hard for one solution to resolve all the forces of the problem it solves. Therefore, different designs may weigh the forces of the problem in a different manner, which is actually the normal expectation. The developer can then make a decision on which pattern to choose, depending on which forces are more critical to the system.

However, choosing the suitable design pattern is not that straightforward nor is it very easy, even with the guidelines described above. There are still many problematic issues, which are associated with making the right decision. For instance, we should remember that patterns aim at helping novice developers, as well as seasoned and skilled professionals. Therefore, with the assumption that the design patterns described only for the novice and average OO developers, it is still difficult to distinguish between different design solutions. Most of the decision-aiding factors require some level of previous experience. It might not be too obvious for a novice developer, how different forces conflict with each other, and which are the most critical forces for the system.

With analysis patterns, the situation is far more complex. Unlike design patterns, where decisions are directed at choosing the appropriate design, the choice between different analysis patterns is a matter of finding the right analysis, instead of the appropriate analysis. Consequently, designers make decisions based on validating the different patterns and examining how accurate they are in modeling the corresponding problem. Without considering time as a critical factor in analysis, validating analysis patterns is not a defined or systematic process to apply. Thus, it is very difficult to guarantee the right type of decision. Therefore, patterns will lose their main advantage of being an easier way to develop systems. Moreover, novice developers will most likely not have the adequate experience for examining different analysis models and choosing the right one.

In conclusion, there should be processes that are more effective or relevant tools that help developers to select the right patterns for the given job. These processes and techniques should not be restricted to specific kinds or groups of patterns as is the case today.

References

1. Vivek Wadhwa, last modified on June 19, 2017, https://en.wikipedia.org/wiki/Vivek_Wadhwa
2. M. Fowler. *Analysis Patterns: Reusable Object Models*, Addison-Wesley, Boston, Massachusetts, 1997.
3. E. Fernandez and Y. Liu. The account analysis pattern, in *Proceeding of the 7th European Conference on Pattern Languages of Programs*, EuroPloPP02, 2002.
4. E. Gamma et al. *Design Patterns: Elements of Reusable Object-Oriented Software*, Addison-Wesley Professional Computing Series, Addison-Wesley, New York, 1995.
5. F. Buschmann et al. *Pattern-Oriented Software Architecture: A System of Patterns*, John Wiley & Sons Ltd, Chichester, 1996.

SIDEBAR 2.4 DRAWING A FINE LINE BETWEEN AN ANALYSIS PATTERN AND A DESIGN PATTERN

There is really no differentiation between the work I make and the world I live in.

Doug Aitken [1]

In spite of very clear theoretical definitions that differentiate between analysis and design aspects of patterns, the resulting fine thin line gets increasingly blurred and invisible, and in some cases it may not even exist at ball. Such confusion may directly lead to more pronounced confusions between analysis and design patterns. In fact, it is quite hard and challenging to decide whether a given pattern is an analysis pattern or a design pattern, unless the patterns presented in a book or an article specifically states their intended usage (GOF [2], Fowler [3], etc.).

In spite of very clear theoretical definitions that differentiate between analysis and design aspects of patterns, the resulting fine this line gets increasingly blurred and invisible, and in some cases it may not even exist at ball!

Strangely, we find strong evidences of confusion during technical lectures, discourses, and presentations that pertain to Coad's patterns [4], while many others consider them to belong to a design-related realm. On the other hand, some skilled developers, such as Fowler, believe in what is called hybrid patterns (which are a mixture of analysis and design) (Fowler's column); however, it becomes very important and critical that they state in clear terms that they are proposing hybrid patterns. For an example, almost all analysis patterns proposed by Fowler in Reference 3 are either design or hybrid patterns or not pure analysis patterns as this book title suggests in a clear term.

References

1. Doug Aitken, 1968, last modified on June 19, 2017, https://en.wikipedia.org/wiki/Doug_Aitken
2. E. Gamma et al. *Design Patterns: Elements of Reusable Object-Oriented Software*, Addison-Wesley Professional Computing Series, Addison-Wesley, Boston, Massachusetts, 1995.
3. M. Fowler. *Analysis Patterns: Reusable Object Models*, Addison-Wesley, Boston, Massachusetts, 1997.
4. P. Coad, D. North, and M. Mayfield. *Object Models: Strategies, Patterns, & Applications*, Second Edition, Yourdon Press, Raleigh, North Carolina, 1997.

SIDEBAR 2.5   KEEPING IT VERY SIMPLE!

Success is nothing more than a few simple disciplines, practiced every day.

Jim Rohn [1]

Current patterns are hard to understand and comprehend. The GOF had earlier mentioned in the preface of their book [2] (page xi): Do not worry if you do not understand this book on the first reading. We did not understand it all on the first writing. Despite the fact that many existing patterns are really worth the time to understand and comprehend, but with such a perceived difficulty, the likelihood of misunderstanding the pattern and not using it correctly in the right manner becomes very high. In addition, some of today's patterns composed very hard to make it understandable.

Describing patterns is hard job and requires careful and calibrated work, since a balance should exist between the pattern's details and pattern's depth.

There are quite a few number of reasons contributed to the difficulty in understanding patterns. One major source of such difficulty is due to the documentation of the patterns. Most of the documents that exist today are confusing and misleading. Choosing the right pattern name, describing the problem in clear way, and presenting the solution without missing any hidden assumptions or details, directly affects understanding the pattern and its reuse.

Describing patterns is a hard job and requires careful and calibrated work since a balance should exist between the pattern's details and pattern's depth. By a pattern's details, we mean the amount of information presented to describe the pattern, while a pattern's depth relates to the technical complexity of the solution the pattern presents. If the ratio of details to depth is either too great or too small, the user will be lost or left adrift and the pattern will be hard to grasp and reuse.

Patterns that are hard and difficult to understand are, most likely, harder to reuse as well!

Patterns that are hard and difficult to understand are, most likely, harder to reuse as well. In addition, it is very important to remember that a pattern is usually a part of an overall project; thus, complex patterns will propagate and induce further complexity to the whole system. Another issue to remember is that the goal of having patterns is to ease the development of systems for both the experienced developer as well as the novice ones. Therefore, complex patterns will serve more as obstacles or impediments to a novice developer, who will likely become frustrated and disappointed, when attempting to read and understand them.

For patterns to preserve their goal of reuse and to ease the development of software, they should be simple, clear, lucid, readable, understandable and easy to grasp by both experts and novices.

For patterns to preserve their goal of reuse and to ease the development of software, they should be simple, clear, lucid, readable, understandable, and easy to grasp by both experts and novices. Developers must develop their patterns with a view to present all necessary pattern details and principles in such a way that everyone will find the reasons, why they need patterns in the first place. Again, I think it is highly debatable that novices should find patterns easy to grasp and comprehend. At the very least, understanding why the pattern was used is not something I would expect out of a novice.

References

1. Emanuel James "Jim" Rohn (September 17, 1930–December 5, 2009), last modified on June 19, 2017, https://en.wikipedia.org/wiki/Jim_Rohn
2. E. Gamma et al. *Design Patterns: Elements of Reusable Object-Oriented Software*, Addison-Wesley Professional Computing Series, Addison-Wesley, Boston, Massachusetts, 1995.

SIDEBAR 2.6   FOCUS! FOCUS! FOCUS! GOLDEN WORDS FOR DEVELOPING MEANINGFUL PATTERNS

Focus on the journey, not the destination. Joy is found not in finishing an activity but in doing it.

Greg Anderson [1]

If a developer wishes to develop meaningful patterns, he or she will need to identify the perimeter of the problems that these patterns model or solve. In many cases, it is quite difficult and cumbersome to identify the boundary of the problems. In the absence of any clear and well-set boundaries for a particular pattern, it will ultimately embody and induce a host of other problems that will be simply out of the scope of the main problem. Consequently, developers end up building large patterns that model or solve a collection of different problems instead of developing patterns for specific problems. It is important to note that this problem dominates almost all existing analysis patterns but rarely exists in current design patterns.

Many design patterns possess very well defined boundaries, such as GoF patterns [1], where each pattern solves a specific and predefined problem, without considering the surrounding environment.

For instance, consider the Account analysis pattern shown previously in Figure SBF2.2 (Pitfall #2). This particular pattern presents a unique situation of modeling a problem related to accounting. Unfortunately, this pattern also includes different types of patterns that lie outside the domain of the problem, such as Institution and Institution branch. On the first glance, it looks as if they are valid classes, but they appear irreverent, when focused solely on the essential idea of an account. In general, more the problems a pattern considers, the less general will be the nature of the problem. Consequently, the issue of reusability will become limited and seriously undermined.

On the other hand, so many design patterns possess very well-defined boundaries, such as GOF patterns [2], where each pattern solves a specific and predefined problem, without considering the surrounding environment.

References

1. Greg F. Anderson (born February 1966), last modified on June 19, 2017, https://en.wikipedia.org/wiki/Greg_Anderson_(trainer)
2. E. Gamma et al. *Design Patterns: Elements of Reusable Object-Oriented Software*, Addison-Wesley Professional Computing Series, Addison-Wesley, Boston, Massachusetts, 1995.

SIDEBAR 2.7   A BRIEF SUMMARY OF PATTERN COMPOSITIONS: UNDERSTANDING THE INSIDER'S SECRETS

Everything we do, every thought we have ever had, is produced by the human brain. However, exactly how it operates remains one of the biggest unsolved mysteries, and it seems the more we probe its secrets, the more surprises we find.

Neil deGrasse Tyson [1]

Pattern composition is the intimate process of integrating and hemming different patterns of the similar type (i.e., design patterns, analysis patterns, etc.) to build and design larger components or different types of patterns that also include process patterns, managerial and organizational patterns, and other types of patterns. Systematic and organized developmental processes that use patterns utilize a well-formed composition mechanism to seam and integrate patterns together at the design level itself. Composition mechanisms display certain inherent characters that are behavioral and structural mechanisms in nature and designs. Right now, there are a few reported successful experiences on composing patterns in software applications, such as the development of highly interactive speech recognition systems [2] and object-oriented cellular communication software [3]; however, such developmental designs are at best only ad hoc and nonsystematic. Structural composition is an innovative approach that uses actual class diagrams to model the designs that would display the patterns in the correct and accurate manner. As such, structural composition is more concrete and robust than the abstract. On the other hand, behavioral compositions rely heavily on composition with objects and the roles they play in fulfilling various other patterns.

However, several critical issues and topics need consideration while considering pattern composition. If a pattern designer wants to develop a meaningful pattern, the following issues need advanced research in a thorough manner:

1. Composition process for integrating analysis, design, implementation, and testing patterns to make the intended design robust and sturdy
2. To assess and quantify the overall impact of software engineering methodologies and environments on the use of composition
3. To assess and redress the impact of programming language and environments on the use of composition
4. To find out the level of tangibility required for creating the composition
5. To find out the impact of generic versus domain-specific patterns on the composition

Thus, when comprised, the software development process will simply does not avoid facing the unique challenge of how these components can fit and integrate together.

Even though the usage and application of different patterns has been successfully demonstrated in system development, these application instances do not provide a clear or explicit process of composing patterns that can be generalized and used by other developers. This crucial and critical issue makes the utilization of patterns in development, a jigsaw puzzle for many software development professionals.

Clearly, pattern composition neither is a natural method nor is it an intuitive process. Even worse, patterns always vary in their behavioral and structural nature and character, according to the nature and scope of the project and/or the developer's innate opinion.

For instance, one of the factors that might hinder integrating two analysis patterns is the fact that each pattern has a different level of abstraction. We simply cannot standardize the level of abstraction to overcome this unique yet common problem; therefore, integrating these two patterns should allow future modifications such that the abstraction level of the resultant pattern is always balanced. In a problem that involves both paying and accounts, an analysis pattern that models payment simply cannot be naturally integrated with a more abstract analysis pattern that attempts to model accounts.

There are many other crucial issues that need to be investigated and solved immediately; otherwise, the utilization of patterns will involve a number of obstacles and hence will diminish their future use. Most of the current texts and journals on pattern composition or systems of patterns do not include concrete or natural examples of methods to build stable systems of patterns but rather include very vague guidelines, inconclusive and hazy suggestions. This makes it extremely difficult for many software development professionals to develop and design systems solely based on patterns and the principles of pattern languages. Although several successful attempts have been made in the past, that demonstrate ways to build small tools and utilities or tiny systems by using patterns, they cannot be used as generalized processes for building systems of patterns.

> Clearly, pattern composition neither is a natural method nor is it an intuitive–process. Even worse, patterns always vary in their behavioral and structural nature and character, according to the nature and scope of the project and/or the developers innate opinion.

References

1. Neil deGrasse Tyson (born October 5, 1958), last modified on June 22, 2017, https://en.wikipedia.org/wiki/Neil_deGrasse_Tyson
2. S. Sirinivasan and J. Vergo. Object-oriented reuse: Experience in developing a framework for speech recognition applications, in *Proceedings of 20th International Conference on Software Engineering, ICSE, 98,* Kyoto, Japan, April 19–25, 1998, pp. 322–330.
3. J. Garlow, C. Holmes, and T. Mowbary. Applying design patterns in UML, *Rose Architect,* 1(2), Winter 1999.

SIDEBAR 2.8 LACK OF PATTERNS CONNECTIVITY ACROSS DEVELOPMENT PHASES

> Greater technological connectivity makes the world wider, and the walls of isolation thinner.
>
> **Tsakhiagiin Elbegdorj [1]**

The point highlighted in the previous pitfall illustrated the challenge of integrating patterns within the same developmental phase. Integrating a given pattern is often a major challenge and a strenuous task for any developer of patterns. However, a different set of challenge emerges when a developer tries to introduce different types of patterns into the developmental phase of a given system. For example, let us assume that a developer uses different phases of development while designing a particular pattern.

> The biggest question here is how does an analysis pattern used in the analysis phase relate to the design pattern used during the design phase? This seems to be a difficult task to many developers of patterns as two or three phases within the same developmental system may pose many riddles and challenges before arriving at a meaningful, yet practical pattern.

For developers, this may pose a crucial question and a critical challenge as they use patterns within a single developmental phase, which is usually the design phase. This is probably attributable to the fact that traditional design patterns are more mature and seasoned than analysis patterns. Successful patterns designed so far always relied on using patterns in a single phase, which is probably the best avenue to exploit the strength and merit of patterns. However, it is also true that final and ultimate strength and maturity of pattern utilization will only result from using them right though the developmental cycle. Overall, improvements and differences will be seen within each phase subsequently resulting in notable improvement as a whole. Right now, there is no known connectivity between the design and analysis patterns. Furthermore, it is still not very clear how this will be tackled and approached in the future, excepting stable software patterns [2–6]. Addressing such issues may become more complicated and challenging as patterns become more mature over a period and there is a burning desire on part of developers to utilize different type of patterns on a concurrent basis.

More obstacles and impediments make the scenario much more difficult for a pattern developer because both the analysis and design patterns developmental process tend to be separate fields of study.

> There is a lack of communication between developers working to enhance and improve analysis patterns and those working on design patterns and vice versa.

The idea and vision of fitting these two different entities seems to be very appealing and attractive as it will assist developers realize and reach the goal of attaining the big picture of pattern utilization. Sadly, right now there is no existing methodology designed to create such connectivity. To make matters worse, patterns of a given type exist at different and varying levels of abstraction, which makes them very difficult to link together. Another negative aspect is the perceived difficulties in implementing the patterns within the software itself. For example, no proven benchmark exists that can accurately measure the effectiveness of the patterns on various issues of software, neither are there any known standards to comply with, when working on patterns or creating them.

> There is no tailor made or exactly fitting mechanisms on both behavioral and structural aspects of many patterns, while there are many critical issues related to the perceived values, benefits, utilization and ROI between the generic and domain specific patterns.

References

1. Tsakhiagiin Elbegdorj (born March 8, 1963), last modified on June 21, 2017, https://en.wikipedia.org/wiki/Tsakhiagiin_Elbegdorj
2. M.E. Fayad. How to deal with software stability, *Communications of the ACM*, 45(3), 2002, 109–112.
3. M.E. Fayad and S. Wu. Merging multiple conventional models in one stable model, *Communications of the ACM*, 45(9), 2002, 102–106.
4. H. Hamza and M.E. Fayad. Applying analysis patterns through analogy: Problems and solutions, *Journal of Object Technology*, 3(3), 2004, 196–208.
5. M.E. Fayad and H.S. Hamza. The trust analysis pattern, in *The Fourth Latin American Conference on Pattern Languages of Programming (SugarLoafPLoP 2004)*, Porto das Dunas, Cearã¡, Brazil, August, 2004.
6. M.E. Fayad, H.S. Hamza, and V. Stanton. The searching analysis pattern, in *The Fourth Latin American Conference on Pattern Languages of Programming (SugarLoafPLoP 2004)*, Porto das Dunas, Cearã¡, Brazil, August, 2004.

SIDEBAR 2.9   UNTRACEABLE PATTERNS: THE PITFALL OF THE PATTERN SYSTEMS

> There are some problems on this planet that seem to be intractable.
>
> **Richard Gere [1]**

Today, developers consider many software patterns, especially those in the area of analysis, as templates. In Reference 2, Peter Code has defined patterns in brief as following: A pattern is a template of interacting objects, one that may be used repeatedly by analogy. Simply speaking, the pattern extracted from a specific project can be integrated into an appropriate abstraction level such that it can be used to model and design the same problem or task in a wide range of applications and domains. Based on this, experts consider the abstracted patterns as a template, which are usable through and by analogy.

Developing patterns as templates, while making them useful as blueprints for a wide variety of applications, nonetheless sacrifices their traceability when used through analogy.

As an example of this approach, Figure 1 shows the class diagram of the Resource Rental pattern [3], which forms the abstract template of the Resource Rental problem. The main objective of the pattern is to provide a model that can be reused to model the problem of renting any resource; therefore, the class diagram does not tie to the renting of a specific resource. Figure 2 shows an example of using the Resource Rental pattern in the application of the library services [3]. Simply through an analogy, one can apply the original abstract pattern into specific applications.

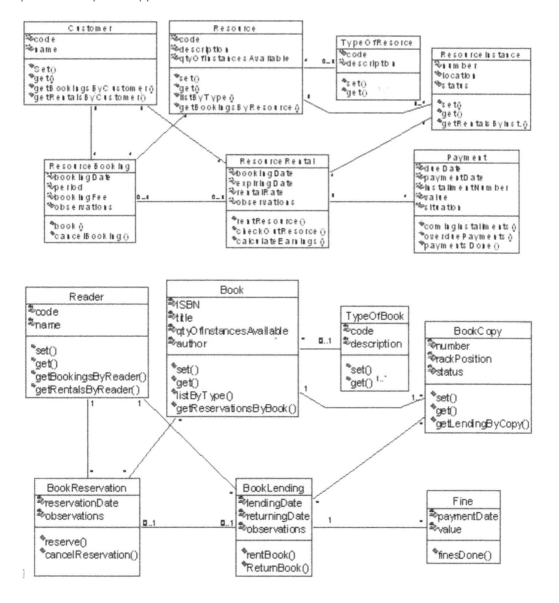

Though using analysis pattern through the process of analogy might appear to be a straightforward and simple technique for maintaining a good level of generality for the pattern, this technique, nevertheless, raises some important and critical problems that should be further investigated. These problems are as follows:

- *It generates untraceable systems*: Once the analysis pattern templates have been instantiated in the developed system, the original patterns will be no longer extractable. For example, consider the instance of the Resource pattern shown in Figure 2 and imagine it as part of a complete library service system. It would be very hard if not impossible to extract the original pattern, shown in Figure 1, after such instantiation.
- *It Complicates system maintainability*: Software maintenance is one of the costliest phases in the system development life cycle; therefore, any design that complicates system maintainability tends to inflate the cost even higher. One can imagine a very simple situation where the developed system documentation needs to be updated, due to some modifications in the system requirements. Since the developed system employs several patterns in its developmental phase, identifying the patterns to which the changes should be applied will always be tedious and time consuming.
- *It Trivializes classes roles of the pattern*: To better illustrate this critical issue, we will use an example from Reference 3 where a class diagram for designing a computer repair shop is used by analogy to build the class diagram of hospital registration project. Instead of a shop that fixes broken computers, we have a hospital that fixes and cures sick people. Therefore, we can simply replace the class named computer in the first project by the new class named "patient" in the next project. Even though such analogy seems achievable, it is highly impractical. There is a perceptible difference between the computer as a machine and the patient as a human. These two classes might look analogous, since they both have problems that need a fix; however, their behavior within the system are completely different. The role of the computer class is completely different from that of the patient. Hence, such analogy is highly inaccurate. There would be even more differences and discrepancies, if we tried to generate the dynamic behaviors of these two systems by using an analogy, as suggested in Reference 3.

References

1. Richard Tiffany Gere (born August 31, 1949), last modified on June 24, 2017, https://en.wikipedia. org/wiki/Richard_Gere
2. P. Coad, D. North, and M. Mayfield. *Object Models Strategies, Patterns, & Applications*, Yourdon Press, Prentice-Hall, Inc., New Jersey, 1995.
3. R.T. Vaccare Braga et al. A confederation of patterns for business resource management, in *Proceedings of Pattern Language of Programs 98 (PLOP, 98)*, September.

SIDEBAR 2.10   NO GUIDELINES FOR EXTRACTING PATTERNS

Some people think we are adrift without any guidelines. I do not. I think we have had instruction on how to live.

Jon Voight [1]

Another major challenge that confronts pattern designers while designing current patterns is the absence of a clear guidance or methodology for already extracting patterns. GOF [2] has stated explicitly in their book on page 355 that finding patterns is much easier than describing them. Creating and documenting patterns that everyone can read, understand, and reuse them in no easy task.

However, extracting a wrong pattern, no matter how perfectly and clearly designed and documented, is USELESS and even very dangerous. Surprisingly, most of pattern developers still do not know the ways and means to extract right and correct patterns.

Theoretically, patterns seem to be obvious and easily located in a given developed system. The GOF recipe for extracting patterns is quite simple: By observing enough number of systems, one can easily identify many patterns. Nevertheless, how much does enough mean in the real sense? Is it two times or a hundred times? On the other hand, is it possible to quantify it?

In practice in reality, extracting patterns is far more difficult, than what GOF has described. The challenge arises from the fact that the boundary of the pattern within the system is neither obvious nor is it easily comprehensible.

The idea is not to merely identify or pinpoint a few classes that do a specific task and later consider them a pattern. Without some clear guidance, such extractions might result in incomplete patterns, which obviously miss some of the classes that make the pattern perform the required function. Alternatively, the system's model can end up extracted itself, instead of the pattern by including too many classes that are outside the boundary or domain of the problem. It also requires dedicated work to know what really constitutes a pattern that lies within a system. The confusion arises, when some of the functional relationships are scattered all over the system model. In this case, extracting a set of classes that form the pattern is very complex and tedious.

References

1. Jonathan Vincent Voight (born December 29, 1938), last modified on June 23, 2017, https://en.wikipedia.org/wiki/Jon_Voight
2. E. Gamma et al. *Design Patterns: Elements of Reusable Object-Oriented Software*, Addison-Wesley Professional Computing Series, Addison-Wesley, Boston, Massachusetts, 1995.

SIDEBAR 2.11   CHASING AN ELUSIVE VOCABULARY: IMPORTANCE OF VOCABULARY IN PATTERN DEVELOPMENT

Loving your language means a command of its vocabulary beyond the level of the everyday.

John McWhorter [1]

One of the specific benefits of using patterns resides in developing a common vocabulary by which software and pattern developers, in any development phase, can easily communicate and discuss their project concepts. Such a common vocabulary simplifies the basic description of complex systems. One can discuss the system at a higher level of abstraction without getting into the details of each pattern.

However, no clear vocabulary currently exists that can allow pattern developers to enjoy these perceived benefits. The main challenge here is the fact that many current patterns try to address the same problem (refer to the discussion of the two problems in the pitfalls #2 and #3 [2]: Multiple Patterns for the Same Problem, and Which Pattern Shall We Choose [3]).

The name of the pattern alone cannot serve as a unique vocabulary; instead, other identifiers need consistent usage to distinguish different patterns for the same problem.

For instance, assume that someone has already developed a new pattern called Observer that does the same thing the Observer pattern as described by Gamma et al., (popularly referred to as the GOF) [4] does. Thus, referring to a pattern as Observer is confusing. The GOF Observer pattern needs thorough reference since there is no reason to assume the GOF pattern by just stating the name Observer. In an analysis, the same problem still exists. The Account pattern discussed previously in the earlier issue must also be qualified to communicate clearly, which Account pattern is used currently.

> It is not entirely practical to expect that a one can achieve a clear pattern vocabulary without first solving the challenges associated with having multiple patterns for the same problem. If the pattern community believes that there is no way to have a single pattern for each problem, then they should not expect to have a common vocabulary.

Choosing the name of the pattern is another immediate challenge that needs consideration. Most of the patterns existing today have some vague names (due to high levels of abstraction), and in most cases, understanding what this pattern does in practice, is not clear until the first few lines in its template have been read. The most useful pattern name provides some useful insight into the problem currently addressed by the pattern.

References

1. John Hamilton McWhorter V (born October 6, 1965), last modified on June 24, 2017, https://en.wikipedia.org/wiki/John_McWhorter
2. M.E. Fayad and G.K. Srikanth. Pitfall #2 Same Problem, but Multiple Patterns: The Common Problem of Duplication, The Software Patterns Blog at pattern.ijop.org.
3. M.E. Fayad and G.K. Srikanth. Pitfall #3 Choosing the Right Pattern—Real Challenges, The Software Patterns Blog at pattern.ijop.org.
4. E. Gamma et al. *Design Patterns: Elements of Reusable Object-Oriented Software*, Addison-Wesley Professional Computing Series, Addison-Wesley, Boston, Massachusetts, 1995.

SIDEBAR 2.12   OUTDATED AND JADED PATTERNS OR FRESH AND EVERLASTING PATTERNS: CHOOSE YOUR PICK!

> The problem of telling contemporary history is that your message gets outdated.
>
> **Salman Rushdie [1]**

> Patterns can die or become useless! According to Reference 1, an existing pattern can die or perish for many causes or reasons: the disappearance of the problem the pattern addresses, the evolution of better alternatives (better patterns appear that deal with the same problem more effectively), and/or changes in current technology (a new paradigm evolves, which makes the existing patterns useless).

As these reasons might provide a useful insight into why patterns die or become outdated, they also provide many ideas and clues about the critical question: Why have not patterns fulfilled the expectation of playing a crucial role in software development?

Even though many of us think of a pattern as a possible solution to a problem, it may or may not really work for every problem, and hence, it is up to the discretion of the developer to either use it or just develop a customized solution. If patterns can die, then is there a risk in heavily utilizing them during development? One obvious and short answer is yes, the patterns may become a legacy system, and developer must deal with upgrading, modifying, or enhancing that system, given that the replacement of the entire system is both costly and time consuming.

Returning to the three reasons cited above, that might cause a pattern to die or become outdated, they can be rephrased to answer the question of why patterns are not as effective as developers may have envisioned during the earlier phases of development.

We can rewrite these reasons as follows: Patterns are not as effective as they should be because they do not present the best possible solution to the problem at hand and there is always a better solution than what the pattern proposes in real. So, what causes a solution to a specific problem be classified as a bad solution or why can a better solution to the same problem evolve?

If patterns are technology interdependent, then the frequency of changes in the technology will drive the frequency of changes in the patterns usage as well. Consequently, rapid changes in technology will render these solutions obsolete and outdated, even before they get the chance to become accepted. This obviously contradicts the original goal of patterns as reusable artifacts.

Buchmann [1] states that "Patterns can become outdated or 'dying patterns'," which contradicts Alexander's notion of "Timeless patterns" [2,3], and conflicts with the reuse principle. Two issues exist here. First, the dying patterns probably were not patterns in the first place. Second, what do you do with those systems using dying patterns? Is it possible to repair or reclaim those system patterns that are already dead or on their way to extinction?

It does not make any sense to use those patterns that are already obsolete and outdated [2–4].

References

1. Salman Rushdie (born 19 June 1947), last modified on June 16, 2017, https://en.wikipedia.org/wiki/Salman_Rushdie
2. F. Buchmann et al. *Pattern-Oriented Software Architecture: A System of Patterns*, John Wiley & Sons Ltd, Chichester, 1996.
3. C. Alexander. *Timeless Way of Building*, Oxford University Press, New York, 1979.
4. A. Christopher, S. Ishikawa, and M. Silverstein. *A Pattern Language: Towns Buildings Construction*, Oxford University Press, Oxford, UK, 1977.

SIDEBAR 2.13   MODELING PROBLEMS ARE NAGGING OBSTACLES TO CREATING MEANINGFUL PATTERNS

If you are trying to achieve, there will be roadblocks. I have had them; everybody has had them. But obstacles do not have to stop you. If you run into a wall, do not turn around and give up. Figure out how to climb it, go through it, or work around it.

Michael Jordan [1]

It is a common fact that the solution for any pattern consists of two different models: static, mainly class diagrams, and dynamic, by using sequence diagrams or state transition diagrams.

The majority of existing patterns, both static and dynamic solutions, are poor in their modeling, and it shows the level of immaturity and lack of knowledge of modeling in general. Most of the majority ignores the behavior models almost completely.

Let us visit Figure 1, in Pitfall #9, the model of the Resource Rental pattern [2,3] that tries to model the common rules that govern any renting, whether the object rented is a DVD or a car. The Figure suffers from several notable deficiencies:

Problem #1: There are dangling classes in the example. Dangling classes in UML diagrams are a clear sign that something is not right or something is very wrong. A dangling class is a class connected to one and the only one class. We have identified two of these: Payment and TypeOfResource.

Problem #2: There are a number of unresolved many-to-many relationships. As noted by several authors, a model is not complete and thorough, until all many-to-many relationships have been decomposed into two one-to-many relations. We have identified two of these: between Resource and ResouorceBooking, and between ResourceRental and ResourceInstance.

Problem #3: This pertains to the use of accessor methods. Accessor methods are operations that have as their only purpose to update the internal state of an object, thus violating one of the main principles of Object-Oriented Design. These methods are called get() and set().

Problem #4: There are no names associated with the relationships. This is an inherent problem because the pattern is incomplete in documentation. If the names are not there then different interpretations of what a given link means may arise.

Problem #5: There is no real behavior associated with the classes. A big portion of the operations described in this model focus just on listing something by a given condition. For example, getRentalsByInstance() of the class ResourceInstance(). This is no real behavior associated with a responsibility for that class but only a listing provided for some other object to operate on it.

Problem #6: Customer, Resource, TypeOfResource, and ResourceInstance have the same assigned responsibility, doing the same thing, get and set, and this a clear violation of the rule of abstractions.

Problem #7: Resource, TypeOfResource, and ResourceInstance are passive objects and they cannot get and set.

Problem #8: What are the differences between Resource and ResourceInstance? There is absolutely no difference.

Problem #9: The model is missing in inheritance and aggregation.

In addition, the figure contains a cross line between two associations.

References

1. Michael Jordan (born February 17, 1963), last modified on June 24, 2017, https://en.wikipedia.org/wiki/Michael_Jordan
2. R.T. Vaccare Braga et al. A confederation of patterns for business resource management, in *Proceedings of Pattern Language of Programs 98 (PLOP, 98)*, September 1998.
3. M.E. Fayad. Pitfall #9: Untraceable Patterns-The Pitfall of the Pattern Systems, The Software Patterns Blog at fayadsblog.vrlsoft.com.

SIDEBAR 2.14   REINVENTING THE WHEEL: AN UNDESIRABLE AND DANGEROUS DEVELOPMENT

What I learned was the quality of continual reinvention.

Merrie Spaeth [1]

A pattern is a plan, rather than a specific implementation [2]. Pattern developers never directly use design patterns to guide programming. They are guides for modeling purposes. A pattern is an attempt to describe successful solutions to common software problems [3]. Design patterns represent solutions to problems that arise when developing software within a particular context [4]. However, how does one follow the patterns to find specific solutions for a given problem? By instantiation, concrete objects from the specific problem will replace the original abstract objects in patterns. The obvious result is the pattern instances, which are the models for the specific problem and a specific guide for programming. The process of applying patterns is in Figure SBF2.3.

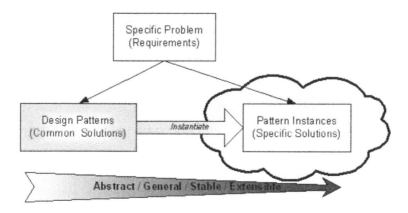

FIGURE SBF2.3 Current design patterns approach [5].

> The basic message of this pitfall is that the implementation (i-model) of design patterns (d-model) leaves the programmer with no trace of what patterns (d-models) were applied and what design decisions were made while designing a pattern. Thus, when changes are necessary, the one needs to reconstruct the entire design. This is the process of reinventing the wheel every time to attempt to implement the same pattern.

In the example of the current Design Patterns Approach, design patterns (d-models) and their instances (i-models) are always separate models. After instantiation, the instances are less abstract, less stable, and inextensible, when compared to their patterns. However, we can see only the instances as the result of the Design Patterns Approach, in general practice. The resulting models will never be displayed in the patterns and the instantiation processes. By instantiating patterns within a particular problem context, the problem at hand is easily manageable; however, the stability of the pattern is totally lost in the final model. This causes a completely new set of problems.

Problems

According to the definition, design patterns plan to facilitate the reuse of software architecture. Therefore, a well-designed pattern should be always stable and provide a strong foundation for reuse. To achieve this ultimate goal, design patterns must be abstract, general, and common in nature. This makes them unsuitable for detailed implementation on a specific problem. Thus, design patterns must be instantiated into pattern instances, specifically based on a specific problem context. This creates a pronounced discrepancy between the design patterns and the final specific pattern instances for the given problem. However, the instantiation process can work as a useful roadmap around this discrepancy. It is an efficient way to migrate from the existing knowledge to the product for a specific problem.

Nevertheless, currently, designers tend to discard this roadmap soon after using it. Generally, the description of this process is unavailable in the final models. This means that it is possible to create a specific model employable for the given problem, but the capability to trace back to its blueprint and later extend the model in the future is missing altogether. The model loses its generality and abstraction soon after instantiation, causing it to be very weak in adaptability and extensibility. When developers want to extend the software at a later stage, they simply cannot directly reuse this model. They will have to reconstruct the whole model directly from the original patterns.

From a utilization perspective, design patterns are the rules of the road. The provided instances are the actual driving and the problem context is supposed to be different driving situations. The rules are always constant and uniform, although the actual driving may change under different circumstances. To drive safely and efficiently, people must learn the basic rules first and then apply them according to the different circumstances. Every time they drive, they must keep the basic rules in mind in order to adapt to different and varying situations. That is the important key to stable driving. However, in software modeling, unlike driving, people tend to discard the patterns they use as soon as they get the desired result. They seldom describe the patterns and the instantiation process in the final models. This makes their results highly unstable and inextensible to different problem contexts. Expecting these models to be stable under future changes is similar to applying past driving sequence actions to different times and locations.

A pattern is definitely not the part of the final software product, though it acts as a powerful tool to find a solution for a specific problem, in a quick and accurate manner. However, the pattern is the only stable part you can apply repeatedly in modeling. The system analyst and designer feed the patterns with a specific problem and get a specific model through instantiation. The final pattern instance is a specific solution to the given problem. In the current Design Patterns Approach, all of the objects in the final model are very concrete and specific. It is usable as a programming guide. However, it is highly unstable and unscalable because of the same set of similar characteristics.

References

1. Merrie Spaeth (born August 23, 1948), last modified on February 12, 2017, https://en.wikipedia.org/wiki/Merrie_Spaeth
2. P. Coad, D. North, and M. Mayfield. *Object Models Strategies, Patterns, & Applications*, Yourdon Press, Prentice-Hall, Inc., New Jersey, 1995.
3. D.C. Schmidt, M.E. Fayad, and R.E. Johnson. Special issue on software patterns, *Communications of the ACM*, 39(10), 1996, 37–39.
4. D.L. Levine and D.C. Schmidt. *Introduction to Patterns and Frameworks*. Department of Computer, Science Washington University, St. Louis. http://www.cs.wustl.edu/~schmidt/PDF/patterns-intro4.pdf
5. S. Wu, H. Hamza, and M.E. Fayad. Implementing pattern languages using stability concepts, in *The Sixth Annual ChiliPLoP 2003*, Carefree, Arizona, March 2003.

SIDEBAR 2.15   ANALYZING PITFALLS TO SEEK A PERMANENT SOLUTION

In the final analysis, a drawing simply is no longer a drawing no matter how self-sufficient its execution may be. It is a symbol, and the more profoundly the imaginary lines of projection meet higher dimensions, the better.

Paul Klee [1]

In the above columns, we have highlighted 14 of the major problems that are believed to diminish the strength of the role of patterns in developing software systems.

Conclusion

Pattern development poses several unique challenges to the pattern developers like the ones clearly highlighted in the previous columns. These challenges tend to diminish the strength of the role of patterns in developing software systems, subsequently contributing to the ineffectiveness of patterns. Probably, nothing can be as damaging as misusing the term experience while attempting to design and write patterns; in fact, this misplaced feeling or perception could severely dent the quality of the developed patterns. Another curious problem occurs in the realm of creating a large number of patterns, when most of them represent and deal with almost the same problem. Frequent duplication of efforts in writing patterns will also make it very cumbersome to devise a common vocabulary for pattern development. Novice and fresh developers of patterns may also get confused while selecting the right and appropriate pattern for a pool of candidate patterns; in many cases, inexperienced pattern developers lack the knowledge of clear guidelines.

Pattern designers also face the piquant situation of differentiating between analysis and design patterns, while there is an immediate need for developing patterns that are easy to understand and comprehend so that one can ensure and ascertain quicker adoption in practice. Another vital string in the earlier series provides criticality of developing patterns to focus on specific patterns defined with clear boundaries. This column also raises another peculiar situation of developing patterns that solve a collection of problems that finally result in large-scale patterns with very limited applicability.

Patterns also display certain specific compositions that integrate and sew together different patterns of the similar type (i.e., design patterns, analysis patterns, etc.) to build and design larger components or different types of patterns that also include process patterns, managerial and organizational patterns, and other types of patterns. In many cases, existing patterns lack proper connectivity across both analysis, design developmental phases, while they can also become untraceable, and lost without providing clear guidelines and rules for traceability. Patterns can also become obsolete because of drastic changes in technology or by the fact that the original problem the pattern devised to sole is no longer in usage. Most of the patterns are modeled very poorly, affected by nagging problems like dangling classes, the use of accessory methods, classes that are useless in performing meaningful functions, classes without inheritance and aggregation, and with the presence of a number of passive objectives.

In addition to the above-mentioned problems, very poor patterns show relationship without any unique names, and under such a scenario, sundry relationship are yet to be fully decomposed as one-to-many relationships. Ultimately, patterns community and design forums suffer from the tendency of reinventing the wheel syndrome, as patterns themselves are not included in the final software product and developers come upon mixed results when attempting to deduce the patterns from the models. These are some of the most important pitfalls while designing a pattern and one may need to conduct advanced research to integrate patterns like designing and testing and integration must encourage a coherent composition of several important parameters. In the future, upcoming columns, we will attempt to introduce a novel thinking of software patterns as a way to confront the problems we have discussed so far in the earlier columns.

In conclusion, before the patterns community can more effectively utilize patterns in practice, the 14 problems or pitfalls discussed in this blog must be confronted. Further research much be done into the process of integrating patterns, such as design and testing, into a coherent composition.

Patterns lack connectivity across the analysis and design development phases, are untraceable and provide no clear guidelines for traceability. Many patterns have no clear and common vocabulary, complicated by the fact that some patterns address the same problem. Patterns can become outdated by changes in technology or by the fact that the original problem the pattern was designed to solve no longer exists. Patterns tend to be modeled poorly, with such obvious problems such as dangling classes, the use of accessory methods, classes that perform no meaningful function, classes without inheritance and aggregation and the presence of passive objects. Additionally, poor pattern models show relationships without unique names, and other relationships have not been fully decomposed into one-to-many relationships. Finally, the patterns community suffers the tendency to reinvent the wheel with software patterns, as patterns themselves are not included in the final software product and developers come upon mixed results when attempting to deduce the patterns from the models.

Reference

1. Paul Klee (December 18, 1879–June 29, 1940), last modified on June 20, 2017, https://en.wikipedia. org/wiki/Paul_Klee

2.3 SUMMARY

Of late, software patterns have emerged as a promising technique for facilitating the development of highly sophisticated software systems. Even with the best of the pattern developmental methodologies, patterns developed are yet to fulfill the expectations software developers always envisioned. In spite of some pronounced glitches in the produced patterns, patterns, as concepts, still hold an important key in developing state-of-the-art software systems in the near future. Broadly termed "pitfalls," these glitches are known to play a vital role in deciding the degree with which one can confidently claim that the designed pattern is foolproof and fail-safe. A deeper study and subsequent evaluation of all pitfalls mentioned in these sidebars suggest that even the best of stable patterns may have some hidden weaknesses that are likely to affect their practicality and reusability.

2.4 REVIEW QUESTIONS

1. Why the factors of simplicity and clarity are essential during the designing phase of patterns?
2. Of all the phases of development of patterns, *diagnostic phase is the most important. Explain why.*
3. Why skills and technical experience is necessary while designing patterns?
4. What are some of the major problems that one will come across while designing patterns?
5. Why do multiple patterns occur for a similar problem?
6. Is it difficult to find an exact pattern vocabulary?
7. Can you write two models for a similar problem?
8. What are the main challenges that crop while writing a good pattern?
9. Define some of the processes for selecting most appropriate pattern for a specific problem.
10. Is it difficult to differentiate between an analysis pattern and design pattern?
11. In the context of previous pattern designing history, will you agree that hybrid pattern or design patterns created by using traditional methods are not pure patterns?
12. Why most of the traditional patterns are difficult to understand and comprehend?
13. Could you name some of the reasons that make reading such patterns difficult and tedious?
14. Do you agree that most of the traditional patterns touch issues that lie outside the problem domain? Explain and give reasons.
15. What is a structural composition?
16. Why structural composition is a unique approach?
17. What are some of the critical issues that one needs to consider while developing patterns?
18. Integrating two different patterns should allow future modifications—True or False.
19. Why introducing different types of patterns into the developmental phase of a given system is always a tedious task?
20. As of today, does any connectivity exist between the design and analysis patterns?
21. Can one use an analysis pattern through the process of analogy to introduce a fair level of generality?
22. If not, list some of the problems that hinder the process which is stated above.
23. Are there any well-defined guidelines for extracting patterns from an existing system?
24. Choosing the right name for a pattern is a big challenge—True or False.

25. "Patterns are not as effective as they should be, because they do not present the best possible solution to the problem at hand and there is always a better solution than what the pattern proposes in real." True or False.
26. "If patterns are technology interdependent, then the frequency of changes in the technology will drive the frequency of changes in the patterns usage" True or False.
27. What are some of the most important modeling problems?
28. Why design patterns are the rules of the road?
29. What is road map to create a solid and stable design pattern? Name some of the properties of a good pattern.
30. Can you summarize the details learned from all these sidebars?
31. Can you suggest improvements to improve the scope of topics that are covered in these sidebars?

2.5   EXERCISES

1. Visit Sidebar 2.2 and create different patterns for a similar problem. Analyze why multiple patterns for a similar pattern is undesirable.
2. Some of the challenges that hinder the development of stable pattern are provided in Sidebar 2.3. Create a project to add some more problems that you think are additional challenges which restrict your ability to create a stable pattern.
3. Create class diagram each for designing an acrobatic dance training shop and an automobile workshop. First, create a diagram for acrobatic dance training shop and use its analogy to create the second one (automobile workshop). Now, explain why such an analogy to create two patterns is highly inaccurate and dangerous.

2.6   PROJECTS

1. Think of more pitfalls that hinder the making of a stable pattern. Describe them in detail, if possible with diagrams and patterns.
2. Try to design a stable pattern that is free of pitfalls and weaknesses that were elaborated in the abovementioned sidebars. Choose a problem of your choice and include class diagrams and notes.

REFERENCE

1. W. M. Gombrowicz (August 4, 1904–July 24, 1969), last modified on June 23, 2017, https://en.wikipedia.org/wiki/Witold_Gombrowicz

3 Engineering Stable Atomic Knowledge Patterns

Software is a great combination between artistry and engineering.

Bill Gates [1]

Domain-independent knowledge is a class of knowledge that deals with notions and concepts that are untied to a specific domain. Hence, it is probable that this knowledge is available in developing systems within different domains. However, reusing domain-independent knowledge might be stalled, when this knowledge is presented as an integral part of domain-specific components. One possible solution to this problem is to identify those portions of knowledge that encapsulate domain-independent concepts and later separate them from the monolithic knowledge component. In this chapter, we will address the problem of reusing domain-independent knowledge by proposing a new class of stable patterns [2,3], called *Stable Atomic Knowledge* (SAK) patterns that includes two kinds of patterns—*Stable Design Patterns and Stable Analysis Patterns* (SAPs) [4]. An SAK pattern encapsulates an atomic notion and presents it in an appropriate abstraction level. Consequently, the SAK can be reused to model this notion whenever it appears, regardless of the domain or the application. We will also present some main properties that characterize these patterns, as well as the high-level engineering process of developing them. In addition, we will demonstrate the proposed concept through numerous, illustrative examples.

3.1 INTRODUCTION

Knowledge artifacts can be broadly classified into two main categories: domain specific and domain independent. Domain-specific knowledge artifacts present the knowledge of notions and concepts that are tied to a specific domain; thus, they are useful in modeling different applications within the same domain. Conversely, domain-independent knowledge artifacts encapsulate the knowledge of notions that are untied to specific application domains and, hence, they can be reused to model the same notions, wherever these notions appear, regardless of the domain or the application they might appear.

The last few decades have witnessed a significant evolution of several software reuse communities, including Model-Based software, Component-Based Software Engineering (CBSE), and software patterns. The ultimate goal of these different reuse communities is to enable the concept of reuse within the different phases of software development.

Different reuse communities convey the reuse message by using different approaches. Nevertheless, the resultant reusable artifacts, whether they are models, components, or patterns, are usually developed for specific applications or domains. They are useful for development within the same applications' domain that they appeared. Consequently, these reusable elements, in most cases, encapsulate a fusion of both domain-specific and domain-independent knowledge. This mixture of knowledge makes it difficult for developers to reuse domain-independent knowledge across domains. By atomizing monolithic knowledge artifacts into smaller atomic components that deal with practical domain-independent problems, we will be able to broaden the reuse of these knowledge components, and hence, enable the knowledge reuse across domains. Achieving such wide reusability is expected to improve both time and cost of software development.

Stable patterns [2,3] is a new approach that has been proposed recently for developing patterns based on software stability concepts [5–8]. The goal of stable patterns is guaranteeing stability.

As a result, these stable patterns could be used to deal with the same problem wherever it appears. A brief background for the basic concepts of software stability and stable patterns is provided in the next section.

This chapter will try to address the problem of reusing domain-independent knowledge across different domains. In particular, it specifically focuses on answering the following question: *How do one present domain-independent notions in atomic reusable knowledge components, such that these components can be used to present the notions they encapsulate regardless of the domain?* This chapter will also try to provide an answer to this question by introducing a new class of stable patterns called SAK patterns. SAK patterns are reusable artifacts that can be used to represent the core knowledge of a concept or notion that appears within a wide spectrum of domains. A comprehensive definition of SAK patterns is given in Section 3.3.

The proposal of atomizing knowledge into small reusable components has motivated other researchers and practitioners. In Reference 9, the concept of a Knowledge Center™ is introduced. A Knowledge Center™ is a network of nodes and relationships that contain available knowledge within a particular system. Each node stores a small component that represents a useful concept or idea. Each system has its own unique network of nodes that are related to the nature of the project. For instance, for a given class its attributes can be broken down and each attribute can then form a node in the network. To instantiate the class that is presented with the nodes, each node (attribute) will receive a value, and the combination of all the nodes with their values can form the instance. However, this work does not address the idea of building knowledge components that can be reused across domains. Instead, it always focuses on small-scale knowledge elements (e.g., class attributes) that are locally reused within a project or system and they not necessary useful across domains.

The remainder of this chapter is organized as below: An overview of software stability concepts and stable patterns is provided in Section 3.2 while the concept of SAK patterns is elucidated in Section 3.3. In Section 3.4, we will present and discuss the essential properties of SAK patterns. Section 3.5 gives a high-level iterative approach for developing SAK patterns. An example of SAK patterns is discussed in Section 3.6 and we will conclude in Section 3.7.

3.2   SOFTWARE STABILITY AND STABLE PATTERNS: AN OVERVIEW

Software stability approach [5–8] is a layered approach for developing software systems. In this approach, the classes of the system are classified into three layers: the Enduring Business Themes (EBTs) layer [8], the Business Objects (BOs) layer, and the Industrial Objects (IOs) layer. Figure 3.1 depicts the layout of the three layers of the stability approach.

Based on its nature, each class in the system is classified into one of these three layers. EBTs are the classes that present the enduring and core knowledge of the underlying industry or business. Therefore, they are extremely stable and form the nucleus of the SSM. BOs are the classes that map the EBTs of the system into more concrete objects. BOs are semiconceptual and externally stable,

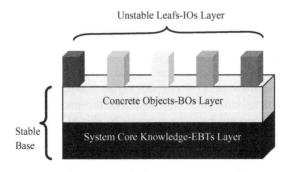

FIGURE 3.1   Software stability approach layers.

but they are internally adaptable. IOs are the classes that map the BOs of the system into physical objects. For instance, the BO "Agreement" can be mapped in real life as a physical "Contract," which is an IO. The detailed properties of EBTs, BOs, and IOs are discussed in References 5–7.

The SAP methodology introduced in References 2 and 3 is an innovative approach for developing patterns by utilizing software stability concepts. The concept of SAPs was proposed as a solution for the limitations of contemporary analysis patterns [2].

The goal of SAPs is to develop models that encapsulate the core knowledge of the problem and present it in terms of the EBTs and the BOs of that problem. Consequently, the resultant pattern will inherit the stability features; hence, they can be reused to analyze the same problem whenever it appears.

Later on, this chapter also tries to generalize the basic theory of SAPs to accommodate design patterns. This generalization will eventually lead to the introduction of the new concept of *Stable Design Patterns*, and later to the more general term of stable software patterns. For the purpose elucidating this chapter, we will not differentiate between analysis and design patterns; instead, we will the generic term stable patterns.

3.3 SAK PATTERNS CONCEPT

A stable pattern, whether it is an analysis or design pattern, can contain one or more EBTs. Based on the number of EBTs they embody, stable patterns can be broadly classified into two main categories: *stable architectural patterns* and *SAK patterns*. An architectural pattern should contain more than one EBT, whereas an SAK pattern should contain just one EBT. An Architectural stable pattern usually addresses a larger problem. On the other hand, an SAK pattern focuses on a smaller problem. Figure 3.2 depicts stable patterns concept and their categories. The dotted line indicates the possibility of using one or more SAK patterns in developing architectural patterns.

To understand the concept of SAK patterns, we might consider a simple example of an account problem. With the emergence of new applications, the word "account" alone has become a vague concept, if it is disjointed with a word that is related to a certain context. For example, besides all of the traditional well-known business and banking accounts, now we have e-mail accounts, online shopping accounts, online learning accounts, subscription accounts, and many others kinds of accounts.

Building a separate account model for different account kinds might not be efficient, especially if we realize that these accounts share some basic aspects and attributes. Encapsulating the common

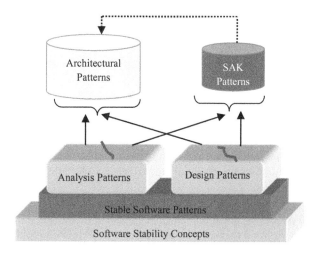

FIGURE 3.2 Stable patterns concept and categories.

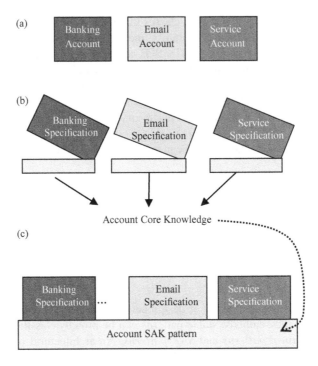

FIGURE 3.3 SAK patterns concept: (a) Different independent account models. (b) Separating the common knowledge of account from specific specifications of each account kind, and present it as account SAK pattern. (c) Using the account SAK pattern in developing different account kinds.

aspects of any account into one SAK pattern, named the account SAK pattern, is beneficial in reducing cost, time, and effort for developing current and other emerging account patterns.

Figure 3.3 illustrates the concept of developing the account *SAK pattern*. Figure 3.3a presents the conventional way of developing different accounts, where each account model is built separately, and serves as a domain-specific model. Figure 3.3b shows how the separation of the account core knowledge from the different independent accounts models. The separated common knowledge can form a solid base for reuse. In Figure 3.3c, the common extracted core knowledge is used to develop the account SAK pattern, which can be utilized as a foundation to model the different account kinds.

Several different patterns that address the account problem have been introduced in the literature [10–12]. Though these patterns offer valuable experience in their prospect, they are restricted to specific applications and domains; thus, they do not allow for a broad reuse, which limits the reusability of the core knowledge they encapsulate.

3.4 ESSENTIAL PROPERTIES OF SAK PATTERNS

An SAK pattern is defined as "A stable pattern that has a single EBT and models a well-defined and focused domain-independent recurring problem such that it can be easily extended, modified, or linked to build complex applications or other architectural stable patterns."

This above definition embodies the essential properties of SAK patterns. An SAK pattern should satisfy the following properties:

1. *Focus*: An SAK pattern is not intended to represent a model for a complete system or combined problem; rather it focuses on a specific and well-defined problem that commonly appears within larger problems. Complex systems may be composite of several SAK patterns, or even several architectural patterns. The generality of SAK patterns is directly

related to the focusing level of the problem. If an SAK patterns an overly broad portion of a system, the generality of resulting model is sacrificed—the maxim holds. The probability of the occurrence of all the problems together is less than the probability of the occurrence of each problem individually.

2. *Stable*: Stability influences pattern reusability. As they consist of the EBTs and BOs of the problem they model, SAK patterns inherent the stability features of these two artifacts. IOs deal with the implementation details and hence they change according to the application and the domain. A key feature in SAK patterns is that they separate the implementation details (IOs) from the core knowledge of the system (EBTs and BOs), which increase the pattern stability.

3. *Natural*: SAK patterns are indented to be used in a broad range of applications, and by different domain expertise. Moreover, the same word might carry different implications even within the same domain. Therefore, it is essential that an SAK pattern is presented and described in a simple and natural language than can reflect the knowledge it embodies.

4. *Domain independent*: From the definition of SAK patterns, they should model domain-independent recurring problems.

5. *Abstract*: To accomplish the adequate level of generality, an SAK pattern should be put in an appropriate level of abstraction. This property is essential to ensure broad model reusability. Generality means that a pattern that models a specific problem is easily used to model the same problem whenever it appears. Abstraction means that the SAK pattern structure and the class naming and convention should be carefully chosen so that the model becomes untied to a specific domain or application.

6. *Easy to use*: As any other reusable components, SAK patterns should be developed and presented in such a way that makes them easy to understand and use.

7. *Single EBT*: Each SAK pattern should contain one EBT. By constraining the number of EBTs to one, this helps in focusing on one problem at a time. Multiple EBTs indicate several problems, which contradicts the defined structure of SAK patterns.

3.5   ENGINEERING PROCESS FOR DEVELOPING SAK PATTERNS

This section gives the high-level details of the process we propose for developing SAK patterns. The process consists of four main phases: *Define the problem*, *Identify the EBT and BOs*, *Construct SAK pattern*, and *verify the SAK pattern*. The high-level engineering process is shown in Figure 3.3. The solid lines in Figure 3.3 indicate the normal directions of the construction process, whereas the dotted lines represent the iterative directions of the process. This process is iterative in the sense that it allows the repetition of one or more phases during the development, as shown in Table 3.1.

In the first phase, the problem is clearly identified. If necessary, the defined problem is further divided into smaller isolated problems. A critical issue in this phase is to ensure that the atomic problem is useful in practical situations, that is, it is important that the identified atomic problem presents a real-life problem that is known to be recurring in different situations. This condition prevents the development of useless and/or meaningless patterns that model unnecessary details. The *ultimate* output of this phase would be a well-defined problem with clear boundaries. (Due to the difficulty of this phase in practice, it is expected that when this phase is first conducted, the resultant problem is not well defined yet; thus several iterations in this phase usually take place.)

The second phase deals with a crucial step in the development process, which is the identification of the EBT in the problem and its BOs. The difficulty of this phase arises from the strict property of SAK patterns in containing only one EBT. Practical experience shows that coming up with this single EBT needs several iterations. The typical procedure of achieving this goal is through identifying a list of few (usually 3–4) candidate EBTs, then conduct several refinement iterations until the most appropriate EBT is chosen, an example of this procedure is given in Reference 3.

TABLE 3.1
Engineering SAK Patterns Phases

Development Phase	Phase Final Production	Heuristics Proposed in Related Work
Define the problem	• Problem statement with well-defined bounders	[6]
Identify EBT and BOs	• EBT	[6,10–12]
	• A list of BOs	
Construct SAK model	• SAK model with defined relationships	[6,10–12]
Verify SAK pattern	• Sample of applications models	[6,7,9–12]
	• Simple stable patterns	
	• Architectural patterns	

In the third phase, all the identified elements from phase two are integrated to form the ultimate SAK Patterns. In this phase, the relationships (associations, aggregation, inheritance, etc.) that depict the interaction between the different elements of the model, along with the roles and multiplicity on each relationship should be defined.

Phase four is concerned with the verification issue for the developed SAK pattern. The main objective of this phase is to validate the constructed SAK patterns to ensure that it satisfies the essential properties discussed in the previous section. One way of doing so is to examine the model into different scenarios, and then verify that the model can represent the core knowledge needed in each scenario. In addition, one can exercise the extension of the model by building on its top to construct more complex models that can be used in specific applications or domains. This extension is expected to be relatively simple with no major modifications in the pattern, and without any changes in the pattern structure (i.e., the extension should not force a change in a relationship between two classes in the SAK pattern). Such major changes indicate something incorrect or inaccurate in the developed pattern (Figure 3.4).

In practice, it is almost impossible to develop an SAK pattern without iterations. This is due to two main challenges. The first challenge is that defining the boundary of the problem that the SAK should deal with is not usually straightforward. The second challenge is due to the identification of the EBTs and BOs of the problem. Our experience reveals that understanding the general characteristics of the EBTs, and BOs might be straightforward to some extend; however, identifying them

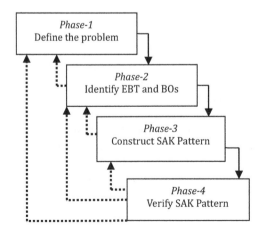

FIGURE 3.4 High-level steps of developing SAK patterns.

in real problems is not so. Usually, group discussion and several iterations are needed to identify these elements in the given problem. For SAK patterns, identifying a single EBT for the problem is even harder and usually involves an extensive group discussion. In References 1, 3, and 7, we highlighted these problems with some examples. Unfortunately, no systematic procedures exist that one can use to identify EBTs, BOs, and IOs. However, several useful heuristics have been developed in References 3, 7, and 8.

3.6   EXAMPLE OF THE SAK PATTERNS: ANYPARTY SAK PATTERNS

In this section, we illustrate the concept of SAK patterns through an example. The SAK pattern we present here is called the *AnyParty* SAK pattern. The objective of this pattern is to provide a generic model that captures the core knowledge about any party or group.

Many factors make the modeling of a party interesting yet, difficult at the same time. In almost every application, some kind of party is usually involved (for instance, banking applications, web applications, and negotiation applications). These all involve an interaction of different kinds of parties. Keeping in mind the very different nature of these applications, the AnyParty SAK pattern should be abstract enough to form a base for modeling any kind of party in any application.

In addition, a party within the same application can play different roles. For instance, the buyer and the seller in a trading application are different parties with different roles. The model should handle whatever role exists in the application.

Moreover, different parties do share common interests, which is the reason why these parties are interacting with each other. For instance, the members of a political party do share the same general view or vision upon which they are grouped into this specific political party (despite the fact that each member has his own beliefs or opinions). The pattern should indicate inappropriate level of abstraction the common interest of specific party to exist.

Figure 3.5 shows the AnyParty SAK pattern. It consists of one EBT; *Orientation*, one BO; *AnyParty*, and *n* IOs. Note that the prefix "Any" is used to indicate the generality of the class and its ability to virtually presents any kind of parties that might be encountered in any application. *Orientation* describes the common motive by which the party's members are grouped. *AnyParty* defines the different possible roles within a specific party.

One key factor that increases the flexibility of the model is the use of the IOs—*Role*. This usage helps in developing a limited size for AnyParty pattern, whilst gives the flexibility of scaling up and down the model according to the application without affecting the structure of the model itself by adding the required number of roles.

The structure of the AnyParty pattern can be examined against each of the challenges discussed early in this section. Moreover, we can verify that this model stratifies all the essential properties of SAK patterns we have discussed in Section 3.3.

To demonstrate the usability of the AnyParty SAK pattern, we applied the pattern in the development of several applications within different domains. Figure 3.6 shows the application of the

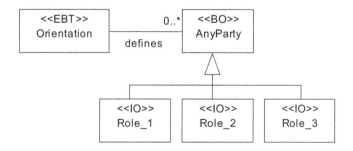

FIGURE 3.5   AnyParty SAK pattern.

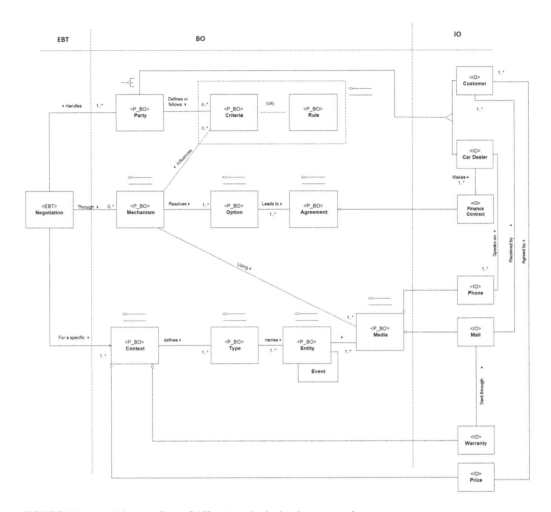

FIGURE 3.6   Applying AnyParty SAK pattern in the buying a car subsystem.

AnyParty pattern into a subsystem of buying an automobile. Figure 3.7 shows the usage of the pattern in another application related to Internet content negotiation [2,13].

Moreover, the AnyParty SAK pattern can be used to construct other stable patterns by incorporating it into a newly developed pattern. We have used the AnyParty pattern in developing the Negotiation SAP show in Figure 3.8 [13,14].

3.7   SUMMARY

In this chapter, we presented the concept of SAK patterns as a solution for enabling the reuse of domain-independent knowledge across domains. An SAK pattern captures the core knowledge of a recurring problem, and it can be used to model this problem wherever it appears, regardless of the application it appears. We also provided and described the high-level engineering process for developing SAK patterns. In order to demonstrate the proposed concepts, we developed the AnyParty SAK pattern, which presents the core knowledge of the party problem. The developed pattern was used in the development of different applications and in the development of other stable patterns as well. The concept of SAK is a promising approach to enable a broad reusability of software patterns within and across domains.

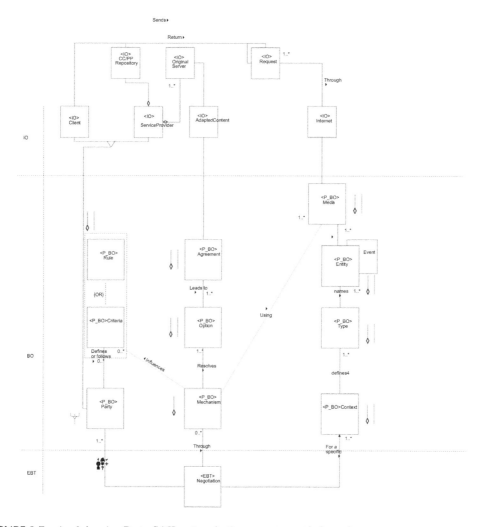

FIGURE 3.7 Applying AnyParty SAK pattern in the content negotiation subsystem.

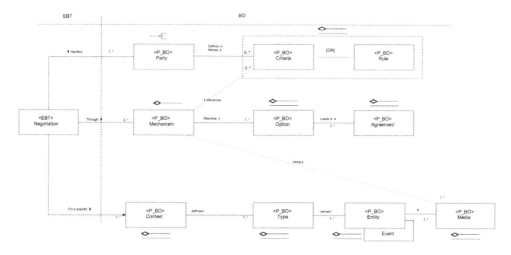

FIGURE 3.8 Negotiation SAK and applying AnyParty SAK pattern in developing the negotiation SAPs.

3.8   REVIEW QUESTIONS

1. What are Stable Atomic Knowledge (SAK) Patterns?
2. How SAK can be reused to model any type of applications?
3. What is a Knowledge Center?
4. What is software stability approach?
5. Can you conceptualize SAK Patterns?
6. Define their advantages.
7. What are some of the essential properties of SAK patterns?
8. What are the different phases of creating an SAK pattern?
9. What are the basic engineering processes for creating SAK patterns?
10. Give three examples of SAK patterns.
11. Can you summarize what you have learnt through this chapter?
12. Can you name two patterns that are included within the domain of SAK patterns?
13. An SAK pattern is not intended to represent a model for a complete system or combined problem—True or False.
14. To accomplish the adequate level of generality, an SAK pattern should be put in an appropriate level of abstraction—do you agree?

3.9   EXERCISES

1. Create three different types of SAK patterns. Provide a schematic diagram for each of them and explain the entire process in detail.
2. Create an SAK pattern use scenario for business negotiation system.

3.10   PROJECTS

1. Define and highlight a problem domain of your choice and create a flow process for creating an SAK pattern.
2. Create a Party SAK model. Here, the word Party could relate to house warming party, bachelor's party, reunion party, or even a cocktail party. Define the word party with respect to each scenario while creating the model.

REFERENCES

1. William Henry Gates III (born October 28, 1955), last modified on June 19, 2017, https://en.wikipedia.org/wiki/Bill_Gates
2. H. Hamza. A foundation for building stable analysis patterns, MS thesis, Department of Computer Science, University of Nebraska-Lincoln, Lincoln, NE, 2002.
3. H. Hamza and M.E. Fayad. A pattern language for building stable analysis patterns. *9th Conference on Pattern Language of Programs (PLoP02)*, Illinois, September 2002.
4. M.E. Fayad. *Stable Analysis Patterns for Software and Systems*, Auerbach Publications, Boca Raton, FL, Taylor & Francis Catalog #: K24627, May 2017, ISBN-13: 978-1-4987-0274-4.
5. M.E. Fayad and A. Altman. Introduction to software stability, *Communications of the ACM*, 44(9), 2001, 95–98.
6. M.E. Fayad. Accomplishing software stability, *Communications of the ACM*, 45(1), 2002, 111–115.
7. M.E. Fayad. How to deal with software stability, *Communications of the ACM*, 45(4), 2002, 109–112.
8. M. Cline and M. Girou. Enduring business themes, *Communications of the ACM*, 43(5), 2000, 101–106.
9. Cagito, Inc. Concept Summary: Atomic Knowledge, available: http://www.cogitoinc.com
10. E. Fernandez and Y. Liu. The account analysis pattern, in *Proceedings of the 7th European Conference on Pattern Languages of Programs (EuroPlop02')*, July 2002.
11. IBM Corp. Patterns for e-business, available: http://www-106.ibm.com/developerworks/patterns/

12. M. Fowler. *Analysis Patterns: Reusable Object Patterns*, Addison-Wesley, Reading, MA, 1997.
13. H. Hamza and M.E. Fayad. The negotiation analysis patterns, *8th European Conference on Pattern Languages of Programs (EuroPLoP 03)*, Irsee, Germany, June 2003.
14. M.E. Fayad. *Stable Analysis Patterns for Software and Systems*, Auerbach Publications, Boca Raton, FL, Taylor & Francis Catalog #: K24627, May 2017, ISBN-13: 978-1-4987-0274-4.

SIDEBAR 3.1 EXTRACTING DOMAIN-SPECIFIC AND DOMAIN-INDEPENDENT PATTERNS

Extracting domain-specific and domain-neutral patterns is a challenge for both expert and novice software engineers. No mature guidelines or methodologies exist for extracting patterns. Software stability model (SSM) proposed in Reference 1 provides a base for extracting the core knowledge of the domain and identifying atomic notions that can be thought of as patterns by extracting another level of abstraction using software stability. This poster presents the concept of extracting both domain-specific and domain-neutral patterns from systems that are built using software stability concepts.

Introduction

Theoretically, patterns seem to be obvious and can be easily located within the developed system. By observing considerable number of systems, developers can spot and extract their patterns [2]. However, in practice this approach does not scale well, since many developers fail to extract the patterns this way.

Domain-specific patterns are defined here as patterns that capture the core knowledge of specific applications; therefore, they can be reused to model applications that share the core knowledge. On the other hand, domain-neutral patterns are those models that capture the core knowledge of atomic notions that are untied to specific application domains; hence, can be reused to model the same notions whenever they appear in any domain.

In SSM, the model of the system is viewed as three layers (Figure SBF3.1): the Enduring Business Themes (EBTs) layer, the Business Objects (BOs) layer, and the Industrial Objects (IOs) layer. EBTs are the classes that present the basic knowledge of the underlying industry or business. Therefore, they are extremely stable. BOs are the classes that map the EBTs of the system into more concrete objects. BOs are tangible and externally stable but they are internally adaptable. IOs are the classes that map the BOs of the system into physical objects. For instance, the BO "Agreement" can be mapped in real life as a physical "Contract," which is actually an IO. The practitioner should consult [3–5] for heuristics to help identify EBTs, BOs, and IOs, and [12,7,5] for examples of building systems using SSM.

This poster presents the concept of extracting both domain-specific and domain-neutral patterns from systems that are built using software stability concepts.

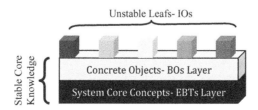

FIGURE SBF3.1 Three layers of SSM.

Extracting Patterns Using Stability Concepts

The concept of extracting stable patterns from SSM is made possible because of the stable characteristic of both the EBTs and BOs. Systems that share the same domain have commonalities in their models. As mapping to software stability concepts, the model that captures the common aspects of some applications is a model that combines both the EBTs and the BOs of these applications. This submodel forms a stable domain-specific pattern for these systems.

To obtain stable domain-neutral patterns from the stability model of one problem, we need to obtain a second level of abstraction for each EBT and BO in the problem stability model. This second level of abstraction is obtained by modeling each EBT and BO using stability concepts. Each new stable model will stand as a pattern by itself. This pattern is focused on a specific problem and can be reused to model this problem whenever it appears; therefore, it is considered a domain-neutral pattern. Stable domain-neutral patterns can be further classified into stable analysis pattern (EBT models) [2,7,9,11,12], and stable design pattern (BOs models). For instance, the stable model that represents the EBT *Satisfaction* is a stable analysis pattern. Similar reasoning can be applied to BOs: The stable model representing the *Account* BO is a stable design pattern.

Examples of Extracting Domain-Specific Patterns

Renting applications have core aspects that are common and independent of the nature of the rented item. Therefore, capturing these core aspects in a single model is beneficial for developing different renting systems without starting from scratch. Figure SBF3.2 gives part of the system for renting a car as modeled using the SSM concept. The EBTs, BOs, and their relationships in this system form a renting pattern that can be reused as a base for modeling the rental of any entity [8]. The extracted renting patterns are given in Figure SBF3.3.

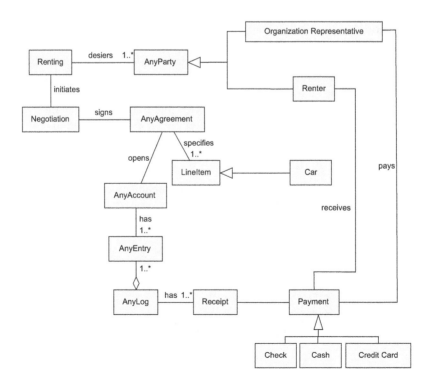

FIGURE SBF3.2   Stable object model for car renting.

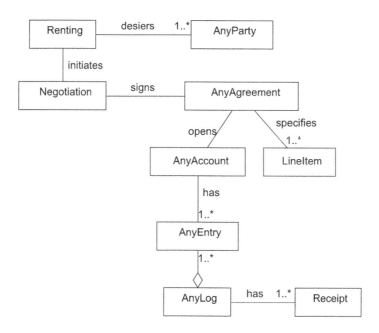

FIGURE SBF3.3 *Renting* pattern object model.

Examples of Extracting Domain-Independent Patterns

The EBT *Negotiation* could be used as a catalyst for building a stable analysis pattern by generating its second level of abstraction. This can be achieved by modeling this EBT further using software stability concepts. The resultant pattern is the *Negotiation* stable pattern (Figure SBF3.4). In this model, AnyParty is a stand-alone stable pattern that models the party notion; hence, it can be used to model any party in any application. The full documentation of this pattern can be found in Reference 10. Extraction of stable design patterns follows the same reasoning and its description can be found in Reference 8.

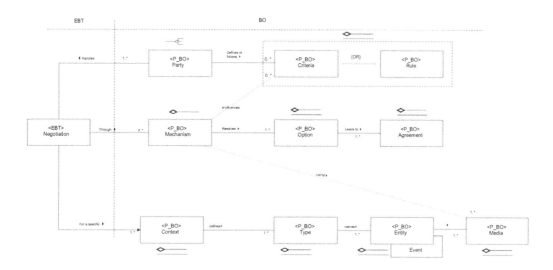

FIGURE SBF3.4 *Negotiation* pattern object model.

References

1. M.E. Fayad. Accomplishing software stability, *Communications of the ACM*, 45(1), 2002, 111–115.
2. E. Gamma et al. *Design Patterns: Elements of Reusable Object-Oriented Software*, Addison-Wesley Professional Computing Series, Addison-Wesley, New York, 1995.
3. M.E. Fayad and A. Altman. Introduction to software stability, *Communications of the ACM*, 44(9), 2001, 95–98.
4. M.E. Fayad. How to deal with software stability, *Communications of the ACM*, 45(4), 2002, 109–112.
5. H. Hamza and M.E. Fayad. A pattern language for building stable analysis patterns, in *Proceedings of Pattern Language of Programs 2002 (PLOP02)*, Monticello, IL, September 2002.
6. M.E. Fayad and S. Wu. Merging multiple conventional models into one stable model, *Communications of the ACM*, 45(9), 2002, 102–106.
7. H. Hamza. A Foundation for building stable analysis patterns, Master thesis, University of Nebraska-Lincoln, 2002.
8. H. Hamza, A. Mahdy, M.E. Fayad, and M. Cline. Extracting domain-specific and domain-neutral patterns using software stability concepts, *9th International Conference on Object-Oriented Information Systems (OOIS 03)*, Geneva, Switzerland, September 2003.
9. H. Hamza. Building stable analysis patterns using software stability, *4th European GCSE Young Researchers Workshop 2002 (GCSE/NoDE YRW 2002)*, Erfurt, Germany, October 2002.
10. H. Hamza and M.E. Fayad. The negotiation analysis pattern, *Eighth European Conference on Pattern Languages of Programs (EuroPLoP 03)*, Irsee, Germany, July 2003.
11. M. Cline and M. Girou. Enduring business themes, *Communications of the ACM*, 43(5), 2000, 101–106.
12. H. Hamza and M.E. Fayad. Model-base software reuse using stable analysis patterns, *ECOOP 2002, Workshop on Model-based Software Reuse*, Malaga, Spain, June 2002.

4 Stable Analysis and Design Patterns
Unified Software Engine

> The way is long if one follows precepts, but short ... if one follows patterns.
>
> **Lucius Annaeus Seneca [1]**

Stable analysis patterns are conceptual models that capture the core knowledge and concepts of the problems they model. In stable analysis patterns, concepts and knowledge are presented in abstraction levels that make them reusable whenever the problem occurs. Driving the design model from analysis and conceptual models is essential to develop the required system. In this chapter, we propose an approach that transforms a stable analysis pattern into a design model by composing the appropriate set of design patterns. The approach can be generalized to transform a collection of analysis patterns (i.e., analysis models) to the system design. We highlight the approach and illustrate its applicability through an example.

4.1 INTRODUCTION

In software development, conceptual models are usually used to develop a good understanding of the main concepts in the problem domain and their relationships. In addition, these models facilitate the communication between developers, stakeholders, and domain experts. Understanding the core concepts of problem domain and the way in which they interact with each other is a key task in developing a system that meets the stakeholders' expectations and requirements.

Unfortunately, developing domain models can be very time consuming and costly. This is because developing conceptual models requires domain knowledge as well as modeling skills; two aspects that novice and even experienced developers may lack. In addition, in many cases, domain experts need to be involved in order to explain, clarify, and verify the core concepts in the domain, which adds to the overall time and cost of the development process.

Domain models, however, should not necessarily be developed from scratch. Indeed, experienced developers naturally use their knowledge about the domain and their experience from similar and related projects to develop conceptual models. If such experience can be documented and communicated among developers, both cost and time of developing conceptual models can be significantly reduced.

Software patterns have emerged as a promising technique for documenting and communicating experience among developers. Patterns can improve quality and condense cost and time of software development [2,3]. In general, a pattern can be defined as "An idea that has been useful in one practical context and will probably be useful in others" [4,5]. Analysis patterns are conceptual models that can be reused to analyze similar and related problems, and hence, they provide a valuable resource for domain modeling.

Stable analysis patterns [6,7] are analysis patterns that separate domain concepts from specific business requirements. This leads to analysis models that are more general, and hence more suitable for modeling a wide range of problems [6]. In addition, stable analysis patterns provide the base for evolving system design and later on for implementation. This alignment between analysis, design, and implementation can improve traceability that has a direct impact on the evolution of the system in the future.

This chapter proposes a new approach to develop design models using stable analysis pattern and design patterns. The approach transforms a stable analysis patterns into several design components. Each design component can be then developed *separately* using existing design patterns.

The reminder of this chapter is organized as follows. Section 4.2 discusses some related work. A brief overview of this concept of stable analysis patterns is given in Section 4.3. The proposed approach is presented in Section 4.4, and an example to demonstrate the approach is given in Section 4.5. Section 4.6 presents the conclusions.

4.2 DEVELOPING SOFTWARE WITH PATTERNS

Despite the existence of analysis and design patterns for more than a decade, less attention has been devoted to integrate the use of analysis patterns into the traditional software development life cycles. Moreover, there has been no attempt to integrate the use of patterns across the different development phases. Existing techniques for using patterns to develop software systems are focused on selecting and integrating architectural and design patterns to evolve systems' models. Analysis phase, on the other hand, is performed using traditional techniques.

To our knowledge, the Pattern-Oriented Software Analysis and Design (POAD) method proposed in Reference 5 is the only approach that emphasizes the use of patterns in both analysis and design. POAD performs analysis by viewing the system as a collection of components and identifying responsibilities and functionalities of each of these components. The design step is conducted by selecting different design patterns to implement the different identified components. The main difference between our approach and the POAD approach is that our approach uses analysis patterns to the analysis and modeling of the problem, while POAD does not.

4.3 STABLE ANALYSIS PATTERNS

Most existing analysis patterns model the knowledge of the domain in a specific business context in which these patterns have been observed. This leads to analysis models that contain a mix of the domain core knowledge and some business requirements that are specific to the current problem. Existing analysis patterns are therefore hard reuse for modeling the same problem but with different business requirements [6].

Stable analysis patterns, proposed in References 6 and 7, try to alleviate this problem by separating domain concepts from any specific business requirements. This can lead to analysis patterns that are more general, and hence, more reusable [6]. To illustrate this, consider the concept of *negotiation*. Conceptually, negotiation can be defined as a decision process that compromises the individual decisions of two or more parties. This definition is untied to a specific domain or certain application. When applying the negotiation concept in a particular application or domain, for example, the e-commerce domain, the conceptual definition of negotiation should be specialized to the nature of the domain. For instance, the two parties in the negotiation definition can be two software agents that automatically perform the negotiation process. Other domains will probably have different nature of parties and negotiation activities; nonetheless, the core concept of negotiation does not change from one domain to another. In this case, negotiation is a *domain-neutral* knowledge that can be used across domains.

To achieve the required abstraction level, stable analysis patterns apply the concept of SSM [8,9]. SSM is a generic modeling approach that stratifies the classes of the system into three layers: the EBTs layer, the BOs layer, and the IOs layer. Each class in the system model is classified into one of these three layers according to its nature.

EBTs are the classes that present the enduring and basic knowledge of the underlying industry or business. Therefore, they are extremely stable and form the nucleus of the SSM. BOs are the classes

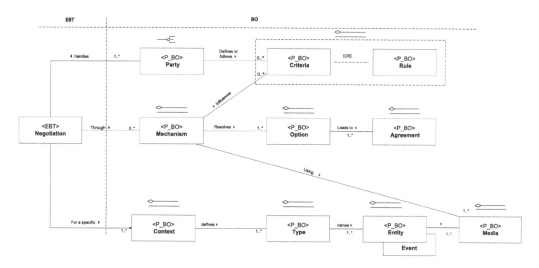

FIGURE 4.1   The Negotiation stable analysis pattern [9].

that map the EBTs of the system into more concrete objects. BOs are tangible and externally stable, but they are internally adaptable. IOs are the classes that map the BOs of the system into physical objects. For instance, the BO "Agreement" can be mapped in real life as a physical "Contract," which is an IO.

Figure 4.1 shows the Negotiation stable analysis patterns [10]. The pattern captures the core concepts that are involved in any kind of negotiation.

4.4   PROPOSED USE

In the proposed approach, we start with a stable analysis pattern (or a model that is composed of several stable analysis patterns), and use the knowledge captured in the analysis pattern to guide the selection of the appropriate design patterns to develop the design model. The USE contains the following main activities (Figure 4.2):

1. Extract association relationships between EBTs and BOs, and identify responsibilities associated with these relationships.
2. Identify the nature of the design pattern that can be used to realize the identified responsibilities.
3. Identify specific design patterns to be used.
4. Construct the identified design patterns to develop a design model.

In the following, we present the detailed steps in the USE:

Step 1 Association Extraction: Stable analysis patterns are modeled using the Unified Modeling Language (UML) [11]. As in any traditional UML class diagram, associations between different classes are annotated with *association names* as shown in Figure 4.1. In our approach, these association names are used to identify the appropriate design that captures the interactions between the different concepts in the domain.

Step 2 Pattern Group Identification: Design patterns can be generally classified into three main categories [3]: *creational patterns*, *structural patterns*, and *behavioral patterns*. Creational patterns provide solutions for class or object instantiation. Structural patterns are design patterns that represent the relationships between entities. Behavioral patterns provide solutions for interactions between entities. In this step, each EBT–BO association in the analysis pattern is analyzed in order to identify which group of design patterns can

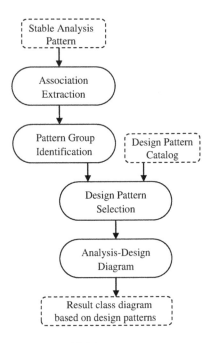

FIGURE 4.2   A process mode of the proposed USE.

be used to realize the semantic of this association. This point will be further elaborated in the case study (Section 4.6).

Step 3 Design Pattern Selection: Once we have identified the appropriate design pattern group for each EBT–BO association, we try to identify the most appropriate pattern within the identified group that can be used in the design.

Step 4 Analysis-Design Diagrams (ADD): The ADD diagram represents the transition diagram between the analysis pattern and the design model. In the ADD, one of the selected design patterns represents each BO that interacts with some EBT in the analysis pattern from Step 3 above. The resultant model has similar structure as the original analysis pattern except that instead of representing each BO as a class, it is now represented as a package that has an internal structure, as we will show in Section 4.4.

4.5   EXAMPLE

To illustrate our approach, we show how to apply it to the design of a negotiation system. For simplicity, we consider only portion of the design that is concerned with the negotiation functionality.

The Negotiation analysis pattern, as shown in Figure 4.1, can be used as part of the conceptual model of a negotiation system in any application. The pattern captures the core components of the negotiation process; however, depending on the application some other components need to be added.

In the following, we illustrate a systematic method to identify and integrate design patterns from the negotiation stable analysis pattern.

Step 1 Association Extraction: The Negotiation analysis pattern contains one EBT: Negotiation, and four BOs: AnyContext, AnyMedia, AnyAgreement, and AnyParty. To extract the associations between the EBT and BOs, we identified the following relationships:

a. AnyParty represents the parties involved in the negotiation process. Each involved party depends on a negotiation environment to conduct negotiation. The relationship between Negotiation and AnyParty can be described as *AnyParty handles Negotiation*.

b. AnyAgreement is the result of Negotiation, and for each negotiation process there is only one agreement that all involved parties agreed on. The relationship between Negotiation and AnyAgreement can be described as *Negotiation generates AnyAgreement*.

c. AnyMedia represents how the communication is conducted for the negotiation. Observe that there are usually multiple options for the same negotiation process. The relationship between *Negotiation* and *AnyMedia* can be described as *Negotiation uses AnyMedia*.

d. AnyContext represents the items to be negotiated. Usually a package includes multiple items. The negotiation parties may focus on the package as a whole, or leverage individual items in the package to obtain the best result. The relationship between Negotiation and AnyContext can be described as *Negotiation negotiates AnyContext*.

Step 2 Pattern Group Identification: After we have identified the association between the different BOs and the EBT in the analysis pattern, we now use this information to identify the type of design patterns that can be used to realize each of the identified relationships. The following are the different design groups for each BO-EBT relationship in our example:

a. *AnyParty handles Negotiation*. The "handles" association implies interactions between entities. Behavioral patterns can be a good candidate.

b. *Negotiation generates AnyAgreement*. The "generate" association strongly suggests creational patterns.

c. *Negotiation uses AnyMedia*. The term "uses" also suggest behavioral patterns.

d. *Negotiation negotiates AnyContext*. This association will not have a strong clue about the candidate for design pattern categories. However, from our analysis, we have identified that the process of negotiation involves the negotiation of a whole package, and the negotiation of individual items within the package. The package-to-individual item relationship suggests structural design patterns.

Step 3 Design Pattern Selection: This step is relatively easier after we have found the candidate design pattern category. For each BO, we select the most appropriate design pattern that realizes the EBT–BO relationship as follows:

a. *Mediator* pattern is a candidate for the association AnyParty handles Negotiation, in the sense that the negotiation parties do not have to have direct contact with each other, and the negotiation process can be mediated by a third party or a negotiation platform that all parties trust.

b. Among creational patterns, *Singleton* pattern fit into the scenario of "Negotiation generates AnyAgreement," since the result of negotiation should be fixed for each items negotiated. Among behavioral patterns, strategy pattern encapsulates interchangeable algorithms. Since the "uses" association suggest there are multiple options, we may apply strategy pattern for this association.

c. In the negotiation process, a group of items can be treated as a whole, or they can be treated individually. In the implementation, both the "set of items" and "individual item" should provide similar interface for the negotiation process. In structural patterns, *Composite pattern* serves the "part-whole" hierarchies and it is a good candidate in this scenario.

Step 4 ADD: Finally, we integrate the identified design patterns to the Negotiation analysis pattern to construct the design model. Figure 4.3 shows the role each class in the Negotiation analysis patterns when the design patterns are applied. We also show some reference IOs to make design patterns easy to understand.

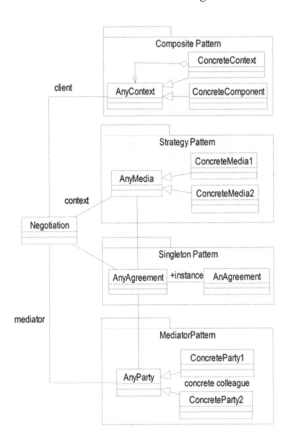

FIGURE 4.3　The ADD of the negotiation example.

4.6　SUMMARY

In this chapter, we propose a new approach to use analysis and design patterns to develop USE and software systems. The approach uses stable analysis patterns to capture the core concepts of the problem, then design patterns are used to transform analysis patterns into design modules. These design modules are integrated to form the final design of the system. We believe that this approach forms a promising step toward a *truly* pattern-driven software development methodology that uses both stable and design patterns to develop the system. Real-life systems, however, are complex and may involve large number of problems. To analyze these problems, several analysis patterns can be used. In such a case, our approach still applies as follows. First, the system can be decomposed into smaller problems, each of which can be modeled using stable analysis patterns. Next, the proposed approach is used to develop the design module of each analysis pattern separately. The relationships between the different analysis patterns are projected to integrate the different design modules (refer to Reference [12]).

4.7　REVIEW QUESTIONS

1. What is USE?
2. Write down some application areas of USE.
3. Give three examples of design patterns.
4. What is a mediator pattern?
5. What is a singleton pattern?
6. What is a composite pattern?

7. What is association extraction?
8. What are Stable Analysis Patterns?
9. How do you develop software with patterns?
10. Define EBT.
11. Define BOs.
12. Explain IOs.
13. What are the main goals of developing USE?
14. What are the preliminary steps in the process of designing a USE?
15. Create a simple process mode of the proposed USE.
16. Pick up a sample scenario and develop an USE model by using the above-mentioned process and explain the entire process in detail.
17. Summarized what you have learnt from this chapter.

4.8 EXERCISES

1. Create a basic stable analysis pattern on the concept of "debate."
2. Extract EBT, BOs, and IOs for "debate."
3. Design a process model for the proposed USE on the concept of "debate."

4.9 PROJECTS

1. Use the model process for creating a USE and create a model case for the concept of "debate."
2. Draw a detailed diagram for the process followed.
3. Discuss and explain the entire process in your own words.

REFERENCES

1. Lucius Annaeus Seneca (c. 4 BC–AD 65), last modified on June 5, 2017, https://en.wikipedia.org/wiki/Seneca_the_Younger
2. D.C. Schmidt, M.E. Fayad, and R. Johnson. Software patterns, *Communications of the ACM*, 39(10), 1996, 37–39.
3. E. Gamma et al. *Design Patterns*, Addison-Wesley, Reading, MA, 1995.
4. M. Fowler. *Analysis Patterns: Reusable Object Patterns*, Addison-Wesley, Reading, MA, 1997.
5. S. Yacoub and H. Ammar. *Pattern-Oriented Analysis and Design*, Addison-Wesley, Reading, MA, 2003.
6. H. Hamza. A foundation for building stable analysis patterns, Master thesis, University of Nebraska-Lincoln, Lincoln, NE, 2002.
7. H. Hamza and M.E. Fayad. Model-based software reuse using stable analysis patterns. *ECOOP 2002, Workshop on Model-Based Software Reuse*, Malaga, Spain, June 2002.
8. M.E. Fayad. Accomplishing software stability, *Communications of the ACM*, 45(1), 2002, 111–115.
9. M.E. Fayad and A. Altman. Introduction to software stability, *Communications of the ACM*, 44(9), 2001, 95–98.
10. M. E. Fayad. *Stable Analysis Patterns for Software and Systems*, Auerbach Publications, Boca Raton, FL, Taylor & Francis Catalog #: K24627, December 2015. ISBN-13: 978-1-4987-0274-4.
11. G. Booch et al. *The Unified Modeling Language User Guide*, Addison-Wesley, Reading, MA, 1999.
12. S. Singh. Unified software engine (USE), Master thesis, San Jose State University (SJSU), June 2013.

SIDEBAR 4.1 UNIFIED SOFTWARE ENGINES

The USE is a unified way of designing and building software solutions. It is the new and improvised, next generation of the solutions that are built on top of any common core infrastructure, for example, of SaaS infrastructures. It facilitates on-demand, highly reusable architectures, as well as applications, with very quick times to market, starting from requirements to final product delivery, while incorporating qualities such as scalability, adaptability, maintainability, and many more, right in the architecture of the software. The USE can be of any "type" domain independent that it can be applied on, for example, USE for Content Management or USE for Maintenance, etc. The term "USE" is a generalization of all these different kinds of USEs and it would refer to the system that satisfies all the core requirements that a system must fulfill. It has all the core capabilities that it must have to qualify as a "USE" and apply them to multiple applicable domains and one or more application contexts, for the kind of USE that it is [1].

Importance of the USE in advancing knowledge and understanding within its own field and across different fields:

1. *Unification, Unlimited, and Stable Reuse*: The USE also advances the field of Software Engineering, moving from an ecosystem of hundreds of different part-solutions and rather quick fixes, to real solutions that are holistic and unified in nature. It enables very high reuse of software over a time, thus saving on costs of maintenance in a very big way (Unified Software Engineering Reuse [USER] [2]).

2. *Domain Independence and Unlimited Applicability*: USE can be utilized to build systems that can be used to support numerous functions in any domain and it not necessarily tied to a single application as such. Based on the principles of SSM [3–5], it is not tied to any specific domain or application context. It is built based on the conceptual knowledge rather than application contexts. Any USE can be divided into two main parts—an inner generic stable core and an outer layer. The core contains the core knowledge for the engine and it is stable over time, that is, it need not be changed repeatedly, such as a search engine [6,7]. Since the core remains stable, it can be reused in multiple scenarios by changing a small percentage of the system, that is, just the application specific outer layer of the system, and to suite it to another context, without reinventing the wheel.

   The USE is designed in a way that makes it extremely adaptable in nature, that is, separation of concerns between functionality sets is seamless and hassle free. Each part of the functionality can be taken out or some other can be plugged into the USE core thus scaling it in whatever manner in which it is required. Moreover, a generic stable core enables it to be it and adaptive to a change in the application context by changing the application context specific outer layer of the software. All of this is possible, since it is designed based on the foundation of SSM. This positive feature adds a number of benefits, such as longevity, high returns on investments, configurability, customizability, and much more.

References

1. S. Singh. Unified software engine (USE), Master thesis, San Jose State University (SJSU), June 2013.
2. C. Flood III. Unified software engineering reuse, Master thesis, San Jose State University (SJSU), March 2017.
3. M.E. Fayad and A. Altman. Introduction to software stability, *Communications of the ACM*, 44(9), 2001, 95–98.
4. M.E. Fayad. Accomplishing software stability, *Communications of the ACM*, 45(1), 2002, 111–115.
5. M.E. Fayad. How to deal with software stability, *Communications of the ACM*, 45(4), 2002, 109–112.
6. Alphaworks.ibm.com. alphaWorks Community. Alphaworks.ibm.com. 2009-10-20, retrieved, 2013-09-17.
7. S. Lawrence and C. Lee Giles. Accessibility of information on the web, *Nature*, 400(6740), 1999, 107–109, doi: 10.1038/21987. PMID 10428673.

Section II

SDPs' Detailed Documentation Template

This part consists of eight chapters and four sidebars.

Each chapter presents the detailed documentation pattern that consists of the following sections: (1) the name and type of the pattern, (2) context, (3) the functional and nonfunctional requirements, (4) challenges and constraints, (5) solution, (6) consequences, (7) applicability, (8) related patterns, quantitative and qualitative measurements, (9) modeling issues, (10) implementation issues, (11) formalizations, (12) testing issues, (13) business issues, and (14) common usages. The chapters presented in this part are

1. AnyActor Stable Design Pattern (Chapter 5)
2. AnyParty Stable Design Pattern (Chapter 6)
3. AnyEntity Stable Design Pattern (Chapter 7)
4. AnyData Stable Design Pattern (Chapter 8)
5. AnyEvidence Stable Design Pattern (Chapter 9)
6. AnyPrecision Stable Design Pattern (Chapter 10)
7. AnyCorrectiveAction Stable Design Pattern (Chapter 11)
8. AnyDebate Stable Design Pattern (Chapter 12)

This part also has four sidebars:

1. Common Stable Design Patterns (SB P2.1)
2. Fayad's Practical Actor's in UML (SB5.1)
3. Introducing Fayad's Legal Actors to UML: Adding a New Dimension to Software Modeling (SB6.1)
4. e-Evidence or Digital Evidence (SB9.1)

SIDEBAR P2.1:   COMMON STABLE DESIGN PATTERNS

This sidebar shows 16 common stable patterns the majority of them are used in Stable Analysis and Design Patterns. Each of the common pattern presents a brief overview of SDPs which play a major and decisive role in reducing the overall cost and in condensing the time duration of software project lifecycles

Detailed Templates
1. AnyActor—It appear in all stable analysis and design patterns—Refer to Chapter 5
2. AnParty—It appear in all stable analysis and design patterns—Refer to Chapter 6
3. AnyEntity—Refer to Chapter 7

Mid-Size Template
4. AnyConstraint—Refer to Chapter 13

Short Templates
5. AnyReason—Refer to Chapter 22
6. AnyConsequence—Refer to Chapter 23
7. AnyImoact—Refer to Chapter 24
8. AnyCause—Refer to Chapter 26
9. AnyCriteria—Refer to Chapter 27
10. AnyRule
11. AnyMechanisim—Refer to Chapter 28
12. AnyOutcome
13. AnyImoact—Refer to Chapter 24
14. AnyReason—Refer to Chapter 22
15. AnyCause—Refer to Chapter 26
16. AnyType
17. AnyEvent—Refer to Chapter 33
18. AnyMedia—Refer to Chapter 29
19. AnyLog—Refer to Chapter 30

5 AnyActor Stable Design Pattern

Actors are agents of change. A film, a piece of theater, a piece of music, or a book can make a difference. It can change the world.

Alan Rickman [1]

The AnyActor design pattern models the concept of actor by using software stability model. AnyActor is used in diverse domains, each domain having a different rationale for the use of this term. Since AnyActor design pattern captures the core knowledge, it is easy to model AnyActor in any application, by just hooking in the dynamic components of the application. The core is highly stable. Again, since software stability model is used for modeling concept of AnyActor, a generic model which can be extended for use by diverse applications is conceptualized. This increases reusability and reduction in repetition of modeling AnyActor for each individual application. It also formalizes the definition and types of actors and reduces the complexity of modeling actors and rules.

5.1 INTRODUCTION

In most common usage, AnyActor is the user of the system. AnyActor can be either an external user of the system or the internal entity or the system and in some cases the system itself.

AnyActor is used for many applications, each system having its own rationale behind using the term Actor. Also, the role played by AnyActor differs from application to application. For example, in context of entertainment, actors are persons, who act in stage plays, motion pictures, television broadcasts, etc. They basically perform in order to entertain public. In other scenarios, actor may mean to be a participant. Whereas in software context, actor is actually a software–hardware combination that performs specific and specialized tasks. Again actor might mean to be an agent who acts mediary.

Since, the function of AnyActor depends on the context in which AnyActor is being used, the traditional approaches to software design will not yield a stable and reusable model. However, by using software stability model, AnyActor can be represented in any context using a single model. The resulting AnyActor pattern is stable, reusable, extendable, and adaptable. Thus, any number of applications can be built by using this common model. The AnyActor design pattern tries to capture the core knowledge of AnyActor that is common to all the application scenarios listed above to come up with a stable design pattern.

The objective in front is to design a stability model for AnyActor by creating knowledge map of AnyActor. This knowledge map or core knowledge can then serve as building block for modeling different applications in diverse domains. The rest of section presents AnyActor, a stable design pattern based on the SSM.

5.2 ANYACTOR DESIGN PATTERN DOCUMENT

5.2.1 PATTERN NAME: ANYACTOR STABLE DESIGN PATTERN

In this pattern, the role of AnyActor is meaningfully generalized. The AnyActor design pattern classifies an actor based on its role in the context of use and thereby achieves a general pattern that can be used across any application. This pattern is required to model the core knowledge of AnyActor without tying the pattern to a specific application or domain; hence, the name AnyActor is chosen.

5.2.2 KNOWN AS

This pattern is similar to AnyUser. AnyActor and AnyUser sound very similar and are used interchangeably in common context. However, when AnyUser is used, it generally refers to a human/person using a system. On the contrary, AnyActor in addition to the human user includes hardware and software that may use an application. AnyActor also includes a creature. Also, AnyUser generally is external to the application, whereas AnyActor can be both external, as well as internal to the application and in many cases the application itself.

Also AnyActor is similar to AnyDoer. AnyDoer is a person or thing that does something, especially a person who gets things done with vigor and efficiency. Thus, AnyDoer satisfies our definition for AnyActor and can be used interchangeably with AnyActor. However, AnyActor conveys the meaning more clearly and hence should be selected for usage over AnyDoer.

Again AnyActor and AnyAdvocate may seem related, but in more general sense an advocate is a person who speaks or writes in support or defense of a person, cause, etc. Since, it is clear from the definition that AnyAdvocate term can be used for specialized cases of AnyActor usage, AnyAdvocate cannot be used interchangeably with AnyActor.

5.2.3 CONTEXT

AnySystem has one or more actors, who coordinate tasks to achieve the goal of the system. Nevertheless, each actor has a specific role to play in any system. In a general context, AnyActor refers to an actor or actress, who can be a comedian, villain, hero, etc. Nevertheless, AnyActor could be any human, hardware, software, and creature, who act in order to achieve a specific goal. To accomplish the process of acting, AnyActor may interact with any system like a movie, television, radio, software application, etc. The AnyActor design pattern deals with the issue of act performed by an actor, based on his, her, or its role, so that the task is completed successfully and in a meaningful manner.

The AnyActor design pattern can be applied to numerous applications and within diverse disciplines such as software systems, entertainment business, and other fields like renting, lending, etc., where AnyActor acts as an intermediary or agent and so on. This pattern can be applied to different applications where AnyActor performs diverse roles. In later sections, application scenarios to illustrate the applicability of AnyActor pattern are shown by flushing AnyActor concept in two distinct scenarios such as a musical theater and a network system to illustrate how an actor can play different roles in different applications.

1. Earthworms (AnyActor) display their unique importance and role in terrestrial farming (AnySystem). They are the beneficial creatures (AnyType) that make (Acting) soil fertile and productive. Farmers (AnyActor) purposefully introduce them in the soil to make it more fertile and grow more crops.
2. In many countries, robots (AnyActor) are used in the field of medical and allied science (AnySystem). Robots are highly efficient hardware devices (AnyType) that have the capability to act on their own will. Scientists (AnyActor) have created these robots to perform many sensitive operations (Acting) and carry out other helpful tasks.

5.2.4 PROBLEM

The AnyActor design pattern covers the concept of acting in the context of many diverse domains that have different functionalities, usage, and nature. Hence, modeling a generic design pattern that can be applied to all these domains makes it more relevant. However, the questions enlisted below need to be answered and elucidated before modeling AnyActor.

- How to build an actor model that can capture the core knowledge of the actor problem and that can be reused to model the actor problem in any application?

- How to find an ultimate goal for AnyActor and what are the other patterns that are used in the design?
- Actor exists to act or/and perform or to do something else?
- Are there types for an actor?
- What are the systems that can interact with an actor?

Given that, AnyActor is used in different context and in different domains, building a generic model is greatly challenging. By using SSM, this problem could be solved and a generic model is modeled for different domains. This model is illustrated and described in the Solution section.

5.2.4.1 Functional Requirements

1. *Acting*: The ultimate goal of AnyActor is to act or to perform acting. This makes Acting as the enduring business theme (EBT) of AnyActor. Acting has a unique ID, type, and property. It also has constraints and requirements and can be performed only in a certain environment. Acting specifies its properties, lists requirements, and constraints and hence meets the demand of the system.
2. *AnyActor*: AnyActor is the actual user of the system. He or she acts on own will and can be of four types: Person/human, creature, hardware device, and software system. AnyActor has a unique ID, name, type, and role. AnyActor can be a member of a group and specifically performs a particular activity. He, she, or it plays one or more roles, can switch roles, can join or leave a group, and can participate in multiple activities. AnyActor has a capability to learn.
3. *AnyType*: AnyType is used to classify AnyActor. AnyType has a unique ID, a name, and a classification property based on which the type is decided. AnyType operates on AnyActor, provides labels, and names them. It can also be used to change the type of AnyActor.
4. *AnySystem*: AnyActor acts within AnySystem. He, she, or it is the actual user of AnySystem. AnySystem has a unique ID, name, type, location, role kind, and role rank. AnySystem has an impact on AnyActor, can have a value, can have a size, and can define what type of actors it need.

5.2.4.2 Nonfunctional Requirements

1. *In Context*: AnyActor should be within the context of the system. For example, in a classroom, there are many creatures in air, but when we consider classroom as a system, creatures do not play a role in the classroom and hence are not in the context of the system.
2. *Interactive*: AnyActor interacts with the system and with other actors of the system. Thus, AnyActor should be interactive. For example, if we consider a scenario of a house, a child interacts with his parents and parents interact with the child. Thus, an actor should be interactive so as to be a part of a system.
3. *Learnability*: AnyActor should have the capability to learn. It should be able to learn from experience. This would help evolve with time and would also help expand skill set that would in turn improve overall performance within the system. AnyActor may also need to learn how to use the system. Therefore, learnability is a very important quality factor of AnyActor.
4. *Relevant*: AnyActor should be relevant to the system. If the system is farming and we need earthworms to improve the fertility of the soil, then a lion would be of no use here and thus would not be relevant for the system.
5. *Motivated*: AnyActor should be motivated to perform given task. If AnyActor would be motivated, he, she, or it would be able to give best performance and thus AnySystem would benefit from the entity.

6. *Informative*: AnyActor should be informative and should have the information of how to use the system and what role he, she, or it has to play in the system. He, she, or it should also know how to play the given role.

7. *Confident*: AnyActor should be confident of self-capabilities. The entity should know the boundary of the role and should perform it with full confidence.

5.2.5 CHALLENGES AND CONSTRAINTS

5.2.5.1 Challenges

Challenge ID	0001
Challenge Title	Multiple Roles Played by One Actor
Scenario	House
Description	In a house, a person (AnyActor) can play a role of a father and a child. He can be a father to his child and a child to his father. He can also play a role of brother, uncle, grandson, etc. Thus, with in a same application or system, a person can play multiple roles. It would not be a problem, if a person were playing 10 different roles. However, consider an application, where a person is playing 100 unique roles. The person may get confused when switching roles and how to act according to a specific one. Thus, managing many roles of the same actor in a particular application can be truly challenging.
Solution	Each role should be defined clearly in the application and there should be a threshold of the number of roles an actor can play.

Challenge ID	0002
Challenge Title	Multiple Actors in a System
Scenario	Music Band
Description	When a band is composing a specific score of music, each band member plays a particular instrument. One member can be a singer, while another member can be a guitarist. Having a close coordination and synchronization between these members is very important to create a fine score. As the number of members increase, it becomes more and more challenging to retain the same level of coordination. Hence, an increase in the number of actors in a particular system makes it demanding to manage the entire system.
Solution	Each actor should have a well-defined role and there should be no redundancy of roles within the system. There should be a threshold on the number of actors a system can support.

5.2.5.2 Constraints

1. AnySystem should have at least one AnyActor.
2. AnyActor should be within the context of AnySystem.
3. AnyActor should belong to AnyType.
4. One or more AnyActor should perform acting.
5. One or more AnyActor can be a part of AnySystem.
6. AnyType of AnyActor is determined by Acting.
7. One or more AnyActor can support other AnyActor in the system.

5.2.6 SOLUTION

The stable design pattern below is the proposed solution for handling the issues in AnyActor and it provides a generic pattern that can be applied to any application by attaching application-specific features (Figure 5.1).

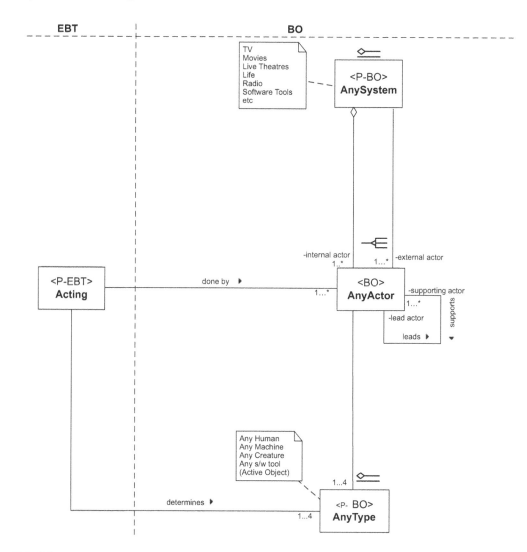

FIGURE 5.1 AnyActor stable design pattern.

5.2.6.1 Pattern Structure and Participants

The participants of the AnyActor design pattern are as described below.

5.2.6.1.1 *Classes*

AnyActor: This class represents the actor itself. It has a type and interacts with a system to do acting.

5.2.6.1.2 *Patterns*

> *AnyType*: This class represents the actor type. Type can be a human, hardware, software, and creatures. Also, it could be determined by acting (EBT).
>
> *AnySystem*: This class represents the system that the actor interacts with. A system can be a television, movie, radio, software application, etc.
>
> *Acting*: This class represents the existence of the actor. Existence of any actor depends on the acting. Without acting, an actor cannot exist.

5.2.6.2 Class Diagram Description

The class diagram visual illustration of all the classes in the model along with their relationships with other classes its description is given below.

- Acting is the EBT of this model and is done by one or more AnyActor (BO).
- Each AnyActor (BO) has a specific role to play and performs different Acting (EBT).
- One or more AnyActor (BO) interact with AnySystem (BO).
- AnyActor (BO) can also be a part of AnySystem (BO).
- AnyActor (BO) may or may not support one or more AnyActor (BO).
- AnyActor (BO) has AnyType (BO) such as hardware, software, human, and creature.
- Acting (EBT) determines the AnyType (BO) of AnyActor (BO).

5.2.6.3 CRC Cards

Responsibility	Collaboration	
	Client	Server
Acting(Acting)(P-EBT)		
The performance/behavior of AnyActor, which is generally different for any actor and based on its role	AnyActor AnyType	nameActingType() specifyActingProperty() meetActingDemand() stateActingConstraint() listActingRequirement() emphasisActingDomain() informActingEntity()
Attributes: id, actingType, actingProperty, constraints, actingRequirement, actingEntity, actingGuide, actingDemand, actingEnvironment		
AnyActor(AnyActor)(BO)		
The one who participates in various activities	Acting AnySystem AnyType AnyActor	learn(), participate(), interact() playRole(), setCriteria() switchRole() monitor()
Attributes: id, actorName, type, role, member, affair, activity, category		
AnyType(AnyType)(P-BO)		
Represent the type of an actor, like hardware, software, human, and creature	AnyActor Acting	name() change() listOperatation() pause() resume() operate() classify() label()
Attributes: id, typeName, property, interfaceList		
AnySystem(AnySystem)(P-BO)		
Acts as a medium through which an actor interacts with each other and accomplishes the goal	AnyActor	status() type() update() new() impact() value() size()
Attributes: id, name, type, location, roleKind, roleRank, vendorName, property, requirement		

5.2.7 Consequences

The AnyActor design pattern is generic enough to serve as building block for applications in diverse domain.

Using the AnyActor design pattern for AnySystem will require interaction among them. However, this does not mean that the pattern is incomplete, as this is the nature of patterns—they need to be used with other components.

A good thing with AnyActor design pattern is that it is more commonly used, because it defines roles in AnySystem. But, the bad thing is that it does not have a large domain.

The AnyActor design pattern is generic enough to serve as building block for applications in diverse domain.

Using the AnyActor design pattern for AnySystem will require an interaction between them. However, this does not mean that the pattern is incomplete, as this is the nature of patterns—they need to be used with other components.

A negative consequence of AnyActor pattern is that merging two different systems with the same actor type by different actors may result in confusion over which actor to use for the integrated system. Some actors may be dropped in favor of others. Even though the pattern is stable, the individual actors may introduce temporary system integration challenges.

The AnyActor pattern has the following benefits.

5.2.7.1 Flexibility

A positive aspect of the AnyActor pattern is that it is highly flexible as per the requirement of an actor for any system and for any given acting type. Many actors of the same type can easily use the system. Replacing one actor with another is easy and it requires lesser amount of effort.

5.2.7.2 Reusability

The AnyActor pattern is a very stable pattern. It can be reused in many different scenarios that are spread across many different fields. A theater AnyActor pattern can be easily reused for movie with a few changes.

5.2.8 Applicability with Illustrated Examples

In this section, two examples to illustrate the use of AnyActor design pattern are depicted by using use case description and behavior model like sequence diagram.

5.2.8.1 Application No. 1: AnyActor in Musical Theater Context

In this example, AnyActor is evaluated in a musical theater scenario. Musical theater is a form of theater that combines music, songs, dance, and spoken dialogue. However, theater is more than just what one sees or hears on stage. A theater involves an entire and complex world behind the scenes that creates costumes, sets, and lighting to make the overall effect very interesting. The table below summarizes the participants in this scenario.

Concept	Description
Acting (EBT)	Systematically control the process of Acting in a musical theater like singing, dancing
AnyActor (BO)	Singer, musician, voice artist
AnySystem (BO)	Represents a musical theater
AnyType (BO)	Identify the type of the actor like human who is a performer

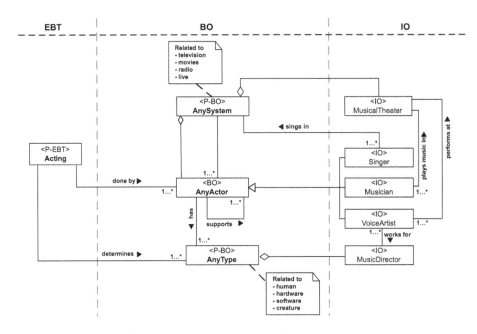

FIGURE 5.2   Class diagram for AnyActor pattern in musical theater context.

5.2.8.1.1   Model

The model for this application is shown below in Figure 5.2.
   Use Case No: 1.0
   Use Case Title: Performs in musical theater (Table 5.1)

5.2.8.1.2   Use Case Description

1. AnyActor (singer/musician/voice artist) does Acting (EBT). What is meant by Acting?
2. Acting (EBT) determines the AnyType (Human) of AnyActor (BO). Is Acting in itself sufficient to determine AnyType (BO)?
3. AnyActor (BO) interacts with AnySystem (MusicalTheater). How does AnyActor (BO) interact with AnySystem (MusicalTheater)?
4. AnyActor (BO) has AnyType (Human). Is AnyType (Human) (IO) the only type of AnyActor (BO)?
5. AnyActor (Singer) sings in AnySystem (MusicalTheater). Does AnyActor (Singer) perform everyday?
6. AnyActor (Musician) plays music in AnySystem (MusicalTheater). When does AnyActor (Musician) play in AnySystem (MusicalTheater)?
7. AnyActor (VoiceArtist) perform in AnySystem (MusicalTheater). Is AnyActor (VoiceArtist) able to mimic well and entertain the audience?
8. AnyActor (Singer/Musician/Voice Artist) is AnyType (Human).

5.2.8.1.2.1   Alternatives

1. Other than AnyType (Human) (BO), there can be other AnyType (BO) say hardware systems for sound.

5.2.8.1.3   Sequence Diagram

The flow diagram in Figure 5.3 represents the flow of messages, when Acting (EBT) is done by AnyActor (BO). The AnyActor (BO) supports Acting (EBT). Acting (EBT) determines AnyType (Human) (IO). The AnyType (human) (IO) is an AnyActor (BO). AnyActor (BO) interacts with

TABLE 5.1

Performs in Musical Theater

Actors	Roles
AnyActor	Singer, musician, voice artist

Classes	Type	Attributes	Operations
Acting	EBT	• id • actingType • actingProperty • constraint • actingRequirement • actingEnvironment	• nameActingType() • specifyActingProperty() • meetActingDemand() • stateActingConstraint() • listActingRequirement()
AnyActor	BO	• id • actorName • type • role • member • affair • activity	• learn() • participate() • playRole() • join() • switchRole()
AnySystem	BO	• id • name • type • location • roleKind • roleRank	• impact() • type() • value() • size()
AnyType	BO	• id • typeName	• name() • change() • operate() • classify() • label()
MusicalTheater	IO	• siteName • siteLocation	• stageShow() • giveNewArtistChanceToPerform() • satisfyGuest()
Singer	IO	• TypeOfSongPlayed • soloPerformance	• perform() • dramatize () • exhibit() • captivateAudience()
Musician	IO	• kindOfMusicPlayed • theaterWherePerform	• play() • enact() • entertain() • cheerUpPublic()
VoiceArtist	IO	• typeOfArtist • performingPlace • performingDate	• emote() • impersonate() • actOut() • mimic()

AnySystem (MusicalTheater) (IO). AnySystem (MusicalTheater) (IO) has AnyActor (Singer) (IO), who sings in AnySystem (MusicalTheater) (IO). AnySystem (MusicalTheater) (IO) also encourages AnyActor (VoiceArtist) (IO), who performs in AnySystem (MusicalTheater) (IO). AnySystem (MusicalTheater) (IO) also hosts AnyActor (Musician) (IO), who plays music in AnySystem (MusicalTheater) (IO).

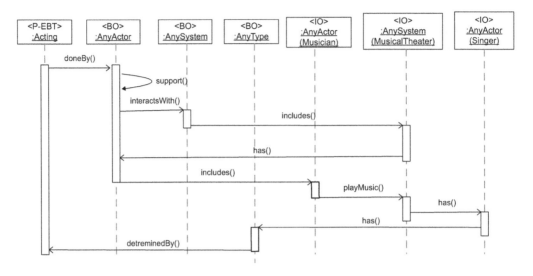

FIGURE 5.3   Sequence diagram for AnyActor pattern in musical theater context.

5.2.8.1.3.1   Sequence Diagram Description
1. AnyActor (Singer/Musician/Voice Artist) does Acting (EBT).
2. AnyActor (Singer/Musician/Voice Artist) supports AnyActor (Singer/Musician/Voice Artist).
3. AnyActor (Singer/Musician/Voice Artist) interacts with AnySystem (MusicalTheater).
4. AnySystem(BO) includes AnySystem (MusicalTheater).
5. AnySystem (MusicalTheater) has AnyActor(BO).
6. AnyActor(BO) includes AnyActor(Musician).
7. AnyActor (Musician) play music in AnySystem (MusicalTheater).
8. AnySystem (MusicalTheater) has AnyActor (Singer).
9. AnyActor (Singer) has AnyType (BO)
10. AnyType (BO) is determined by Acting (EBT).

5.2.8.2   Application No. 2: AnyActor in the Network System
AnyActor in a network scenario involves Acting by AnyActor on the network side. Server/client is involved in the Acting process. The table below summarizes the participants in this scenario (Table 5.2).

5.2.8.2.1   Model
The model for the application is as shown in Figure 5.4.
   Use Case No: 1.0
   Use Case Title: Act by AnyActor in network system (Table 5.3)

TABLE 5.2
Participants in AnyActor in the Network System Scenario

Concept	Description
Acting (EBT)	Systematic control of the process of Acting in a network.
AnyActor (BO)	Server, client
AnySystem (BO)	Represent a list of systems like network that interacts with AnyActor
AnyType (BO)	Identify the type of the actor like hardware

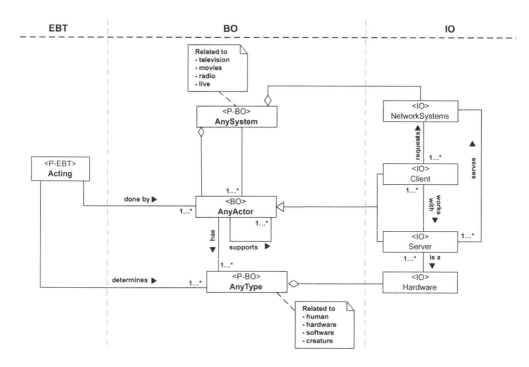

FIGURE 5.4 Class diagram of Anyactor pattern in the network system.

5.2.8.2.2 Class Diagram Description

1. AnyActor (server/client) has the responsibility of Acting (EBT). What is meant by Acting?
2. Acting (EBT) determines AnyType (hardware). Is Acting in itself sufficient to determine AnyType (BO)?
3. AnyActor (BO) has AnyType (BO).
4. AnyActor (BO) interacts with AnySystem (NetworkSystem). How does AnyActor (BO) interact with AnySystem (NetworkSystem)?
5. AnyActor (Client) sends requests to AnySystem (NetworkSystem). Are requests received by AnySystem (NetworkSystem)? If they are received, then how are they processed?
6. AnyActor (Server) serves AnySystem (NetworkSystem). Is AnyActor (Server) able to service the request sent by AnyActor (Client)?

5.2.8.2.3 Use Case Description

1. Acting (EBT) is done by AnyActor (Server/Client). What is meant by Acting?
2. AnyActor (BO) support AnyActor (BO).
3. AnyActor (BO) interacts with AnySystem (BO). How does AnyActor (BO) interact with AnySystem (BO)?
4. AnySystem (BO) includes AnySystem (NetworkSystem).
5. AnySystem (NetworkSystem) has AnyActor (BO).
6. AnyActor (BO) includes AnyActor (Client).
7. AnyActor (Client) sends requests to AnySystem (NetworkSystem). Are requests received by AnySystem (NetworkSystem)? If they are received, then how are they processed?
8. AnySystem (NetworkSystem) sends message to AnyActor (Server).
9. AnyActor (Server) has AnyType (BO).
10. AnyType (BO) is determined by Acting (EBT).

TABLE 5.3

Act by AnyActor in Network System Use Case

Actors	Roles		
AnyActor	Client, server		

Classes	Type	Attributes	Operations
Acting	EBT	• id • actingProperty • constraint • actingEntity • actingRule	• specifyActingProperty() • stateActingConstraint() • emphasisActingDomain() • informActingEntity() • actingRule()
AnyActor	BO	• id • actorName • type • role • activity	• learn() • interact() • playRole() • setCriteria() • collectData() • monitor()
AnySystem	BO	• id • name • type • location • vendorName • property • requirement	• status() • type() • update() • impact() • value() • size()
AnyType	BO	• id • typeName • property • interfaceList	• name() • attach() • pause() • resume() • operate() • detached()
NetworkSystem	IO	• typeOfSystem • noOfTerminal	• provideEnvironmentForInteraction() • serviceUserRequest() • maintainSystem()
Server	IO	• id • name • function • configuration • numberOfCPU • numberOfClientServiced	• serveClient() • processRequest() • receiveRequest() • sendResponse()
Client	IO	• id • name • serverAttachedTo • specification	• sendRequest() • receiveResponse() • displayResult() • takeUserQuery()

5.2.8.2.3.1   Alternatives
1. Other than AnyType (Hardware) (BO), there can be other AnyType (BO) such as humans using the network systems.
2. AnyActor (server) is not able to receive requests from AnyActor (client).

5.2.8.2.4   Sequence Diagram

The flow diagram in Figure 5.5 represents the flow of messages, when Acting (EBT) is done by AnyActor (BO). The AnyActor (BO) supports Acting (EBT). Acting (EBT) determines AnyType (Hardware)

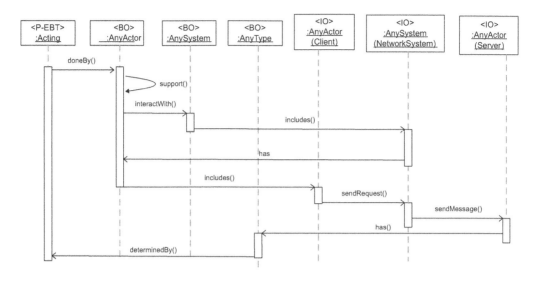

FIGURE 5.5 Sequence diagram of AnyActor pattern in the network system.

(IO). The AnyType (Hardware) (IO) is an AnyActor (BO). AnyActor (BO) interacts with AnySystem (NetworkSystem) (IO). AnySystem (NetworkSystem) (IO) has AnyActor (Client) (IO), which is sends request to AnySystem (NetworkSystem) (IO). AnyActor (Client) (IO) is served by AnySystem (NetworkSystem) (IO). AnySystem (NetworkSystem) (IO) sends message to AnyActor (Server) (IO). AnyActor (Server) (IO) processes the request from AnyActor (Client) (IO) send in form of message by AnySystem (NetworkSystem) (IO) and sends response back to AnySystem (NetworkSystem) (IO).

5.2.8.2.4.1 Sequence Diagram Description
1. Acting (EBT) is done by AnyActor (Server/Client).
2. AnyActor (BO) supports AnyActor (BO).
3. AnyActor (BO) interacts with AnySystem (BO).
4. AnySystem (BO) includes AnySystem (NetworkSystem).
5. AnySystem (NetworkSystem) has AnyActor (BO).
6. AnyActor (BO) includes AnyActor (Client).
7. AnyActor (Client) sends requests to AnySystem (NetworkSystem).
8. AnySystem (NetworkSystem) sends message to AnyActor (Server).
9. AnyActor (Server) has AnyType (BO).
10. AnyType (BO) is determined by Acting (EBT).

5.2.9 Design and Implementation Issues

5.2.9.1 Design Issues (Only Two of the Following Issues)
1. **Type/Subtypes: Situation is AnyType and Actor/Roles**
2. **AnyActor reduces the complexity of Actor/Roles in the Traditional Model:** In a traditional model, for every type of actor, we will need to create a separate model. For example, Meta model illustrated in Figure 5.6 under related patterns and measurability section is specific and special to the human. For any application that has no human influence, but software or some other application as the driver (actor), the Meta model illustrated in Figure 5.6 cannot be used. We will need to have another Meta model for such applications. Similarly for hardware or creature as an actor in any application, we will need different Meta model.

Also, any application that has multiple actors cannot use the Meta model would need different Meta model for such scenarios. However, as seen from the two applications illustrated under applicability section, the AnyActor design pattern can be used for any kind of application because it is general enough into fit to any application that uses an actor.

The type of actor in the pattern makes it possible to plug in any kind of actor being used in the application as an Industrial Object (IO). Again, since the core of the pattern is stable over time, multiple actors can inherit from AnyActor (BO), when application-specific IOs are to be attached, thus making the pattern useable in any scenario and context. Since, single pattern can be used for all cases, AnyActor pattern would reduce the complexity. This can be seen in 2 and 4, where all roles of actor and functionality are specified with business objects (BOs). Only application-specific IO's have to be attached to the corresponding BOs. As a result, any application can be designed quickly. This will eventually reduce cost of development too.

3. **AnyActor and/or AnyParty:** In some applications, both actor and party may be applicable. Actor refers to human, hardware, software, and creature, while party refers to human, political party, organization, or country. Using AnyActor and AnyParty stable design patterns together for such applications will suffice the requirement. However, with traditional models such flexibility is not available, as there is no distinction between actor and party, everything is categorized as person. This inflexibility in traditional approaches proves a big constraint in describing the application accurately.

5.2.9.2 Implementation Issues (Two of the Following Issues)

5.2.9.2.1 Type/Subtypes

Acting (EBT) has just a relationship with AnyType (BO) and AnyActor (BO); they are not inherited. AnySystem (BO) will have IO systems in the implementation phase. AnyActor (BO) will also inherit any kind of actor in the implementation phase. Similarly, AnyType (BO) will also inherit any kind of type in the implementation phase.

5.2.9.2.2 Hooking

BOs provide hooks to which application-specific IOs can be attached. Because the pattern provides hooks, the core design remains same for all the applications that has an actor.

BOs are the capabilities of the system which are used to achieve a goal (EBT). BOs are actually work objects because they result in design patterns. BOs are stable too, but differ from EBTs in the fact that BOs are externally adaptable and internally stable. BOs are sometimes explicitly stated in the problem statement; however, in rest cases intuition has to be used to identify them. For acquiring BOs for the problem statement, one needs to get an answer to what you do with the concept question.

5.2.9.3 Implementation Delegation Pattern

In the below diagram (Figure 5.6), AnyActor can have many roles like singer, musician, or a voice artist. A person (AnyActor), who is a singer, can also be a voice artist. In the above scenario, we

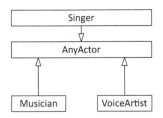

FIGURE 5.6 AnyActor with three roles.

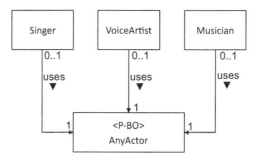

FIGURE 5.7 The use of delegation pattern of AnyActor with three roles.

also see that the same person at different times can play various roles. So, if we need to represent the same person for different roles, then we need to create multiple instances, which can represent different changes for different roles. In order to avoid this problem, delegation pattern can be used. Delegation pattern uses delegation to represent persons for different roles. The diagram above (Figure 5.7) explains how a delegation pattern can be used to delegate various roles to different objects at run time.

Java Code example:

```java
class AnyActor{
      void function1(){
             //implementation
      }
         void function2(){
                //implementation
      }
}

class Singer {
      //delegation
      AnyActor actor = new AnyActor();

         void function1() {
         actor.function1();
      }

         void function2() {
         actor.function2();
      }
}

public class Main {
      public static void main(String [] args){
         Singer singer = new Singer();
                singer.function1();
                singer.function2();
      }
}
```

5.2.9.3.1 Nonhuman Intervention

There may be different applications in which no human exists as a controller, while another application or hardware controls the execution. Such nonhuman intervention applications can easily be implemented by using AnyActor design pattern. This is possible because of the presence of

BO—AnyType in design pattern (refer to Figure 5.1). This extensibility is one of the main advantages of using AnyActor pattern.

AnyActor design pattern is designed, so that it can be adapted in diverse domains that need actors in their application. This section describes some of the issues faced in designing and implementing a stable AnyActor design pattern.

EBTs are the goals of the system and they result in an analysis pattern. EBTs are stable over time. Since EBTs are not stated in the problem statement, identifying EBTs for the problem domain is challenging and requires thorough understanding of problem domain as well as user experience. Using intuition is a common way of finding EBT. An answer to question what is this concept used for helps us to cull EBTs for the problem being modeled.

BOs are the capabilities of the system, which are used to achieve goal (EBT). BOs are actually work objects because they result in design patterns. BOs are stable too, but differ from EBTs in the fact that BOs are externally adaptable and internally stable. BOs are sometimes explicitly stated in the problem statement; however, in other cases, intuition has to be used to identify them. For getting BOs for the problem statement, you will need to get answer to what you do with the concept question.

Some issues in implementation of AnyActor design pattern are described next. Acting (EBT) has just a nominal relationship with AnyType (BO) and AnyActor (BO), and they are not inherited. AnySystem (BO) will have IO systems in the implementation phase. AnyActor (BO) will also inherit any kind of actor in the implementation phase. AnyType (BO) will inherit any kind of type in the implementation phase.

Designing generalized pattern that is applicable to all possible business domains is bit time consuming, as it requires identification and finding interrelatedness of all possible BOs for the corresponding EBT. The pattern can be designed only through a thorough knowledge of the problem domain and it requires identification of EBTs, BOs, and IOs to accommodate all concepts of stability model.

5.2.10 Related Patterns and Measurability

Actor can be defined as a person, who acts and gets things done; "he's a principal actor in this affair"; "when you want something done get a doer"; "he's a miracle worker" [5]. Based on this definition, following Meta model can be modeled for actor problem (Figure 5.8).

5.2.10.1 Traditional Model (Meta Model) versus Stable Model (Pattern)

Traditional modeling (Meta model) generally results in an unstable model which is not adaptable, scalable, and extensible. On the other hand, analysis/design pattern created by using SSM is adaptable to changing requirements and can be scaled easily with minimum effort.

In traditional modeling, since we only design as much is needed for a specific application in question, and by not thinking of its applicability in other domain, the resulting application is only very specific. As a result for each application, there is a specific Meta model resulting in multiple Meta model for the same concept. Within the case of stability, we use general enduring concepts and hence the resulting pattern can be used for building numerous applications. In short, the resulting pattern can serve as the building block for diverse application domains. Thus, with stable patterns, we do not need multiple models as is the case with traditional models; single pattern suffices for any kind of application using the concept.

With traditional modeling, the resulting design is tangible enough to be comprehended; while with stability model, pattern is intangible. Designing aspects in traditional modeling come from the problem statement; while with stability modeling, we need experience to derive solution by intuition and by thinking out of the box. However, implementing successful pattern requires expertise and meticulous work and detailed understanding of the problem domain.

It is always easier to test the pattern derived with stability in mind than a traditional model. As traditional model is specific to a particular application, it becomes very difficult to test this model

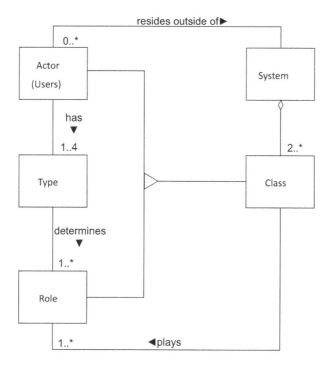

FIGURE 5.8 Actor Meta model.

without building the application. Even though a developer is successful in building the application from the traditional model, to test its suitability is very difficult, as there is no way to be sure of that. On the other hand, as described in the later section under testability, stable design patterns can be easily tested by just trying to find out an application that uses the concept, but the design pattern does not satisfy it.

Traditional model is just the first design step in the software life cycle. When using a traditional model, one then, needs to create a high-level and low-level design. Later, one needs to implement and test the application. Thus, each step is time consuming. Again, it cannot be generalized, as we have numerous application-specific traditional models. On the other hand, in the case of stability, the pattern created covers all the three steps of design, implementation, and testing. Eventually, this process reduces effort and energy. Also, it can be considered to satisfy maintenance stage of the software too, as the pattern is easily extendable.

The services provided by both models are as follows.

Traditional Model: Are created with IOs, which are unstable over period. This model is difficult to change and expand too, as compared with the stable model. When creating a new model by using traditional model, a lot of effort is required. Each application has a specific model while following the approach of traditional model, so reusing the existing model is not possible.

Stability Model: Overcomes some of the disadvantages of the traditional analysis pattern. It provides some of the new approaches for designing software with the use of software stability concept in mind. The stable analysis pattern is used to analyze any problem, and then classify the problem into EBT and BO. With this novel approach, the stability and reusability of the software increases. The result will be a stable pattern, which can be reused in any situation. Hence, with the help of stability model, it is now possible to build stable and reusable software [2–4].

The Life Cycle Coverage: Traditional model life cycle is quite long because the design and implementation is context specific. Stable model life cycle is short, as you leverage EBTs. Traditional model needs tweaking and remodeling, when entities change, but stable model does not need changes during the life cycle. All traditional models need more maintenance and testing effort when compared with a stable model.

Testability: If AnyActor design pattern can be used, as it is without changing the core design and by only plugging IOs for infinite number of applications, then AnyActor pattern can be said to be testable. In the applicability section, two widely different applications are illustrated and they do not require changing the core design of the pattern. Using the scenarios listed in this paper, many such scenarios can be deduced and proved that AnyActor pattern is indeed testable.

The overall testing methodology for pattern must satisfy the following four parts [6]:

- *Literature comparisons*: Compare the pattern to existing literature on pattern in the greater patterns community. Is the pattern well formed and does the guidance it provides agree with established patterns?
- *Comparison against other implementations*: Compares the implementation described against past implementations discussed in previously existing patterns. This includes implementations in other technologies.
- *Implementation testing*: Validates the implementation against best practices for the technologies discussed. This is close to traditional testing, except that one may need to create the code to be tested.
- *Brainstorming and proof by contradiction*: Applies aims of testing for exception conditions to the content.

Patterns must be tested from two different viewpoints—namely testing each pattern independently and then testing the pattern within the context of other patterns, or according to pattern clusters. Some of the test cases can be

1. Is the business problem that the pattern defines relevant to the problem and forces driving the pattern?
2. Is the recurring problem inherent to the application, technology, or programming language?
3. Are the key decisions required for making architecture and design choices evident?
4. Does the solution consist of elements that can be used in the architecture or design for the problem being addressed?
5. Does the solution consist of implementation (programmatic) sequences to be used for addressing the problem being addressed?
6. Is the pattern name representative of what the pattern intends to achieve?
7. Is the pattern name ambiguous or verbose (longer than a few words)?
8. Does the pattern have an alias?
9. Is the alias name descriptive without being verbose?
10. Is the solution generic enough to enable it to be applied to specific business application scenarios having the relevant context?
11. Does this pattern, or its context, provides a solution that could form a part of the solution to a larger problem?
12. Does this pattern, while helping to solve a larger problem, excludes the usage of any other patterns?
13. Does this pattern work with other patterns to provide a solution to a larger problem, apart from solving the problem it is intended for?
14. Is the usage and applicability of the pattern explained clearly?
15. Does it address the case, if multiple inheritances are not supported?

5.2.11 BUSINESS ISSUES (TWO OF THE FOLLOWING ISSUES)

5.2.11.1 Business Rules

The business rules for AnyActor pattern are

- An actor must act.
- The acting is dependent on the role performed by the actor.
- Actor is either external or internal to the system.

Whereas according to meta model, the business rules are

- An actor is hired, when there is a need/reason.
- An actor must carry out a task and is thus identified as a worker.

5.2.11.2 Business Models

The stable AnyActor design model is generic and so can be extended by plugging in specific IOs, which are specific to the application domain, eventually resulting in numerous stable and adaptable business models [7,8].

The term business model describes a broad range of informal and formal models that are used by enterprises to represent various aspects of business, such as operational processes, organizational structures, and financial forecasts [9]. Business models are targeted to generate revenues by building customer base and strong customer relationships. Since customers represent the concept of actor, using AnyActor design pattern is quite useful and profitable for designing good business models.

5.2.11.3 Business Integration

Since AnyActor pattern provides hooks, the integration of the AnyActor design pattern to existing business model is quite easy. Integrating AnyActor pattern to a particular application domain will require the plugging of application-specific IOs to the hooks.

5.2.11.4 Business Transformation

Use of AnyActor design pattern by a company in development of any application strengthens and cements its business transformation process because AnyActor pattern quickens the development process by providing the framework for the application. Also, since AnyActor pattern is tested and verified, chances of getting nagging bugs are minimal.

In a business transformation process, humans are involved and, AnyActor pattern can be used to depict the whole process of transformation. As a result, this use will speeds up the entire process by tying the transformation process together and making people more accountable.

5.2.12 KNOWN USAGE

AnyActor is commonly used in an Ecommerce, health management, and takeoff airplane systems. It is also used in network systems and all kinds of software applications. It is very commonly used or known as an actress in a movie.

AnyActor is used widely in applications. Some of them are listed below.

- Ecommerce Applications like eBay, Amazon, etc. In these systems, AnyActor refers to the customers, who make transactions for buying or selling goods such as books, electronic items, cosmetics, etc. over the Internet.
- Health Management Systems: These are widely used by health management organizations (HMOs) such as Kaiser Permanente, WellPoint, etc. In these systems, AnyActor refers to general public availing medical facilities, as well as healthcare practitioners. Also, various equipments used to diagnose and treat patients like x-ray, ventilator, etc. are actors.

- Network systems, where computers/nodes in the network, software that runs on this nodes, and users of these terminals are actors.
- All kinds of software applications in which the software as well as the person using it are actors.
- In entertainment industry like movies, theater, and arts, where the performers who entertain audience are actors.

5.3   TIPS AND HEURISTICS

- Describing patterns is a difficult job and it requires careful and calibrated work.
- Meta model is totally different than stable model, it is a traditional model.
- AnyActor should do acting.
- Pattern design must be generic, so that it can be applied to applications that are spread across various domains.
- EBT must represent the goal of the pattern.
- Intuition and experience are required in order to find correct EBT for the pattern.
- BOs provide capabilities to achieve the goal of the pattern. Identification of BO requires spending some time in thinking and finding correct BOs.
- BOs provide hooks to which specific IOs can be plugged and getting varied applications in diverse domains. This reduces the cost by encouraging reusability.

5.4   SUMMARY

Although, building a stable design pattern for AnyActor that can be reused and reapplied across diverse domain is difficult and requires thorough understanding of the problem, it is worth the effort and time. Creating AnyActor pattern by using SSM results in reusable, extensible, and stable pattern.

The correct identification of EBT and BOs for AnyActor is the most challenging task and requires knowledge and experience. Once EBT and BOs are correctly identified, next challenge is to determine the relationship between EBT and BOs, so that AnyActor pattern can hold true in any context of usage for an actor. Once this is done, depending on the application, the IOs are attached to the hooks which are provided by BOs. Hence, using AnyActor pattern as a basis, infinite number of applications can be built by just plugging in the application-specific IOs to the pattern. This results in reduced cost, effort, and stable solution. Hence, AnyActor design pattern is very useful.

5.5   REVIEW QUESTIONS

1. Explain why the traditional approach of software design fails to create a stable and reusable model?
2. What are the limitations of using AnyUser in place of AnyActor?
3. Describe AnyDoer.
4. Why should we use AnyActor instead of AnyDoer?
5. Why can we not use AnyAdvocate in place of AnyActor?
6. Describe a few properties of AnyActor.
7. Describe four scenarios where AnyActor can be applied from day-to-day life.
8. List and explain the functional requirements of AnyActor stable design pattern.
9. List and explain the nonfunctional requirements of AnyActor stable design pattern.
10. List two challenges of AnyActor stable design pattern.
11. List five constraints of AnyActor stable design pattern.
12. Explain AnyActor stable design pattern with the help of a class diagram.
13. List the participants in AnyActor stable design pattern and explain them.
14. Write CRC cards for stable design patterns.

15. Describe the negative consequences of AnyActor pattern.
16. What are the benefits of AnyActor pattern?
17. Describe the application of AnyActor in musical theater context with the help of a class diagram.
18. Write a use case for the application of AnyActor in musical theater context and draw a sequence diagram for that use case. Describe both of them.
19. Describe the application of AnyActor in networking system context with the help of a class diagram.
20. Write a use case for the application of AnyActor in networking system context and draw a sequence diagram for that use case. Describe both of them.
21. Mention a few design and implementation issues related to AnyActor stable design pattern.
22. Define Actor.
23. Draw Meta model for actor.
24. Compare the traditional model of AnyActor with the stable model.
25. What are the advantages of stable model over traditional model?
26. What are the services provided by both the models?
27. What is the life cycle coverage for both the models?
28. Compare both the models on testability.
29. Write a few test cases for AnyActor stable model.
30. Write 10 business rules for AnyActor stable model.
31. Describe 10 usages of AnyActor stable model.
32. What have you learnt from this chapter? List a few tips and heuristics.
33. (T/F) The AnyActor pattern can only represent Hollywood movie stars.

5.6   EXERCISES

1. Describe a scenario of an application of actor as a mythological creature.
   a. Draw a class diagram of the application.
   b. Document a detailed and significant use case.
   c. Create a sequence diagram of the created use case of b.
2. Describe a scenario of an application of actor as a hardware device.
   a. Draw a class diagram of the application.
   b. Document a detailed and significant use case.
   c. Create a sequence diagram of the created use case of b.
3. Describe a scenario of an application of actor as a software system.
   a. Draw a class diagram of the application.
   b. Document a detailed and significant use case.
   c. Create a sequence diagram of the created use case of b.
4. Research on a virus.
   a. Draw a class diagram of the application.
   b. Document a detailed and significant use case.
   c. Create a sequence diagram of the created use case of b.
5. Actors in a classroom.
   a. Draw a class diagram of the application.
   b. Document a detailed and significant use case.
   c. Create a sequence diagram of the created use case of b.
6. Application for Ocean.
   a. Identify actors in an application for Ocean.
   b. Draw a class diagram of the application.
   c. Document a detailed and significant use case.
   d. Create a sequence diagram of the created use case of c.

7. Illustrate with a class diagram the applicability of the AnyActor SAP to an ecommerce site such as eBay.
8. Now, adapt the AnyActor to an Internet forum discussing software engineering.

5.7   PROJECTS

1. Model you own house. Identify actors. Draw a class diagram by using stable pattern and a Meta model. Which model do you prefer more? Write CRC cards and use cases. Draw sequence diagram for the same.
2. You are a part of a team that is developing an application like YouTube, where people can view, upload, watch, and like videos. They can also subscribe to channels and can make their own channel. They can also comment on a video. Identify actors for such an application and illustrate with a class diagram the applicability of AnyActor SDP.
3. Select any animated movie that you like. Identify actors for that movie and illustrate with a class diagram the applicability of AnyActor SDP.
4. University Student and Employee System—A university would like to develop a system for tracking all the employees and students at the university. Each person in the system has an associated ID number. Students and employees are distinguished by their roles in the system and on campus. Design a corresponding class diagram, use case, and sequence diagram as shown in the solutions in this chapter.

REFERENCES

1. Alan Sidney Patrick Rickman (February 21, 1946–March 14, 2016), last modified on June 25, 2017, https://en.wikipedia.org/wiki/Alan_Rickman
2. M.E. Fayad and A. Altman. Introduction to software stability, *Communications of the ACM*, 44(9), 2001, 95–98.
3. M.E. Fayad. Accomplishing software stability, *Communications of the ACM*, 45(1), 2002, 111–115.
4. M.E. Fayad. How to deal with software stability, *Communications of the ACM*, 45(3), 2002, 109–112.
5. K. Körner and V. Neale. *An Introduction to Proof by Contradiction. Stage: 4 and 5*, Cambridge, England, December 2005, February 2011. http://nrich.maths.org/4717
6. M.E. Fayad. *Stable Analysis Patterns for Software and Systems*, Auerbach Publications, Boca Raton, FL, Taylor & Francis Catalog #: K24627, May 2017, ISBN-13: 978-1-4987-0274-4.
7. M.M. Al-Debei, R. El-Haddadeh, and D. Avison. Defining the business model in the new world of digital business, in *Proceedings of the Americas Conference on Information Systems (AMCIS)* (vol.), 2008, pp. 1–11.
8. G. George and A.J. Bock. The business model in practice and its implications for entrepreneurship research, *Entrepreneurship Theory and Practice*, 35(1), 2011, 83–111.
9. M.M. Al-Debei and D. Avison. Developing a unified framework of the business model concept, *European Journal of Information Systems*, 19(3), 2010, 359–376.

SIDEBAR 5.1   FAYAD'S PRACTICAL ACTORS IN UML

Movies are a fad. Audiences really want to see live actors on a stage.

Charlie Chaplin [1]

The main actors in Unified Modeling Language (UML) are humans and machines. After examining thoroughly, we find that this is quite arbitrary and it may actually limit the types and the numbers of applications that that could be modeled. It is most challenging to symbolize the interaction of UML actors with a system, when the actors are not classified and defined properly. Hence, a conventional UML is not suitable for some domains and it will not model the requirements in an adequate manner.

This work proposes the introduction of practical enhancements, and extension of actors within the UML. By reclassifying actors into four types, instead of two, we will demonstrate introduction of practical improvements to UML class diagrams, which can yield accuracy in the UML modeling of any system.

We will extend the actors into four different types:

1. Human or person
2. Hardware devices or machines
3. Software packages
4. Creatures, which include all animated, mythological, and fictional characters

This reclassification of actors into four different types provides numerous benefits, such as

1. Increasing the applicability of UML [1] to different domains, such as animation, the making of movies, fiction, and mythology.
2. All the possible users fall under one of the four types of actors defined above. Hence, the system built will satisfy all kind of users.
3. In modeling, when the actor classification is abstracted correctly, it gives an exact and broader picture and good understanding of the system.
4. If the actors are classified accurately, the system can be built with better quality.

Introduction

Actors are external entities that interact with a system. Each actor plays one or more roles within the system represented by one or more use cases and a class diagram, where each of the use cases is an instance of behavior when actors interact. Use cases provide information about the roles played by the actors in the system. Class diagrams show the interaction and relationship between the actor/role classes with the system classes.

Section 2 discusses the UML Classification [2–4]. Section 3 discusses the Practical Actors Classification. Section 4 shows examples of the Practical Actors Classification. This paper concludes with Section 5.

UML classification [4] is the broader classification of actors in UML and it is categorized under human actors and nonhuman actors. The human actors are the actual users, who play one or more roles, which interact outside the scope of the system. These actors are always external to the system and they define the boundary of scope of the system. The nonhuman actors (machines) are the robots that are automatic systems initiating the use case.

Actors may represent roles played by human users, external hardware, or other subjects. Note that an actor may not necessarily represent a specific physical entity, but merely a particular facet (i.e., "role") of some entity that is relevant to the specification of its associated use cases. Thus, a single physical instance may play the role of several different actors and, conversely, a given actor may be played by multiple different instances [4].

"UML does not permit associations between actors" [2,3]. Moreover [4] the use of generalization/specialization relationship between actors is useful in modeling overlapping behaviors between actors, and it does not violate this constraint, since a generalization relation is not a type of association [2].

Actors interact with use cases.

Example: Project Management System

The following schema shows:

Human or person actors such as Project Manager, Human Resource Representative, Team Leader, and System Administrator (refer to Figure SBF5.1)

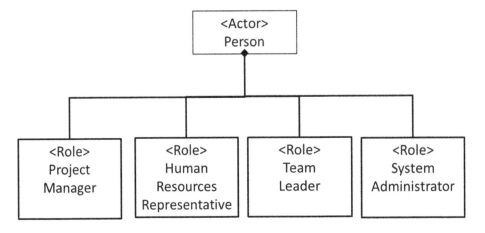

FIGURE SBF5.1   Human actors.

Nonhuman actor (machines): Another class of actors is hardware actors. This class
    includes actors like Honda's Asimo Robot, Roomba, an automobile and a traffic light
    as depicted in Figure SBF5.2.

Through a concise way of representing hardware actors, we can empower designers
to model a broader range of solutions with far greater accuracy. For instance, the
abovementioned hardware actors can be used to model a housekeeping system driven by
Asimo and Roomba or a traffic management system involving actors such as automobiles
and traffic lights.

In the UML actors' classification system, actors are classified into persons and hardware
devices or machines [1]. It is very broad and does not satisfy actors in different type of
applications. In fact, not all the actors interacting with the system fit into one of these
categories. Hence, this type of classification gives an incomplete picture of the system.

Let us consider a system, where a robot and a cat are interacting with the system. These
two actors are unrelated and hence cannot fall under one actor category. There is no clear
classification of the actor types. This classification is considered to problematic to the UML
users and many of them do not distinguish the differences between a robotic machine and
no-robotic one [Problem 01].

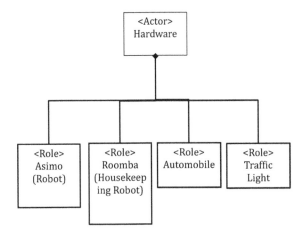

FIGURE SBF5.2   Hardware actors.

Likewise, if there are few animals such as cows, goats, tigers, and lions playing their roles in the system, these are broadly classified as animals. Although they are animals, they have major differences among them. For example, cows and goats are domestic and herbivorous animals, whereas lions and tigers are wild and carnivorous. Therefore, it is not wise to categorize them as animals [Problem 02].

In some systems, animated characters play a vital role. They are not assigned any proper actor classification. Animated characters have a life and they interact with the system intelligently. Hence, it is essential to have a classification to fit all types of actors. Overall, the classification is incomplete to represent all kinds of use case scenarios [Problem 03].

Fayad's Practical Actors Classification

Reference 5 influence Fayad's actor classification. And in Reference 6, there are four types of actors:

1. Human or person
2. Hardware devices or machines
3. Software packages
4. Creatures

Human or person: Person actors are the active objects that play one or more roles within the system that it is interacting.

Hardware devices or machines: These actors are smart robotic and other similar devices.

Software packages: Sometimes, external software systems interact with the targeted system in the use case. Software packages come in different architectures and formats and most of them are not equal to each other. Each one of them has its own role to play, while two or more such systems existing within one big ecosystem may show signs of conflicts with each other. Hence, they cannot be categorized into one single category and their classification is incomplete.

Creatures: Creatures are living entities or animated, mythology, and fiction characters. In a national animal reserve like Serengeti, hundreds of different animals and birds live together in perfect coexistence. However, the entire ecosystem displays many contradictions and differences. Most of these animals and birds follow a rigid food chain system, where the weaker ones become preys to the stronger ones in the evolutionary ladder. For example, a deer is the natural prey for carnivorous animal like cheetah or a lion. The dead carcass of deer will provide food source to many other animals and birds. Hence, it is difficult to classify these creatures into a common category. Similarly, a cartoon movie will involve many cartoon characters playing a myriad of roles. Not all characters could be classified into one single category.

Applicability

Scenario [1]: Soils are alive forever! A wide variety of soil organisms lives and coexists in the soil. These include bacteria, fungi, micro arthropods, nematodes, earthworms, and insects. These organisms sustain by feeding on organic matter or other organisms, and they perform a number of vital processes in soil, while other organisms are involved in the transformation of inorganic molecules. Very few soil organisms are pests and parasites.

Scenario [2]: Conflict is inevitable in workplace settings, and conflicts can arise between coworkers, supervisors, and subordinates or between employees and external stakeholders, such as customers, suppliers, and regulatory agencies. Managing conflicts is a key management competency and all small business owners should study and practice effective conflict management skills to maintain a positive workplace environment. Reviewing examples of conflicts and resolutions in the workplace can give one an idea of what to expect when conflicts arise [7].

Scenario [3]: In many countries, robots are used in the field of medical science and manufacturing. Robots are smart hardware devices that have the capability to act on their own will. Scientists have made these robots to perform operations and do other helpful tasks.

Conclusions

The main goal of this work is to suggest a modest refinement to enhance conventional UML without radically changing the notation, while simultaneously moving UML in a more practical direction. This improvement might address some of the fundamental reasons that UML is not employed with greater consistently across a broader set of domains which are likely to benefit from UML. By freeing practitioners of an unreasonable constraint ascribed to UML, systems engineers and architects will be better equipped to model a variety of solutions that heretofore is not easily modeled with UML.

References

1. C. Chaplin and D. Robinson. *(Introduction) My Autobiography*. Melville House Publishing, New York, December 2012, 528 pp., Reprint, ISBN-13-9781612191928.
2. OMG Unified Modeling Language (OMG UML), Superstructure, V2.1.2, pp. 586–588, archived from the original on September 23, 2010, retrieved November 7, 2010.
3. Problems and Deficiencies of UML as a Requirements Specification, s.3.2. (PDF), archived (PDF) from the original on October 17, 2010, retrieved November 7, 2010.
4. UML 2 Specification, retrieved July 4, 2012.
5. M. E. Fayad. *Stable Analysis Patterns for Software and Systems*, Auerbach Publications, Boca Raton, FL,Taylor & Francis Catalog #: K24627, May 2017, ISBN-13: 978-1-4987-0274-4.
6. M. Fayad. Flood III, Charles, Unified software engineering reuse (USER), *Master thesis*, San Jose State University, CA, November 2016.
7. David Ingram. Examples of Conflicts & Resolutions in the Workplace. *Demand Media*.

6 AnyParty Stable Design Pattern

A peace is of the nature of a conquest; for then both parties nobly are subdued, and neither party loser.

William Shakespeare [1]

The AnyParty design pattern molds the concept of party by using software stability model (SSM). AnyParty is used in diverse domains, where each domain has a different rationale for the use of this term. Since AnyParty design pattern captures the core knowledge, it is easy to model AnyParty in AnyApplication just by hooking in the dynamic components of the application. The core is stable. Since, SSM is used for modeling concept of AnyParty, we can conceptualize a generic model which can be extended for further use by diverse applications. This approach increases reusability and reduction in repetition of modeling AnyParty for each individual application. It also formalizes the definition and types of parties thereby reducing the complexity of modeling humans or groups of human with a particular structure like organizations, political parties, etc. and rules to use them.

6.1 INTRODUCTION

AnyParty is the most commonly used entity in the entire system. Generally, AnyParty refers to an external legal user of the system. However, AnyParty can also be an internal entity, when it is part of the system or the system itself.

AnyParty is used in various scenarios, each having its own context and rationale for using the term party. AnyParty can represent a human, any organization, any country, or even any political party depending on the context, where the term AnyParty is being used. For example, in context of politics, AnyParty refers to a group of persons with common political opinions, beliefs, and purposes, whose goal is to gain political influence, seize initiative, and governmental control for directing future government policy. Examples of political parties include the Republican Party, the Democratic Party, etc. In other scenarios like peace treaty signing, AnyParty may represent a country. Example of such scenario includes Camp David Accord. Whereas a party may represent a social gathering of humans to celebrate some events or occasion like birthday, promotion, etc. Again, party might also mean a person or group of persons who are working as an organization having specific goals like United Nations, NGOs, etc. The AnyParty design pattern tries to capture the core knowledge of AnyParty that is common to all these application scenarios to create stable, extendable, adaptable, and reusable pattern.

Since the function of AnyParty depends on the context in which AnyParty is being used, the traditional approaches to software design may not yield a stable and reusable model. However, by using SSM, AnyParty can be represented in any context by using a single model. The SSM requires creation of knowledge map by identifying underlying EBTs and BOs. By hooking IOs, that is specific to each application, the model can be applied to AnyApplication domain. The resulting AnyParty pattern is stable, reusable, extendable, and adaptable. Thus, any number of applications can be built by using this common model. The AnyParty design pattern tries to capture the core knowledge of AnyParty that is common to all the application scenarios listed above to come up with a stable design pattern.

The objective of this chapter is to design a robust and meaningful stability model for AnyParty by creating knowledge map of AnyParty. This knowledge map or core knowledge can then serve as building block for modeling different applications in diverse domains. The rest of this chapter presents AnyParty, a stable design pattern based on the SSM.

6.2 ANYPARTY DESIGN PATTERN DOCUMENT

6.2.1 PATTERN NAME: ANYPARTY STABLE DESIGN PATTERN

The AnyParty design pattern classifies a party based on its role in the context of use and thereby achieves a general pattern that can be used across any application. This pattern is required to sculpt the core knowledge of AnyParty, without tying the pattern to a specific application or domain; hence the name AnyParty is chosen. Again, since Party is semitangible and externally stable, AnyParty is chosen and is a BO.

6.2.2 KNOWN AS

This pattern is similar to AnyActor, AnyUser, and AnyDoer patterns.

AnyParty and AnyActor sound almost similar and are used interchangeably in a common context. AnyActor, in addition to the human user, includes hardware and software that may use an application. AnyActor also includes a creature. While, AnyParty may be human, organization, country, or political party depending on the context, where AnyParty is being used [2].

In addition, AnyParty and AnyUser seem related to each other. Nevertheless, when AnyUser is used, it generally refers to a human/person that is using a system. On the contrary, AnyParty in addition to the human user includes organization or a political party, as well as country that may use an application. Moreover, AnyUser is generally external to the application, whereas AnyParty can be both external as well as internal to the application, and in many cases the application itself. Thus, AnyParty and AnyUser should not be used interchangeably.

Furthermore, AnyParty is quite similar to AnyDoer. AnyDoer is a person, who gets things done with vigor and efficiency. Thus, AnyDoer satisfies our definition for AnyParty and can be used interchangeably with it. On the contrary, AnyParty conveys the meaning more clearly and hence should be selected for usage over AnyDoer.

6.2.3 CONTEXT

A given system might have one or more parties participating in coordination to achieve the goal of the system. However, each party plays a specific role in AnySystem. In general, AnyParty refers to a group of people who are sharing common interests. However, AnyParty can be classified as any human, organization, country, and political party, who have a specific role to play in AnyApplication. To accomplish the Classification, AnyParty supports any system like political system, Camp David Accord, social gathering, etc. The AnyParty design pattern deals with the issue of classifying various parties that work together in an application to achieve the task.

In practice, the AnyParty design pattern can be applied to numerous applications in diverse disciplines such as political scenario, socializing, and scenarios where two countries interact like signing peace treaty, and so on. This pattern can also be applied to different applications in which AnyParty is classified under various headings and interact together. In the later section, application scenarios that illustrate the applicability of AnyParty pattern are shown in finer detail. AnyParty concept is illustrated in two distinct scenarios like political scenario and communication between two countries like signing of peace treaty to illustrate how a party can be classified in different applications.

1. *Renting an apartment*: When a potential tenant (AnyParty) wishes to rent an apartment (AnyApplication), he signs a contract with the leasing department (AnyParty) of the organization (AnyType) that either owns the apartment or manages the apartment (Classification).

2. *Admission in college*: A student (AnyParty) may want to take admission (AnyApplication) in a college (AnyParty) for graduate studies. The colleges are of two types (AnyType): graduate college and community college (Classification). The student will have fill an application and provide it to the department (AnyParty) of admissions.

6.2.4 PROBLEM

The AnyParty design pattern involves the concept of Classification in the context of numerous diverse domains that have different functionality, usage, and nature. Hence, modeling a generic design pattern that can be applied to all these domains makes some sense. However, listed points should be considered before modeling AnyParty pattern.

- The model should capture the core knowledge of the party problem.
- The model can be reused to model the party problem in AnyApplication.
- The ultimate goal for AnyParty must be established along with the other patterns used in the design pattern of AnyParty.
- Identifying types of parties and the systems that can interact with a party.

Since, AnyParty is used in different context in different domains, building a generic model is intricately challenging. By using SSM [3–5], this problem could be solved and a generic model modeled for different domains. This model is illustrated and described in the solution section.

6.2.4.1 Functional Requirements

1. *Classification*: Classification is the EBT of this model and is used for classifying AnyParty. Classification has a unique ID, a grouping factor, criteria, and a Classification property. It distinguishes one type of party from another based on how it acts. It also assigns roles and responsibilities to the parties and classifies them based on that.
2. *AnyParty*: AnyParty is the legal user of the system. He, it, or she acts on own will and can be of four types: countries, political parties, organizations, and person. AnyParty has a unique ID, name, type, and role. AnyParty can be a member of a group and performs a particular activity. The entity plays one or more roles, can switch roles, can join or leave a group, and can participate in multiple activities. AnyParty has a capability to learn.
3. *AnyType*: AnyType is used to classify AnyParty. AnyType has a unique ID, a name, and a Classification property based on which the type is decided. AnyType operates on AnyParty, provides labels and names them. It can also be used to change the type of AnyParty.
4. *AnyApplication*: AnyParty acts within the context of AnyApplication. AnyParty is the legal user of AnyApplication. AnyApplication has a unique ID, name, type, location, role kind, and role rank. AnyApplication has an impact on AnyParty, can have a value, a size and can define what type of party it needs.

6.2.4.2 Nonfunctional Requirements

1. *In Context*: AnyParty should be within the context of the application. For example, a chef in a restaurant is a party, but when considering a scenario of voting for a political party, the chef acts as a citizen and hence the citizen is in context of the application, but not the chef.
2. *Interactive*: AnyParty interacts with the application and with other actors and parties of the system. Thus, AnyParty should be interactive. For example, if we consider a scenario of a classroom, professor interacts with students and students interact with professor and both of them with each other. Thus, a party should be interactive so as to be a part of an application.

3. *Relevant*: AnyParty should be relevant to the application. If the application is a movie and we need actors to play hero and heroine, then a politician would be of no use here and thus would not be relevant for the application.
4. *Legality*: AnyParty is bound in legal terms and has some legal duties and responsibilities. Hence, legality is a key quality factor of AnyParty. Whenever an actor or user of a system is legally bound with a contract, it acts as a party and not as an actor in UML.
5. *Binding*: Binding is another important quality factor of AnyParty. AnyParty is legally bound to perform some duties and to fulfill some responsibilities, either because they have signed a contract/agreement or at times it can be a verbal promise.

6.2.5   CHALLENGES AND CONSTRAINTS

6.2.5.1   Challenges

Challenge ID	0001
Challenge Title	Multiple Roles Played by One Party
Scenario	An Employee of a Software Company
Description	In a company, an employee can play multiple roles like he or she can be owner of a company, as well as a director. He or she can have to think like a customer and work like an employee. If there are a few roles that an employee has to play, then it would not be a problem. But, it would be very challenging when the employee or AnyParty has to play a myriad of roles.
Solution	Each role should be defined clearly in the application and there should be a threshold of the number of roles a party can play.
Challenge ID	0002
Challenge Title	Multiple Parties in an Application
Scenario	Music Band
Description	When a band is composing a piece of music, each band member plays a particular role. One member can be a singer, while another member can be a guitarist. Ensuring perfect coordination and synchronization between these members is a very important task so as to make a fine piece of music. As the number of members increase, it becomes more and more challenging to maintain the required level of coordination. Thus, an increase in number of actors in a particular system makes it taxing to manage the system.
Solution	Each actor should have a well-defined role and there should be no redundancy of roles in the system. There should be a threshold on the number of actors a system can support.

6.2.5.2   Constraints

1. AnyParty should belong to one or more AnyApplication.
2. Classification classifies one or more AnyParty.
3. AnyParty may or may not follow another AnyParty.
4. AnyParty may or may not lead another AnyParty.
5. AnyParty can be of one out of four types.
6. Classification classifies AnyParty into one of four types.

6.2.6   SOLUTION

The stable design pattern given below is the proposed solution for handling issues in AnyParty and provides a generic pattern that can be applied to AnyApplication by attaching application-specific features (Figure 6.1).

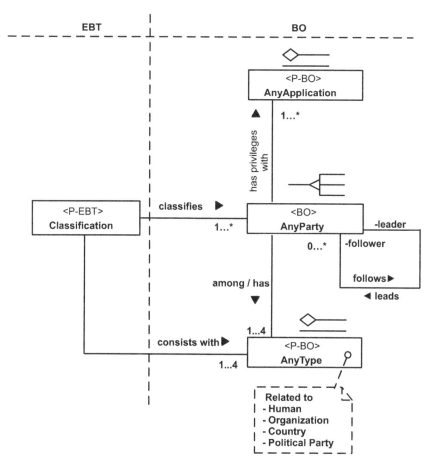

FIGURE 6.1 AnyParty design pattern.

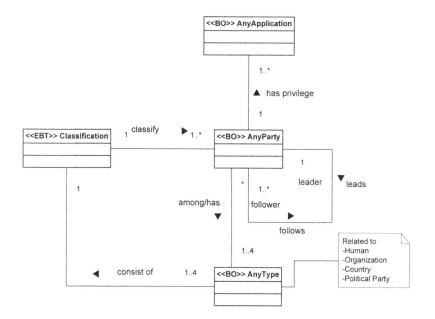

6.2.6.1 Pattern Structure and Participants

The participants of the AnyParty design pattern are as described below.

6.2.6.1.1 Classes

AnyParty: This class represents a party itself. It has a type and supports a system to make Classification.

6.2.6.1.2 Patterns

AnyType: This class represents the party type. Type can be a human, organization, country, and political party. Also, it consists of Classification (EBT).

AnyApplication: This class represents the application that AnyParty uses or supports. In short, AnyParty has the associated privileges with it.

Classification: This class classifies AnyParty. Classification of AnyParty is done with the help of AnyType.

6.2.6.2 Class Diagram Description

The class diagram visual illustration of all the classes in the model along with their relationships with other classes. Description of the class diagram is given as below.

- Classification is the EBT of this model and is used for classifying AnyParty (BO).
- AnyParty (BO) has AnyType (BO). AnyType (BO) can be related to human, organization, country, or a political party.
- AnyType (BO) consists with Classification (EBT).
- One or more AnyParty (BO) has privileges with AnyApplication (BO).
- AnyParty (BO) uses or supports AnyApplication (BO) to help the application achieve its goal.
- AnyParty (BO) can either play a role of leader by leading other parties or the role of a follower by following another party.

6.2.6.3 CRC Cards

	Collaboration	
Responsibility	**Client**	**Server**
Classification (Classification) (P-EBT)		
The categorization of AnyParty which is generally different for different parties	AnyParty AnyType	distinguish() assign() classify()
Attributes: id, grouping, criteria, classificationProperty		
AnyParty (AnyParty) (BO)		
The one who is classified	Classification AnyApplication AnyType AnyParty	use() lead() follow() hasPrivilege()
Attributes: id, partyName, type, role, activity		
AnyType (AnyType) (P-BO)		
Represent the type of party like human, organization, country, and political party	AnyParty Classification	consistWith() has() typeOfParty()
Attributes: id, type, actorName, property		

Continued

Responsibility	Collaboration	
	Client	Server
AnyApplication (AnyApplication) (P-BO)		
A party has privileges with it and uses it to help achieve goal of the application	AnyParty	function() achieveGoal() produceResult()
Attributes: id, type, applicationName, partyInvolved		

6.2.7 CONSEQUENCES

The AnyParty design pattern is generic enough to serve as a building block for applications in diverse domain.

Using an AnyParty design pattern for AnyApplication will require that the AnyParty has sufficient privileges to use the application. However, this does not mean that the pattern is incomplete, as this is the nature of patterns—they need to be used with other components.

One good thing with AnyParty design pattern is that the pattern has been derived with the factor of stability in mind. It has captured the enduring knowledge of business and its capabilities and will stand the test of time. However, the unpleasant thing with it is that it has a very small domain of application.

6.2.8 APPLICABILITY WITH ILLUSTRATED EXAMPLES

In this section, following two examples illustrate the use of AnyParty design pattern by using use case description and a behavior model like sequence diagram.

6.2.8.1 Application No. 1: AnyParty in Political Scenario

Here, AnyParty is evaluated in a political scenario. American politics has been dominated by two major parties, the Democratic Party and the Republican Party, ever since the American Civil War, though other parties of marginal political significance has also always played their minor role. The politics of the United States assumed the framework of a presidential republic, where the President of the United States is the head of state, head of government, and of a two-party legislative and electoral system. The federal government shares sovereignty and federal coexistence with all state governments, while the Supreme Court acts a force to balance the rights of each [6–8].

Concept	Description
Classification (EBT)	To arrange and organize the AnyParty based on its type and role
AnyParty	Named political parties
AnyApplication	Represents the U.S. political system
AnyType	Identify the type of the party like political parties which participate in the political system

6.2.8.1.1 Model
The model for this application is shown below (Figure 6.2).

6.2.8.1.2 Use Case Description
Use Case No: 1.0
Use Case Title: Classify AnyParty in Political Scenario

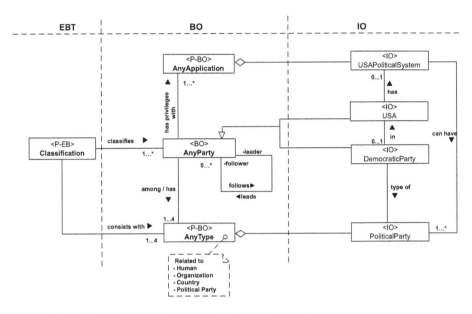

FIGURE 6.2   Class diagram for AnyParty pattern in political scenario.

Actors	Roles		
AnyParty	NamedPoliticalParties		

Classes	Type	Attributes	Operations
Classification	EBT	• id • classificationProperty • grouping • criteria • classificationRule	• stateClassifyingCriteria() • classify() • emphasisClassificationProperty() • classificationRule()
AnyParty	BO	• id • partyName • type • role • activity	• use() • lead() • follow() • hasPrivilege()
AnyApplication	BO	• id • name • type • partyInvolved	• function() • achieveGoal() • produceResult()
AnyType	BO	• id • typeName • property	• consistWith() • has() • typeOfParty()
USAPoliticalSystem	IO	• id • name • noOfParties • typeOfSystem	• maintainDemocracy() • grantCitizensRight() • createSafeEnvironment()
NamedPoliticalParties	IO	• id • name • function • noOfMembers	• govern() • layLaw() • protect() • formPolicy()
Political party	IO	• id • name • role	• negotiate() • communicate() • spreadInformation()

1. Classification (EBT) classifies AnyParty (NamedPoliticalParties). On what basis is Classification (EBT) carried out? Is Classification (EBT) reliable and accurate?
2. AnyParty (NamedPoliticalParties) has AnyType (PoliticalParty). Does AnyParty (BO) always have AnyType (BO)? If no, then what is the scenario?
3. Classification (EBT) consists of AnyType (PoliticalParty) of AnyParty (BO). Is this information correct?
4. Two or more AnyParty (BO) uses/supports AnyApplication (USAPoliticalSystem). What if there is only one AnyParty (BO)? In that case, does single AnyParty (BO) still support AnyApplication (USAPoliticalSystem)?
5. AnyParty (BO) enjoys privileges for AnyApplication (USAPoliticalSystem). How are privileges assigned?
6. AnyParty (BO) might act as a follower or leader depending on its role in AnyApplication (BO). How to find out whether AnyParty (BO) is a follower or leader?

6.2.8.1.2.1 Alternatives
1. Other than AnyType (PoliticalParty) (BO), there can be other AnyType (BO) say organizations that support these political parties.
2. Classification (EBT) of AnyParty (BO) is not accurate (Figure 6.3).

6.2.8.1.3 Sequence Diagram

6.2.8.1.3.1 Sequence Diagram Description The flow diagram in Figure 6.3 represents the flow of messages, when Classification (EBT) classifies AnyParty (NamedPoliticalParty) (IO). The AnyParty (NamedPoliticalParty) (IO) can either be a leader or follower depending on its role. AnyParty (BO) reflects on the act of Classification (EBT). Classification (EBT) consists with AnyType (PoliticalParty) (IO). The AnyType (PoliticalParty) (IO) has AnyParty (NamedPoliticalParty) (IO). AnyParty (NamedPoliticalParty) (IO) has privilege with AnyApplication (USAPoliticalSystem) (IO). AnyParty (NamedPoliticalParty) (IO) uses AnyApplication (USAPoliticalSystem) (IO).

6.2.8.2 Application No. 2: AnyParty in the Context of Interaction between Two Countries

AnyParty in the well-known Camp David Accord Application involves participation of the Prime Ministers of Egypt and Israel. These two countries negotiated at Camp David, which led to the creation of peace treaty. The table below summarizes the participants in this scenario.

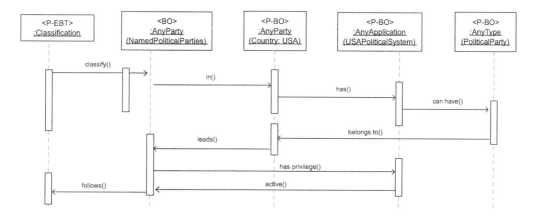

FIGURE 6.3 Sequence diagram for AnyParty political scenario.

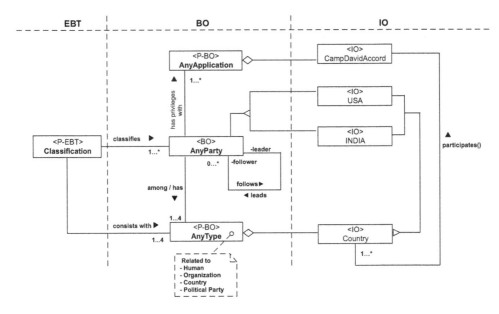

FIGURE 6.4 Class diagram of AnyParty pattern in the context of interaction between two countries.

Concept	Description
Classification (EBT)	To arrange and organize the AnyParty based on its type and role
AnyParty	Named country
AnyApplication	Represents Camp David Accord—the treaty between two or more countries
AnyType	Identify the type of the Party like countries that communicate with each other to discuss and solve various issues

6.2.8.2.1 Model

The model for the application is as shown in Figure 6.4.

6.2.8.2.2 Use Case Description

Use Case No: 1.0
Use Case Title: Interact within countries

Actors	Roles		
AnyParty	NamedCountry		

Classes	Type	Attributes	Operations
Classification	EBT	• id • classificationProperty • grouping • criteria • classificationRule	• stateClassifyingCriteria() • classify() • emphasisClassificationProperty() • classificationRule()
AnyParty	BO	• id • partyName • type • role • activity	• use() • lead() • follow() • hasPrivilege()

Continued

Classes	Type	Attributes	Operations
AnyApplication	BO	• id • name • type • partyInvolved	• function() • achieveGoal() • produceResult()
AnyType	BO	• id • typeName • property	• consistWith() • has() • typeOfParty()
CampDavidAccord	IO	• id • name • noOfParties • typeOfSystem	• formTreaty() • communicate() • negotiate()
NamedCountry	IO	• id • name • policy • agreement • rule	• participate() • initiateTalk() • respond() • signAgreement() • follow()
Country	IO	• id • name • location • capital	• protect() • trade() • exportGood() • importGood()

1. Classification (EBT) classifies AnyParty (NamedCountry). On what basis is Classification (EBT) carried out? Is Classification (EBT) reliable and accurate?
2. AnyParty (NamedCountry) has AnyType (Country). Does AnyParty (BO) always have AnyType (BO)? If no, then what is the scenario?
3. Classification (EBT) consists of AnyType (Country) of AnyParty (BO). Is this information correct?
4. AnyParty (BO) uses/supports AnyApplication (CampDavidAccord). How does AnyParty (BO) support AnyApplication (CampDavidAccord)?
5. AnyParty (BO) has privileges for AnyApplication (CampDavidAccord). How are privileges assigned?
6. Two or more AnyParty (NamedCountry) participate in AnyApplication (CampDavidAccord). What is the criterion for AnyParty (NamedCountry) to participate?
7. AnyParty (BO) might act as a follower or leader depending on its role in AnyApplication (BO). How to find out whether AnyParty (BO) is a follower or leader?

6.2.8.2.2.1 Alternatives

1. Other than AnyType (Country) (BO), there can be other AnyType (BO), say organizations like United Nations, that mediate these communications between two or more countries.
2. Classification (EBT) of AnyParty (BO) is not accurate.

6.2.8.2.3 Sequence Diagram

The flow diagram in Figure 6.5 represents the flow of messages when Classification (EBT) classifies AnyParty (NamedCountry) (IO). The AnyParty (NamedCountry) (IO) can either be a leader or follower depending on its role. AnyParty (BO) reflects on the act of Classification (EBT). Classification (EBT) consists with AnyType (Country) (IO). The AnyType (Country) (IO) has AnyParty (NamedCountry) (IO). AnyParty (NamedCountry) (IO) has privilege with AnyApplication (CampDavidAccord) (IO). AnyParty (NamedCountry) (IO) uses AnyApplication (CampDavidAccord) (IO) and is thus a participant.

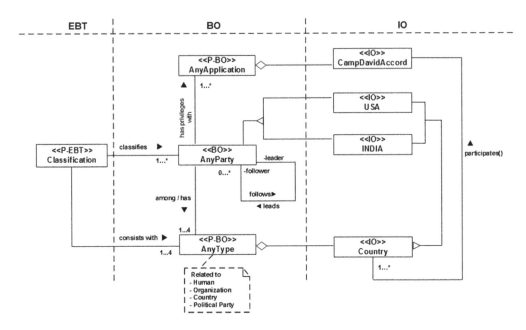

FIGURE 6.5 Sequence diagram of AnyParty pattern in the context of interaction between two countries.

6.2.9 Design and Implementation Issues

6.2.9.1 Design Issues (Two or Three of the Following Issues)

6.2.9.1.1 Design of AnyParty Pattern: Concurrent Development Cycle

AnyParty design pattern is designed in such a way that it can be adapted and reused in diverse domains and under different scenarios. This section describes some of the issues that are faced in designing and implementing a stable AnyParty design pattern.

To implement AnyParty design pattern, one needs to come up with Enduring (EBTs) and BOs. The EBT for AnyParty is Classification. Identifying EBTs is the hardest and challenging part while designing a pattern because EBTs are conceptual and one needs some skill and intuition to identify them.

Another challenge is identifying correct supporting BOs and then connecting them correctly to EBTs in the design pattern. An important issue in implementing design pattern is to ensure that there is no hook up between EBTs and IOs.

Designing generalized pattern that is applicable to all possible business domains is bit time consuming because it requires identification and finding interrelatedness of all possible BOs for the corresponding EBT. The pattern can be designed only through a thorough knowledge of the problem domain and requires identification of EBTs, BOs and IOs to accommodate all concepts of stability model.

EBTs represent analysis, BOs represent design objects and IOs depict implementation. Identification of EBT, BOs, and IOs for modeling stable patterns can happen simultaneously and thus it represents a concurrent development cycle.

6.2.9.1.2 Type/Subtypes Design Specifications: AnyParty/Roles

For some types of applications, there might be a need for a party has to perform multiple roles in order to achieve the desired goal. By using traditional model, it is not possible to achieve this kind of flexibility. As illustrated in Figure 6.6 under related patterns and measurability section, the party Meta model is too rigid and inflexible to accommodate multiple roles of a party. However, with stability design pattern, any number of parties with diverse roles can be easily incorporated. As

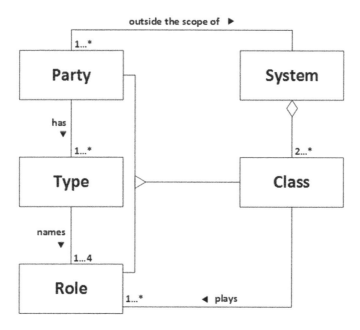

FIGURE 6.6 Party Meta model.

illustrated in Figure 6.1 and later on under applicability section, the pattern is very flexible because of the presence of AnyType as a BO. Roles can be inherited from AnyType.

Implementation Issues (two or three of the following issues): Some issues in implementation of AnyParty design pattern are described in the next section. Classification (EBT) is not inherited and thus has an associative relationship with AnyType (BO) and AnyParty (BO). AnyApplication (BO) will have IO applications in the implementation phase. AnyParty (BO) will also inherit any kind of party in the implementation phase.

6.2.9.1.2.1 How to Use AnyParty and AnyActor   The Meta model illustrated in Figure 6.6 is not applicable for applications, where individuals as well as groups of organizations are involved. This is because in Meta models, there is no clear distinction between actor and parties. Traditional patterns do not allow one to represent organizations. For using stability modeling for such applications, we can use AnyParty and AnyActor stable design patterns in combination. Since, design patterns are extensible, AnyParty and AnyActor patterns can be easily plugged together to satisfy the requirements of an application.

6.2.9.1.2.2 Role Delegation   Role delegation is an important part of the patterns, as it has to be designed in such a way that replacement becomes very easy. One actor can perform multiple roles in a system and can be given more while some roles can be removed too. Thus, the system must still not be affected, as this is a volatile issue.

6.2.9.1.2.3 Pluggable Roles   The system must be designed to accommodate many roles in the future. Pluggable roles are special kind of roles, which when inserted in to the system, does not affect the stability of the system. These are unique and flexible objects to make the system more efficient.

6.2.10 RELATED PATTERN AND MEASURABILITY

Party can be defined as a group that is gathered for a special purpose or task: *a fishing party, a search party*. Based on this definition, following Meta model can be modeled for party problem.

6.2.10.1 Traditional Model (Meta Model) versus Stable Model (Pattern)

Traditional modeling (Meta model) generally results in a unstable model which is not adaptable, scalable, and extensible. On the other hand, analysis/design pattern created by using SSM is adaptable to changing requirements and can be scaled easily with minimum most effort.

In traditional modeling, because we only design as much is needed for a specific application in question and not thinking of its applicability in other domain, the resulting application is only very specific. On the contrary, in case of stability, we tend to use general enduring concepts, and hence the resulting pattern can be used for building numerous applications. In nutshell, the resulting pattern can serve as a building block for diverse application domains.

With traditional modeling, the resulting design is tangible enough to be comprehended, while with stability model, pattern is intangible. Designing traditional modeling comes from the problem statement, while stability approach needs experience to derive solution that usually demands application of intuition and thinking outside the box. However, implementing successful pattern requires expertise and meticulous work and detailed understanding of the problem domain.

6.2.10.2 Testability

If AnyParty design pattern can be used (as it is and without changing the core design) and by only plugging IOs for infinite number of applications, then AnyParty pattern can be said to be testable. In the applicability section, two widely different applications are illustrated and they do not require changing the core design of the pattern. By using the scenarios listed in this chapter, many such scenarios can be deduced and proved that AnyParty pattern is indeed testable.

6.2.11 BUSINESS ISSUES

6.2.11.1 Business Rules

The business rules for AnyParty pattern are

- A party must be classified.
- The Classification is dependent on the role performed by the party.
- Party is either external or internal to the system.

Whereas according to meta model, the business rules are

- A party is thrown, when there is a purpose/reason.
- Person, who throws party, is called host and those who attend the party are called guests.

6.2.11.2 Business Models

The stable AnyParty design model is generic and so can be extended by plugging in specific IOs specific to the application domain thereby resulting in numerous stable and adaptable business models.

6.2.11.3 Enduring Business Theme

The EBT of this pattern is Classification and it is the generic goal that remains same for the variety of the applications in varied fields. By using this common goal for designing pattern, stability and extensibility are achieved.

6.2.11.4 Business Integration

As AnyParty pattern provides hooks, the integration of the AnyParty design pattern to existing business model becomes easier. Integrating AnyParty pattern to a particular application domain will also require the plugging of application-specific IOs to the hooks.

6.2.11.5 Business Transformation

Use of AnyParty design pattern by a company in development of AnyApplication strengthens its business transformation process. This is due to AnyParty pattern's ability to quicken the development process by providing the framework for the application. Also, since AnyParty pattern is tested and verified, chances of getting bugs are minimal.

6.2.12 KNOWN USAGE

The AnyParty design pattern can be applied in any scenario such as

- Political context. In this context, the various political parties involved in the scenario are referred to as AnyParty.
- Communication between countries scenario like signing a peace treaty. Here, AnyParty refers to various countries that are involved in the communication process.
- Applications involving organizations like United Nations, etc. Some examples are striving for world peace, etc. In such scenarios, AnyParty is referred to organizations such as WHO, UNO, etc. and countries which are involved.
- Software scenarios, where any parties like humans, organizations, political parties, and governments are involved. Together, they constitute AnyParty.
- Social gathering and events like birthday parties, etc. Here, AnyParty are generally humans, though sometimes organizations may throw a social gathering say on occasion of their inauguration.

The generalized stability pattern has its own consequences, which are discussed below:

Good	Bad	Ugly
• The generalized pattern allows wide openness and applicability	• Difficult to come with common attributes for the four types of AnyParty	• Nothing as of now
• Is based on legality and helps to solve any issues	• Some legality cannot be solved at all	•
• Allows humans and creatures to be part of it	• AnyActor is not the replacement for AnyParty.	•

6.3 TIPS AND HEURISTICS

- Describing patterns is a hard job and requires careful and calibrated work.
- Meta model is totally different than the stable model, and it is a traditional model.
- Pattern design must be generic, so that it can be applied to applications spread across various domains.
- EBT must represent the goal of the pattern.
- Intuition and experience are required in order to find correct EBT for the pattern.
- BOs provide capabilities to achieve the goal of the pattern. Identification of BO requires skill and times, and culling out correct BOs.
- BOs provide hooks to which specific IOs can be plugged and creating varied applications in diverse domains. This will reduce the cost by encouraging reusability.

6.4 SUMMARY

Although, building a stable design pattern for AnyParty that can be reused and reapplied across diverse domain is a difficult task and requires thorough understanding of the problem, it is worth the effort and time to design and create it. Modeling a problem by using SSM usually results in a reusable, extensible, and stable pattern.

The correct identification of EBT and BOs for AnyParty is the most challenging task and it requires some experience and skill. Once EBT and BOs are correctly identified, the next immediate challenge is to determine the relationship between EBT and BOs, so that AnyParty pattern can hold true in any context of usage for party. Once this is done, depending on the application, the IOs are attached to the hooks provided by BOs. Thus, using AnyParty pattern as a basis, infinite number of applications can be built by just plugging in the application-specific IOs to the pattern. This results in reduced cost, effort, and stable solution. Hence, AnyParty design pattern is very useful.

6.5 REVIEW QUESTIONS

1. What are the four types of AnyParty?
2. Define AnyParty.
3. Why traditional approaches of software design do not yield a stable and reusable model of AnyParty?
4. What are the properties of AnyParty SDP?
5. How does SSM help in making pattern stable, reusable, extensible, and adaptable?
6. How is AnyParty different than AnyActor?
7. Is it possible to use AnyUser in place of AnyParty?
8. Why do we use AnyParty more often than AnyDoer?
9. Identify four scenarios from day-to-day life, where AnyParty SDP can be applied.
10. List and explain the functional requirements of AnyParty SDP.
11. List three nonfunctional requirements of AnyParty that has not been discussed in this chapter. Justify your findings.
12. Write two challenges of AnyParty SDP using the challenge template.
13. Identify the constraints of AnyParty SDP.
14. Draw the class diagram from AnyParty SDP.
15. Discuss the participants in the class diagram of AnyParty SDP and justify their existence.
16. Write down the CRC cards for AnyParty SDP.
17. What are the consequences of AnyParty SDP?
18. Discuss the applicability of AnyParty SDP in a political scenario with the help of a class diagram. Write a use case for it and draw a sequence diagram for that use case.
19. Discuss the applicability of AnyParty SDP in the context of interaction between two countries with the help of a class diagram. Write a use case for it and draw a sequence diagram for that use case.
20. Discuss the design issues of AnyParty SDP.
21. Discuss the implementation issues of AnyParty SDP in terms of how to use AnyParty, Role delegation, and Pluggable roles.
22. Draw and explain Meta model of AnyParty.
23. Compare the Meta model of AnyParty with AnyParty SDP.
24. Discuss testability of AnyParty.
25. Discuss the business issues of AnyParty in terms of:
    a. Business Rules
    b. Business Models
    c. EBT
    d. Business Integration
    e. Business Transformation
26. Describe some of the known usages of AnyParty SDP.
27. Discuss the good, bad, and ugly consequences of generalized stability patterns.
28. Write some tips and heuristics that you have learnt from this chapter.

6.6   EXERCISES

1. Describe a scenario of an application of AnyParty in creation of a job search web site.
   a.   Draw a class diagram of the application.
   b.   Document a detailed and significant use case.
   c.   Create a sequence diagram of the created use case of b.
2. Describe a scenario of an application of AnyParty in apartment renting system.
   a.   Draw a class diagram of the application.
   b.   Document a detailed and significant use case.
   c.   Create a sequence diagram of the created use case of b.
3. Describe a scenario of an application of AnyParty in university admission application for graduate program.
   a.   Draw a class diagram of the application.
   b.   Document a detailed and significant use case.
   c.   Create a sequence diagram of the created use case of b.
4. Describe a scenario of an application of AnyParty in a project team in a company.
   a.   Draw a class diagram of the application.
   b.   Document a detailed and significant use case.
   c.   Create a sequence diagram of the created use case of b.

6.7   PROJECTS

1. Model a customer service department. Identify all parties. Draw a class diagram by using stable pattern and a Meta model. Which model do you prefer more? Write CRC cards and use cases. Draw sequence diagram for the same.
2. You are a part of a team that is developing an e-commerce web site. Identify parties for such an application and illustrate with a class diagram, applicability of AnyParty SDP.

REFERENCES

1. William Shakespeare (April 26, 1564–April 23, 1616), last modified on June 12, 2017, https://en.wikipedia.org/wiki/William_Shakespeare
2. M.E. Fayad. Introducing Fayad's legal actors to UML – Adding a new dimension to software modeling. *International Journal of Advanced Research and Studies (IJARS)*, II(3), September–October Issue 2016.
3. M.E. Fayad and A. Altman. Introduction to software stability, *Communications of the ACM*, 44(9), 2001, 95–98.
4. M.E. Fayad. Accomplishing software stability, *Communications of the ACM*, 45(1), 2002, 95–98, 111–115.
5. M.E. Fayad. How to deal with software stability, *Communications of the ACM*, 45(4), 2002, 109–112.
6. S.P. Huntington. *American Politics: The Promise of Disharmony*, Cambridge, MA: Harvard University Press (HUP), September 1983, 320 pp., ISBN-10: 0674030214 and ISBN-13: 978-0674030213.
7. P.U. Bonomi. *A Factious People: Politics and Society in Colonial New York*, Cornell University Press, Ithaca, New York, 1st Edition, September 2014, 360 pp.
8. A.H. Chroust. *The Rise of the Legal Profession in America* (2 vols), vol. 1, P.U. Bonomi, A Factious People, pp. 281–286, 1965.

SIDEBAR 6.1   INTRODUCING FAYAD'S LEGAL ACTORS TO UML: ADDING A NEW DIMENSION TO SOFTWARE MODELING

Legal actors do not exist in UML, which prevent us from using it in numerous legal applications such as binding contracts. The goal of this chapter is to utilize the services of legal users in UML to model numerous legal applications. The main idea is to introduce legal users in modeling and law applications and in what manner one can use them in an UML construct.

Reduction in the legal cost is one of the benefits apart from simplicity and convenience.

A clear understanding of what a legal user is and in what manner one can use it in a professional manner will help developers in creating a superior software modeling system by using UML. Legal parties are different from actual actors, because humans as actual actors have no obligations, but the human in legal parties have numerous obligations.

Introduction

Actual Users

These are actual users (role players) of the system are called actors in UML. AnyActor has two types: human (person) and machine or hardware [B]. Professor Fayad, in his literature, extended these to two new types that are software packages such as browsers and creatures like other living creatures, animated and fiction characters [A].

Legal Users

Legal users are called parties. Party, defined by law as a party, is a *person* or group of persons that compose a single *entity* that can be identified as one for the purposes of the law. Parties include *plaintiff* (person filing suit), *defendant* (person sued or charged with a crime), *petitioner* (files a petition asking for a court ruling), *respondent* (usually in opposition to a petition or an appeal), cross-complainant (a defendant, who sues someone else in the same lawsuit), or cross-defendant (a person sued by a cross-complainant). References 1 and 2 have four types namely human (person), country, organization, and political party which cover a lot more applications than actual actors.

Both actors and parties are outside the scope of the system.

Roles

Actual users (actors) and legal users (parties) play roles. Roles are part of the system. Traditionally, one has to name the type to name the role.

Section 2 discusses the traditional UML modeling.

Section 3 discusses software stability model as a new approach to design of a more durable and long-lasting system.

Section 4 includes a comparative study of legal users with other systems with respect to criteria such as scalability, reusability, etc.

Section 5 provides an analysis of the legal users based on comparative study results.

We finally try to conclude in *Section 6* followed by references in *Section 7*.

Traditional Model

In UML, an actor/party specifies roles played by a person, organization, country, or a political party.

Examples:

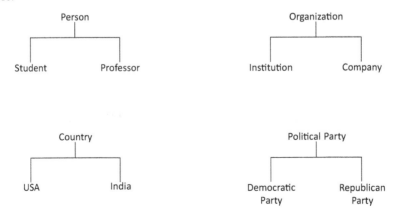

Stable Model

Dr. M.E. Fayad introduced an approach in software engineering that targets software stability model [3–6], and it is used to model a generic design pattern that can be used in the various application domains. The software stability model is used to identify the elements which represent the stable aspects of the problem domain. To identify these elements, the concepts of enduring business themes (EBTs) and business objects (BOs) have been introduced. EBTs are the enduring knowledge areas of the concept and which represent the goal of the concept. BOs map the EBTs to tangible objects. The EBT and BOs compose the design pattern and the design pattern can then be applied as a solution to different domains by plugging in the Industrial Objects (IOs), which are domain-specific objects. Applying these concepts results in a stable core design and changes to the software will change only the IOs while the EBTs and BOs remain stable (see Figure SBF3.1).

Utilization and Comparison

S. No.	Property	Actual Users	Legal Users
1	UML constructs	Exists in UML and are called Actors	An addition to UML and is called party
2	Type	i. Person ii. Hardware	i. Human or persons ii. Country iii. Organization iv. Political parties
3	Applications		**Property legal users:** It governs ownership and possession. It refers to ownership of land, stocks, money, notes, Patents, and copyrights, etc.
4	Example	Professor, Root	Professor IBM, Web, Dog, Cat, Jerry, and Spark

Conclusions

Legal actors do not exist in UML and it is an enhancement to UNL applicability, which prevents us from using UML in legal applications such as binding contracts, etc. The goal is the utilization of legal users in UML to model legal applications. The idea is to introduce legal users in modeling and law applications and how to use them.

References

1. M.E. Fayad. Introducing Fayad's legal actors to UML – Adding a new dimension to software modeling. *International Journal of Advanced Research and Studies (IJARS)*, II(3), September-October Issue 2016.
2. M.E. Fayad. Software System Engineering Courses, CmpE 202, Lecture Notes, Fall and Spring 2000–2013.
3. M. Fayad and A. Altman. An introduction to software stability, *Communications of the ACM*, 44(9), 2001, 95–98.
4. M.E. Fayad. Accomplishing software stability, *Communications of the ACM*, 45(1), 2002, 111–115.
5. M.E. Fayad. How to deal with software stability, *Communications of ACM*, 45(4), 2002, 109–112.
6. M.E. Fayad, H. Sanchez, S. Hegde, A. Basia, and A. Vakil. *Software Patterns, Knowledge Maps, and Domain Analysis*, Boca Raton, FL: Auerbach Publications, Taylor & Francis Catalog #: K16540, December 2014. ISBN-13: 978-1466571433.

7 AnyEntity Stable Design Pattern

The true knowledge or science which exists nowhere but in the mind itself, has no other entity at all besides intelligibility; and therefore whatsoever is clearly intelligible, is absolutely true.

Ralph Cudworth [1]

The AnyEntity design pattern molds the concept of an entity by using software stability model (SSM) concepts [2–4]. AnyEntity is used in diverse domains, where each domain will have a different rationale for the use of this term. Since AnyEntity design pattern captures the core knowledge, it is quite easy to model AnyEntity in any application by simply hooking different dynamic components of the application. Its core is stable too. Again, since SSM [2–4] is used for sculpting the concept of AnyEntity, a generic model, that can be extended for use by diverse applications, is conceptualized. This is known to increase reusability and reduction in repetition of modeling AnyEntity for each individual application. In this chapter, we will introduce a general AnyEntity design pattern based on stability model and later introduce few scenarios where this pattern can be used.

7.1 INTRODUCTION

In a general context, an entity is something that has a distinct and separate existence; however, it need not be a material existence. In particular, abstractions and legal fictions are known as entities. In general, it is not presumed that an entity is animate [5–8].

Entity is used in various contexts and scenarios, where each of them has its own meaning and rationale behind using the term. The word *entity* is often useful, when one wants to refer to something that could be a human being, a nonhuman animal, a nonthinking life form such as a plant or fungus, a lifeless object, or even a belief for instance [5–8]. In each application scenario, this basic idea about an entity is similar, but the context in which entity is being used differs widely. For example, entities are used in system development as models that display communications and internal processing of, say, documents compared to order processing [5–8]. In philosophy, an entity could be viewed as a set containing subsets and such sets are said to be abstract objects. In law, an entity is something capable of bearing legal rights and obligations. It also means "legal entity" or "artificial person," but it may include a "natural person."

Since the function of AnyEntity depends on the context in which AnyEntity is used, the traditional approaches to software design will not yield a stable and reusable model. However, by using SSM, AnyEntity can be represented in any context by using a single model. The SSM requires creation of knowledge map by identifying underlying enduring business themes (EBT) and business objects (BO). By hooking industrial objects (IO) that is specific to each application, the model can be applied to any application domain. The resulting AnyEntity pattern is stable, resusable, extendable, and adaptable. Thus, any number of applications can be built by using this common model. The AnyEntity design pattern tries to capture the core knowledge of AnyEntity that is common to all the application scenarios that are listed above to create a stable design pattern.

The main objective of this chapter is to design a stability model for AnyEntity by creating knowledge map of AnyEntity. This knowledge map or core knowledge can then serve as building blocks for modeling different applications in diverse domains. The rest of this chapter presents AnyEntity, a stable design pattern based on the SSM.

What are the different perspectives of entity, and its impact on software development and management?

The term entity can be used in many frameworks. It is something that has a distinct and a unique existence. Any things that can be referred to are usually entities. The reference could be any living thing or a nonliving thing. Living things could be human beings, animals, and plants. Nonliving things could be any lifeless objects.

AnyEntity design pattern may have an immense impact on software development. However, traditional approaches to software design will not yield a stable and reusable model because the function of AnyEntity depends on the context in which AnyEntity is being used. Nonetheless, this instability can be avoided by using SSM, whereas AnyEntity can be represented in any context by using a single model. In addition, SSM requires identification of underlying EBTs and BOs, which helps in creating a knowledge map. Then, by hooking IOs, that are specific to each application, the model can be applied to any other application domain. Hence, a stable, reusable, extendable, and adaptable application is obtained.

What are the consequences of entity on software modeling, architecture, software development, and more?

The AnyEntity design pattern attempts to capture the core knowledge of AnyEntity that is common to all the application scenarios to design a stable design pattern. By using the principle of the SSM, AnyEntity design pattern will shape the concept of an entity. Each domain has a different rationale for the use of this term; hence, AnyEntity is used in diverse domains and in different perspectives. AnyEntity design will also capture the core knowledge of the term entity; it is easy to model AnyEntity in any application by simply hooking the dynamic components of the application. Because, the software stability pattern is used for modeling the concept of AnyEntity, a generic model can be extended for use by diverse applications and this idea could be conceptualized to include all those domains that relate to the definition of an entity. This will increase the reusability factor and it can result in lessening in the instances of repetition of modeling AnyEntity for each individual application.

Different contexts of using an entity have different processes and operations. Moreover, an entity can be used in different domains, as it is described as any real world object. The reference could relate to any living thing or a nonliving thing. The AnyEntity design pattern covers the concept of existence of an entity in the context of many diverse domains that have different functionalities, usage, and type of entities. The pattern designed here focuses mainly on depicting entity based on its state and type, which presents it in a readable format. It is necessary to model a pattern to target a generic domain that might contain different types of entities. Thus, modeling a generic design pattern that can be applied to all these domains makes perfect sense. A few numbers of things should be taken into consideration while modeling AnyEntity pattern. First, the model should capture the core knowledge of the entity problem. Second, the model should be designed in such a way that it can be reused to model the entity in any application without losing the basic functionality. Third, the ultimate goal for AnyEntity must be established along with other patterns used in the design pattern of AnyEntity. By using software stability concepts, it is possible to build AnyEntity that can be used in different context in different domains.

In addition, report about having AnyEntity stable design pattern is a very useful thing.

7.2 ANYENTITY DESIGN PATTERN DOCUMENT

7.2.1 Pattern Name: AnyEntity Stable Design Pattern

Entity is very essential in representing an object. The objective of pattern is to generalize the idea of entity, so that it can be used as a foundation for representing different things. The AnyEntity design pattern abstracts the existence of an entity based on certain properties inherent to the context in which entity is used, and thereby it helps create a general pattern that can be used across any application. This pattern is required to model the core knowledge of AnyEntity without binding the pattern to a specific application or domain; hence, the name AnyEntity is chosen. Again, AnyEntity name is also chosen because it is semitangible and externally stable, and it is a BO. A BO is considered as a capability for an EBT. Moreover, all BOs have a beginning and an end. We always use the word 'Any' with the name of the pattern because the entity has a beginning and an end. The EBT is existent as it is the goal of AnyEntity.

Why did we choose that specific name?

The name is right and the most appropriate, for an entity is something that has a distinct and separate existence, though it need not be a material existence. Entities are used in system developments, as models that display communications and internal processing of, say, documents compared to order processing. Thus, this is an excellent choice of title. The name is fitting and suitable to the analysis on how an entity can be stabilized, so that the wheel is not reinvented every time.

7.2.2 KNOWN As

An entity is also an object, article, creature, and individual. An entity is something that has a real existence. Therefore, entity is a living or a nonliving thing. Hence, the all synonyms to an entity will fall under this category; therefore, it can be used interchangeably within a common context.

7.2.3 CONTEXT

As highlighted before, an entity is something that has a distinct, separate existence; it need not be a material existence. In particular, abstractions are also considered entities. An entity is a thing of significance also, either real or conceptual, about which the business or system being modeled needs to hold information.

AnyEntity pattern can be applied to various domains of daily importance, which include diverse areas such as philosophy, law, healthcare, and computing. AnyEntity pattern is applied to different domain to achieve different set of goals. The AnyEntity design pattern deals with the issue of abstracting various kinds of entities with different characteristics and types.

In the later section, many application scenarios to illustrate the applicability of AnyEntity pattern are described. AnyEntity concept is illustrated with two distinct scenarios, like database and healthcare, to demonstrate how an existence of entity can be abstracted in such applications.

Within the context of an entity as the main theme, developers can visualize many scenarios where it can be used as a tool to highlight different BOs and the main EBT. Here are two scenarios that work in the context of usage of entity theme.

7.2.3.1 Scout Movement

Scout is a fairly well-known global organization that works under the principle of universal friendship, peace, and a conflict-free brotherhood. Scout is a global entity (AnyEntity) that seeks to promote unity and the understanding of Scouting's purpose and principles (AnyResponsibility), while facilitating its expansion and development (AnyRole). Scouts, guides, cubs, and bulbul (AnyActor) are a part of the system and they interact with the system. On the other hand, World Scout Movement (AnyActor) represents different parties involved in the process of abstracting an entity. Scouting is an exciting entity that allows its pupils to participate in several on-field activities like camps and jamborees (AnyEvent) and it helps provide a humanitarian perspective to the entire movement (AnyView).

7.2.3.2 Rotary International as a Universal Entity

Rotary International is a legendary nonvoluntary organization that works as a universal organization with an established motto of "Service Above Self" and "One profits most who serves best." Rotarian (AnyActor) are the heart and soul of Rotary International (AnyEntity) and it seeks to advance peace, friendship, goodwill, and freedom from hunger among different nations (AnyRole). Rotarians are expected to understand and follow these guidelines as an essential rule to serve as Rotarians (AnyResponsibility). The Board of Governors of Rotary International (AnyParty) formulates and recommends guidelines, action plans, and global schedules to serve humanity while Rotarians carry out their responsibilities by following these packages of practices (AnyRole). Rotary International is viewed as a facilitator (AnyView) of peace and conflict-free world that is free from strife, wars, conflicts, and disturbances.

7.2.4 PROBLEM

AnyEntity highlights the problem over which the pattern concentrates. Entity can be used in many contexts and scenarios and among different operations. Since entity is described as any real world object, it is used everywhere. Almost everything in the world falls under the category of entity. A car, person, and a creature are a few examples of entities. The AnyEntity design pattern covers the concept of existence of entity in the context of diverse domains that have different functionalities, usage, and type of entities. The pattern focuses on depicting entity, based on the state and type of entity, which presents entity in a readable format. It is necessary to model a pattern to target a generic domain that might contain different types of entities. Thus, modeling a generic design pattern that is applicable to all these domains makes a good sense. However, pattern developers should note down below listed points before modeling AnyEntity pattern.

- The model should capture the core knowledge of the entity problem.
- The model can be reused to shape the entity in any application without losing the basic functionality.
- The ultimate goal for AnyEntity must be established along with the other patterns used during the process of designing pattern of AnyEntity.

Since AnyEntity is used in different context and in different domains, building a generic model that depicts an entity in a lucid manner is truly challenging; however, developers could rectify this problem by creating a generic model. This model is illustrated and described in the solution section.
Entity properties are

1. Existence
2. Wide and Multiple Occurrences in a single application
3. Common Abstraction
4. Structural Adequacy
5. Descriptive Adequacy
6. BO

7.2.4.1 Existence

Attributes: Description, type, needs, when, how long, duration, range, existenceRequirements, existenceStates, existenceLimitations, and others.

7.2.4.1.1 Operations

existenceRequirements(): List the existence requirements or allow user to impose his/her own requirements.
find existenceMechanism(): Locate existence mechanism.
satisfyexistenceRequirements(): Which mechanism would satisfy the existence requirements.
existenceGuide(): List of criteria for existence, existence guides—user defined type and allow user to select criteria from the guide or impose his/her own criteria.
rule(): Define the rules of existence or allow user to impose the adaptability rules.

II. Capabilities (BOs):

7.2.4.2 AnyEntity

Attributes: id, entityName, entityType, status, position, states, and type.

7.2.4.2.1 Operations

performFunction(): Name the behavior of an entity and show the ability of performing the behavior.

status(): Return the status of the entity.

type(): Determine the entity type.

update(): Report any update related to the entity.

new(): Create a new entity.

relationship(): List and name the relationships of the entity.

7.2.4.3 AnyParty

This party participates in various activities.

Attributes: id, partyName, type, role, member, affair, activity, partiesInvolved, id, activity, category (or orientation), and purpose.

Operations: participate(), playRole(), interact(), group(), associate(), organize(), request(), setCriteria(), switchRole(), partake(), join(), monitor(), explore(), receive(), collectData(), integrate(), agree(), disagree(), leave().

AnyParty in XXX Context:

participate(): Actual participation, proactive, and involvement. Defined role. When one participates, he/she has a role.

playRole(): Participates in an activity or group. This is like subscriber. It links the party to an activity.

Interact(): One needs to interact with other member through activities. Timed interaction set up the mechanism for doing it.

join(): Join members to the party. One needs the parties to join, and the idea supporting that join. The group was already there and this new player joins.

leave(): A party leaves the group, with an excuse or idea.

Expell(): Force members to leave the party. This keeps a bad record on the expelled party.

group(): Groups the various parties in the system. One creates the party; this is different from join because there was no group before it was grouped. If there had been a group formed, we would have joined that group.

Associate: It is a noninteracting association, perhaps no actual work developed as a group, but, for example, with marketing purposes. One just needs to associate the company name with a bigger one.

organize(): Organize the activity, do planning, scheduling, and who is going to report what and when. Receives an activity and will return a plan for that activity.

Request(): Request an entity. Returns is the entity, for example, a librarian that returns a specific book

setCriteria(): Set by the member to any entity. This is the criteria to, for example, pick a woman,

switchRole(fromRole, toRole): You need the fromRole because you might be playing several roles at the same time.

Partake(): Take part in an activity, which is an entity; here, one will just take the share. It is divided in several parts, not necessarily equal, and then this takes its part and does it.

Monitor(): Observe; one monitors activities, collect data, and take notes.

Explore(): This receives an entity, and one travels through it or examines it with a purpose. It is a research and it will return some kind of information.

Receive(): Receives an entity and plays a passive role, in contrast to the Request(), that is actually active. For example, if one receives an email, he/she may forward it or reply.

collectData(): Receives an entity, for example, one collects data from a girl that lives around the corner, and then he/she organizes and stores it or you clean it up.

agree(): Gets an entity, they look at it and they decide to say yes or no.

7.2.4.4   AnyObject

Attributes: id, entityName, entityType, status, position, states, and type.
Operations: Similar as AnyEntity

7.2.4.5   AnyState

Attributes: stateId, stateProperty, type, name, description, status, conditions, and stateActions.

7.2.4.5.1   Operations

visualize(): Visualizes the state graphically.
show(): Display the state of the entity.
hide(): Hide the state of an entity.
exhibit(): Explore the state.
representState(): Represent the state.
simpleState(): Simplify the state.
expandedState(): Expand the state.
defineStateProperties(): Define the state properties.

7.2.4.6   AnyType

Attributes: id, typeName, properties, interfaceList, methodList, attributeList, and clientList.

7.2.4.6.1   Operations

change(): Changeable from one type to another. What are the changes between different types in use?
operateOn(): operateOn what?
pass(): Is it passable? Then how?
resume(): It renames itself from one name to the name.
label(): Name a label.
classify(): Is it classifiable? Then, what is class of this type?
functionality(): List type functionality.
attached(): Is it attachable? If so, how?
detach(): Is it detachable? If so, how?
nameAttribute(): List type attributes.
subtype(): What its own subtype?
Examples:
MediaType: This is the interface to create various media types.
change(): Change between media types.
operates(): Operate a media.

7.2.4.7   AnyResponsibility

Attributes: id, type, name, description, status, conditions, and responsibilityActions.

7.2.4.7.1   Operations

hide(): Hide the responsibility of an entity.
name(): Name the responsibility of an entity.
expandedResponsibility(): Expand the responsibility.

7.2.4.8   AnyView

Attributes: Visual, property, type, id, name, description, state, and status.

7.2.4.8.1   Operations

   visualize(): Visualizes the graphic view.
   show(): Display a view of the entity.
   hide(): Hide a view of an entity.
   resize(): Resize the view of an entity.
   exhibit(): Explore the view.
   representView(): Represent the view.
   simpleView(): simplify the view.
   expandedView(): Expand the view.
   defineViewProperties(): define the view properties.

7.2.4.9   AnyData/AnyProperty

   Attributes: id, name, valueRange, uperLimit, lowerLimit, type, domain, constraints, and
      description.

7.2.4.9.1   Operations

   exhibitedByAnyState(): Explore the data views
   displayedByAnyState(): Display the data views
   type()

7.2.4.10   Functional Requirements

   1. *AnyEntity*: This class represents entity itself and it is used to abstract any object. It also
      represents the behavior and attributes, which summarizes any object.
   2. *Existence*: This class represents the EBT. It characterizes the existence of any entity.
      Existence is what is asserted by the verb "exist." Existence is the state or fact of having
      independently of human consciousness and as contrasted with nonexistence.
   3. *AnyActor*: Any person, who is a part of the system and who interacts with the system, is
      known as an actor. It can be a creature, hardware, software, or any person. In this context,
      we can label a person, a start-up, or the organization, which develops context-based soft-
      ware as an actor.
   4. *AnyParty*: This class represents the parties involved in the process of abstracting an entity.
      It represents a person, organization, political party, country, or anyone who are involved in
      the process of abstracting an entity.
   5. *AnyState*: This class represents the state in which the entity exists. This class represents the
      state of an entity in any point of time during the process of abstracting an entity.
   6. *AnyType*: This class represents the type of entity. In addition, a class or group can be identi-
      fied in type software development. In this case, the entity identified for software develop-
      ment forms a legitimate example of AnyType.
   7. *AnyData*: This class represents the characteristics of the entity. It summarizes all innate
      features and concepts of the entity. It also tells us the composition of parameters that form
      the available data.
   8. *AnyView*: A view can be something that is seen, believed, outlined, or just spectacle. Other
      names for view are aspect, glimpse, outlook, perspective, or picture. Generality is the main
      reason for choosing this term, as this term is appropriate for all the possible scenarios of
      AnyView.
   9. *AnyResponsibility*: An existence of an entity also defines its mandated responsibilities and
      this is justified by different roles played by the entity itself. An entity performs some spe-
      cific functions in its domain to meet a preset goal or target.
   10. *AnyRole*: AnyEvent: This class represents any viewable occurrence. An event is something
      that happens. It has attributes such as *name, occasion, type, and outcome*. It has operations
      such as *appear(), happen(), and arrange()*.

7.2.4.11 Nonfunctional Requirements

1. *Useful*: An entity should be useful for the cause for which it is used. In other words, it should also extend all its benefits to the user who wants to create a stable pattern. Useful entities will not only save time and energy, it will also help developers bring down the cost of developing a stable pattern that is useful for a wide variety of purposes.

2. *Needed*: A developer, who is keen on developing a pattern, should find an entity that is invariably "needed" for a specific usage or cause. He or she should find an entity for immediate deployment whenever its utility value is needed for a specific task.

3. *Abstraction*: Any entity that is used for abstraction purpose should be named in a way that makes practical sense and that it should have all the appropriate aspects included and no one of the external ones. In other words, an entity should reflect an abstraction principle that reduces complexity and allows a seamless design and implementation of an efficient stable pattern. Independence: An entity should be independent of all existing anomalies, errors, and shortfalls. It should also be free of needless interference from other entities that might exist along with an entity that is used by a developer. An independent entity is something that exists by itself and it is something that is separate from other entities.

4. *Unique*: An entity should be unique in its existence and performance parameters. It should not mimic other entities in any aspects and attributes. A stable pattern designed after an entity should be unique to its main theme and it should not conflict with any other entities that are likely to arise during the process of pattern development.

5. *Stable*: Any stable pattern designed and developed after an entity should be stable, robust, and repeatable. A good entity stable pattern is easily extendible to any scenarios and contexts, while the factor of stability will emerge only when the main business theme (EBT) is identified in the most appropriate manner.

7.2.5 SOLUTION

The stable design pattern in Figure 7.1 is the proposed solution for handling the issues in AnyEntity and provides a generic pattern that can be applied to any application by attaching application specific features.

7.2.5.1 Class Diagram Description

Each class diagram provides a visual illustration of all the classes in the model along with their relationships with other classes. A brief description of the class diagram is given below.

- Existence is the EBT of this model. AnyEntity (BO) has existence.
- AnyEntity (BO) is classified into AnyType (BO) and has one or more AnyState (BO).
- AnyState (BO) exhibits AnyProperty (BO). Again, AnyProperty (BO) can be a part of any other property.
- AnyEntity (BO) is a part of AnyDomain (BO).
- AnyParty (BO) chooses AnyDomain (BO). In addition, AnyType (BO) binds AnyDomain (BO).
- Existence is the EBT of this model and helps one or more AnyEntity (BO) to exist and live.
- AnyEntity (BO) exhibits existence.
- AnyEntity (BO) has one or more AnyState (BO).
- AnyEntity (BO) is of one or more AnyType (BO).
- AnyEntity (BO) is a part of AnyDomain (BO).
- AnyType (BO) binds one or more AnyDomain (BO).
- AnyParty (BO) chooses AnyDomain (BO).
- AnyProperty (BO) exhibits one or more AnyState (BO).
- AnyProperty (BO) can be part of any other property.

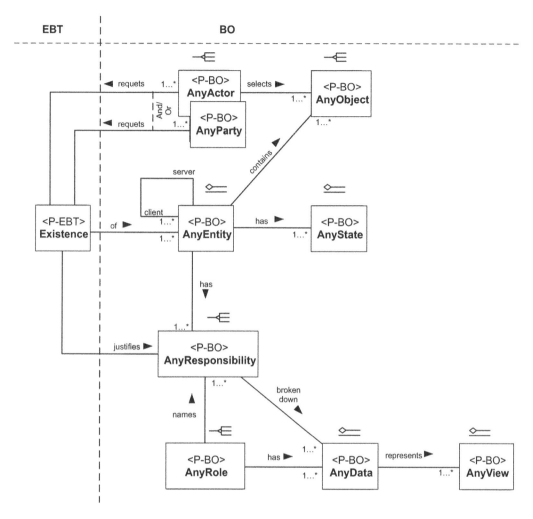

FIGURE 7.1 AnyEntity design pattern.

7.2.5.2 CRC Cards

	Collaboration	
Responsibility	**Client**	**Server**
	Existence (P-EBT)	
Represents the existence of an entity	• AnyDomain	• presentInAnyDomain()
	• AnyEntity	• existAnyEntity()
		• abstract()
Attributes: type		
	AnyEntity(AnyEntity)(BO)	
To perform some functions in its domain	• Existence	• exists()
	• AnyType	• hasAnyState()
	• AnyState	• isOfAnyType()
	• AnyDomain	• exhibits()

Continued

	Collaboration	
Responsibility	**Client**	**Server**
Attributes: name, type, state, domain		
	AnyDomain(AnyDomain)(P-BO)	
Represents a domain where any entity is present	• AnyType • AnyParty	• belongToAnyParty() • boundedByAnyType() • numberOfEntity()
Attributes: name, features, entities		
	AnyParty(AnyParty)(BO)	
Chooses the domain of the entity	• AnyType • AnyProperty • AnyEntity • AnyExistence	• displayedByAnyProperty() • includeAnyType() • chooseAnyDomain()
Attributes: name, address, domain		
	AnyState(AnyState)(P-BO)	
Represents the state of an entity	• AnyEntity • AnyProperty	• exhibitAnyProperty() • has()
Attributes: name		
	AnyProperty(AnyProperty)(P-BO)	
Describes the properties/characteristics of any entity	• AnyProperty • AnyState	• exhibitedByAnyState() • displayedByAnyState()
Attributes: name, state		
	AnyType(AnyType)(P-BO)	
Represents the type of an entity	• AnyDomain • AnyEntity	• boundedByAnyDomain() • includedByAnyParty()
Attributes: type, name		

7.2.6 Consequences

Generic Pattern: The AnyEntity design pattern is generic enough to serve as building block for applications in diverse domains.

Easy to Integrate: By using the AnyEntity design pattern for any application will require that the AnyEntity exists in a correct and appropriate state. In addition, the AnyParty so involved, must choose correct domain. However, this does not mean that the pattern is incomplete, as this is the nature of patterns; they will need to be used with other components.

Stable Pattern: An advantage with AnyEntity design pattern is that the pattern has been derived with stability in mind. It is known to capture the enduring knowledge of business and its capabilities, and it will stand the test of time. Nevertheless, the main disadvantage is that it might not represent the entity correctly, if type and property in the pattern are not properly selected.

The AnyEntity design pattern is generic enough to serve as building blocks for creating applications in diverse domains and diverse types. To use AnyEntity design pattern for AnyDomain and AnyParty, the entire process will require close interaction between them. However, this does not mean that the pattern is incomplete, as this is the nature of patterns; they need to be used with other components.

Sometimes entity, party, and domain may have strong couplings and so any changes to one of them may be difficult to incorporate; incidentally, this seems to be a negative consequence for AnyEntity design pattern.

The AnyEntity design pattern provides following benefits:

7.2.6.1   Flexibility

AnyEntity design pattern is more flexible. Many different entities that follow the same protocol for interacting within the domain can be easily interchanged. Different parties can be added and removed without any issues.

7.2.6.2   Reusability

AnyEntity design pattern is scalable. It does not matter how many entities, parties, and states are supported in a system as long as they follow the same predefined protocol.

7.2.6.3   Generic

The AnyEntity design pattern is generic enough to serve as building block for applications in diverse domains.

7.2.6.4   Easy to Integrate

Using the AnyEntity design pattern for any application will require AnyEntity existing in a correct state. In addition, the AnyParty so involved, must choose the correct domain. However, it does not imply that the pattern is not complete because it is the nature of patterns that they are used with other components.

7.2.6.5   Stable Pattern

With a fair amount of stability in mind, AnyEntity design pattern provides a position of advantage to a pattern developer. It successfully captures the enduring knowledge of business and its inherent capabilities and it will stand the test of time and prove its merit in the years to come. However, it might not represent the entity in a correct manner, if type and property in the pattern are improperly selected.

7.2.7   Applicability with Illustrated Examples

This section provides two examples to illustrate the use of AnyEntity design pattern that uses a use case description and behavior model like a sequence diagram.

7.2.7.1   Application No. 1: Database

In this example, database is depicted as an entity. The administrator maintains the database. Database can be of any type like relational, object oriented, etc. In addition, databases can have state like clustered, linked, etc. The following table summarizes the database as an entity and the manner in which it fits into the class diagram depicted in Figure 7.1.

Concept	Description
Existence (EBT)	The phenomenon of existing
AnyEntity	Database
AnyParty	Administrator
AnyType	Relational, object-oriented
AnyState	Linked, clustered
AnyDomain	Finance
AnyProperty	It is the characteristic of the database that distinguishes one database from another

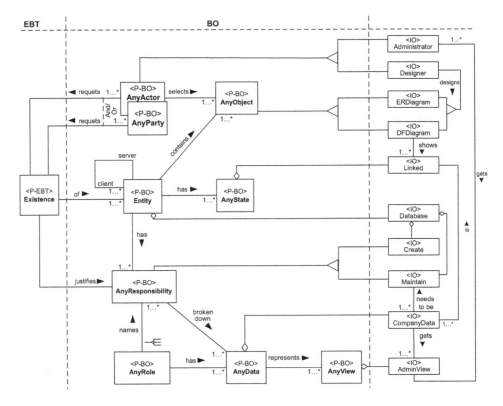

FIGURE 7.2 Class diagram for database as an entity.

7.2.7.1.1   Model

The model for this application is shown below in Figure 7.2.

7.2.7.1.2   Use Case No: 1.0

Use Case Title: Use Database as an Entity

Actors	Roles
AnyParty	Administrator
	Maintainer
	PersonalOrganizer
	DataStorer

Classes	Type	Attributes	Operations
Existence	EBT	• type	• duration()
AnyEntity	BO	• name	• exists()
		• type	• hasAnyState()
			• isOfAnyType()
AnyParty	BO	• name	• chooseDomain()
		• role	
		• address	
AnyType	BO	• type	• categorizeEntity()
			Continued

Classes	Type	Attributes	Operations
AnyProperty	BO	• type • name • characteristic	• depictEntity()
AnyDomain	BO	• name • feature	• environement()
AnyState	BO	• name • type	• isStateValid()
Administrator	IO	• id • name • function • responsibility	• maintainDatabase()
Maintainer	IO	• id • name • function • responsibility	• maintainData()
PersonalOrganizer	IO	• id • name • function • responsibility	• organizePersonalData()
DataStorer	IO	• id • name • function • responsibility	• storeData()
DataBank	IO	• type • size • use	• store()
Finance	IO	• utility	• getStored() • isSafe()
RelationalData	IO	• type	• isRelational() • correlated()
LinkedData	IO	• master • slave	• isLinked()

7.2.7.1.3 Use Case Description

1. AnyEntity (Database) must have an existence (EBT). How is the existence of AnyEntity (Database) established?
2. Existence must exist in AnyDomain (Finance). How is AnyDomain (Finance) established?
3. AnyParty (Administrator) chooses AnyDomain (Finance). On what basis does AnyParty (Administrator) choose AnyDomain (Finance)?
4. AnyEntity (Database) has AnyType (RelationalData). How is AnyEntity (Database) classified into AnyType (BO)?
5. AnyEntity (Database) has AnyState (LinkedData). Is the AnyState (LinkedData) valid for AnyEntity (Database)?
6. AnyState (LinkedData) exhibits specific AnyProperty (BO). How are the properties of AnyState identified and tracked?
7. AnyDomain (Finance) binds AnyType (Relational). How is it verified that AnyType (relational) binds AnyDomain (Finance)?

7.2.7.1.3.1 Alternatives

1. AnyEntity (Database) does not have existence (EBT)?

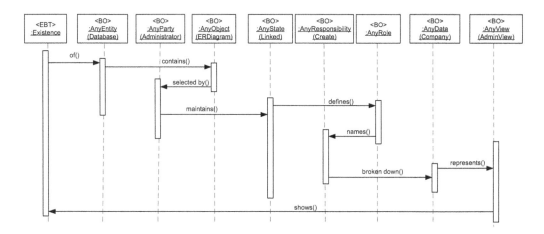

FIGURE 7.3   Sequence diagram for database as an entity.

7.2.7.1.4   Sequence Diagram

7.2.7.1.4.1   Sequence Diagram Description   The flow diagram in Figure 7.3 represents the flow of messages when AnyEntity (Database) has existence (EBT) in AnyDomain (Finance AnyParty [Administrator] chooses AnyDomain [Finance]). In addition, AnyParty (Administrator) has the responsibility of maintaining AnyEntity (Database). AnyEntity (Database) is of AnyType (Relational) and has AnyState (Linked). Certain specific AnyProperty identifies AnyEntity (Database). Again, AnyEntity (Database) needs to have existence (EBT).

1. Existence must exist in AnyDomain (BO).
2. AnyParty (BO) chooses AnyDomain (BO).
3. AnyParty (BO) maintains AnyEntity (BO).
4. AnyEntity (BO) is of AnyType (BO).
5. AnyProperty (BO) identifies AnyType (BO).
6. AnyEntity (DataBank) is a part of AnyEntity (BO).
7. AnyEntity (BO) has to have existence (EBT).

7.2.7.2   Application No. 2: Healthcare

In this example, CoveredEntity is depicted as an entity. An entity that is one or more of the following types of entities is referred to as a "CoveredEntity."

* Patient who is a primary healthcare user
* Patient's family who are secondary healthcare users
* Health plan

A healthcare provider takes care of AnyEntity—CoveredEntity. In order for a person to avail healthcare facilities, the person or his dependent must have a valid state, that is, a valid healthcare policy. The following table summarizes the database as an entity and in what manner it fits into the class diagram depicted in Figure 7.1.

Concept	Description
Existence(EBT)	The phenomenon of existing
AnyEntity	CoveredEntity

Continued

Concept	Description
AnyParty	HealthcareProvider
AnyType	Various categories by which the CoveredEntity can be recognized
AnyState	ValidState
AnyDomain	HealthCare
AnyProperty	It is the characteristic of the database that distinguishes one database from another

7.2.7.2.1   Model
The model for this application is shown below in Figure 7.4.

7.2.7.2.2   Use Case No. 10
Use Case Title: HealthCare as an Entity

Actors	Roles
AnyParty	HealthCareProvider
	AvailHealthCareFacility

Classes	Type	Attributes	Operations
Existence	EBT	type	duration()
AnyEntity	BO	name	exists()
		type	hasAnyState()
			isOfAnyType()
AnyParty	BO	name	chooseDomain()
		role	
		address	
AnyType	BO	type	categorizeEntity()
AnyProperty	BO	type	depictEntity()
		name	
		characteristic	
AnyDomain	BO	name	environement()
		feature	
AnyState	BO	name	isStateValid()
		type	
HealthCareProvider	IO	id	TakeCareOfPatient()
		name	
		function	
		responsibility	
CoveredEntity	IO	type	availMedicalFacility()
		coverage	
HealthCare	IO		
ValidState	IO	isCorrectState	describeCorrectState()

7.2.7.2.3   Use Case Description
1. AnyEntity (CoveredEntity) must have an existence (EBT). How is existence of AnyEntity (CoveredEntity) established?
2. Existence must exist in AnyDomain (HealthCare). How is AnyDomain (HealthCare) established?

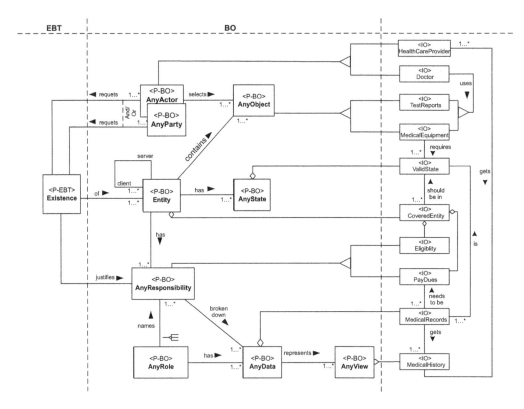

FIGURE 7.4   Class diagram for healthcare as an entity.

3. AnyParty (HealthCareProvider) chooses AnyDomain (HealthCare). On what basis does AnyParty (HealthCareProvider) choose AnyDomain (HealthCare)?
4. AnyEntity (CoveredEntity) has AnyType (BO). How is AnyEntity (CoveredEntity) classified into AnyType (BO)?
5. AnyEntity (CoveredEntity) has AnyState (ValidState). Is the AnyState (ValidState) valid for AnyEntity (CoveredEntity)?
6. AnyState (ValidState) exhibits specific AnyProperty (BO). How are the properties of AnyState identified and tracked?
7. AnyDomain (HealthCare) binds AnyType (BO). How is it verified that AnyType (BO) binds AnyDomain (HealthCare)?

7.2.7.2.3.1   Alternatives
1. AnyEntity (CoveredEntity) does not have existence (EBT).

7.2.7.2.4   Sequence Diagram
7.2.7.2.4.1   Sequence Diagram Description   The flow diagram in Figure 7.5 represents the flow of messages when AnyEntity (CoveredEntity) has existence (EBT) in AnyDomain (Finance). AnyParty (HealthcareProvider) chooses AnyDomain (Healthcare). In addition, AnyParty (HealthcareProvider) has the responsibility of maintaining AnyEntity (CoveredEntity). AnyEntity (CoveredEntity) is of AnyType and has AnyState (ValidState). A specific AnyProperty identifies AnyEntity (CoveredEntity). Again, AnyEntity (CoveredEntity) needs to have existence (EBT).

1. Existence must exist in AnyDomain (BO). How is AnyDomain (BO) established?

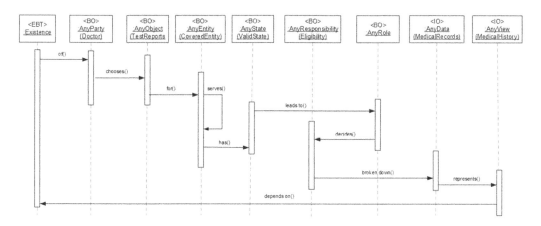

FIGURE 7.5   Sequence diagram for healthcare as an entity.

2. AnyParty (BO) chooses AnyDomain (BO). On what basis does AnyParty (BO) choose AnyDomain (BO)?
3. AnyParty (BO) maintains AnyEntity (BO).
4. AnyEntity (BO) is of AnyType (BO). How is AnyEntity (BO) classified into AnyType (BO)?
5. AnyProperty (BO) identifies AnyType (BO).
6. AnyEntity (CoveredEntity) is a part of AnyEntity (BO).
7. AnyEntity (BO) has to have existence (EBT). How is existence of AnyEntity (CoveredEntity) established?

7.2.8   MODELING ISSUES

7.2.8.1   Abstraction

Software stability stratifies different classes of the system into three layers: the EBTs layer, the BOs layer, and the IOs layer. Each class in the system model is classified into one of these three layers according to its nature and attributes.

EBTs are the classes that present the enduring and basic knowledge of the underlying industry or business. Therefore, they are extremely stable and form the nucleus of the stable model. BOs are the classes that map the EBTs of the system into more concrete objects. BOs are tangible and internally stable, but they are externally adaptable through hooks. IOs are the classes that map the BOs of the system into physical objects. For instance, the BO "Agreement" can be mapped in real life as a physical "Contract," which is an IO.

Even though the properties of each layer are clear, identifying these layers in practice is not always obvious. Stable analysis pattern is a new approach for developing patterns by utilizing software stability concepts. The concept of stable analysis patterns was proposed as a solution for the limitations of contemporary analysis patterns.

The goal of stable analysis patterns was to develop models that capture the core knowledge of the problem and present it in terms of the EBTs and the BOs of that problem. Consequently, the resultant pattern will inherit the stability features and hence it can be reused to analyze the same problem whenever it appears.

Within the AnyEntity stable analysis pattern:

• Existence is an EBT.
• AnyDomain, AnyParty, AnyState, AnyEntity, AnyType, and AnyProperty are BOs.

Existence of AnyEntity is an inherent property of any object we experience and any idea we form, and thus it is an enduring theme. In business entities such as databases, programmers, applications, etc., are entities and a common theme for all of them is *existence*. Therefore, existence is an EBT.

AnyDomain of existence is a domain of operational coherences entailed by the distinction of AnyEnity by an observer. The distinction of AnyEntity entails its AnyDomain of existence. Hence, there can be many AnyDomains of existence as distinguished by the observer. For example, if we distinguish molecules as composite AnyEntity, they reside in the AnyDomain of existence of their components, and their existence implies the distinction by the observer. Since the observer as a living system is a composite AnyEntity, the observer makes distinctions in his or her interactions, as a living system through the operation of AnyProperty of his or her components.

7.2.9 Design and Implementation Issues

AnyEntity design pattern is designed in such a way that it can be adapted and reused in diverse domains. This section describes some of the issues faced while designing and implementing a stable AnyEntity design pattern.

To implement AnyEntity design pattern, EBTs and BOs must be identified. The EBT for AnyEntity is existence. Identifying EBTs is the hardest and challenging part in designing a pattern, because EBTs are conceptual and one needs intuition to identify them. Another challenge is identifying correct supporting BOs and then connecting them correctly to EBTs in the design pattern. An important issue in implementing design pattern is to ensure that there is no hook up between EBTs and IOs.

Designing generalized pattern that is applicable to all possible business domains is bit time consuming, as it requires identification and finding interrelatedness of all possible BO for the corresponding EBT. The pattern can be designed only through a thorough knowledge of the problem domain and requires identification of EBT, BO, and IO to accommodate all concepts of stability model.

Some issues that are noticeable in implementation of AnyEntity design pattern are described next. Existence (EBT) is not inherited and has an associative relationship with AnyEntity (BO) and AnyDomain (BO). AnyParty (BO) will inherit various parties involved in application context during the implementation phase.

7.2.10 Delegation Pattern

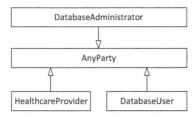

In the above diagram, AnyParty can have many roles such as Database Administrator, HealthCare provider, or a Database User. A person (AnyParty) who is a Database Administrator can also be a Database User. In the above scenario, we also see that the same person at different times can play various roles. Therefore, if we need to represent the same person for different roles, then we need to create multiple instances that can represent different changes for different roles. In order to avoid the above problem, delegation pattern can be used. Delegation pattern uses delegation to represent persons for different roles. The diagram given below explains how a delegation pattern can be used to delegate various roles to different objects at run time.

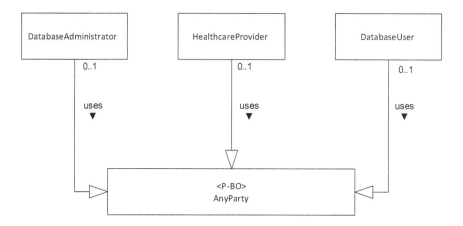

7.2.11   Java Code example

```
class AnyEntity {
        void function1(){
             //implementation
        }
        void function2(){
             //implementation
        }
}
class DatabaseAdministrator {
        //delegation
        AnyEntity entity = new AnyEntity ();

        void function1() {
             entity.function1();
        }

        void function2() {
             entity.function2();
        }
}

public class Main {
        public static void main(String [] args){
             DatabaseAdministrator admin = new DatabaseAdministrator();
                  admin.function1();
                  admin.function2();
        }
}
```

7.2.12   Testability

If AnyEntity design pattern can be used as it is and without changing the core design and by only plugging IO for infinite number of applications, then AnyEntity pattern can be said to be testable. In the applicability section, two widely different applications are illustrated and they do not require changing the core design of the pattern. Using the scenarios those are listed in this chapter, many such applications could be deduced and proved that AnyEntity pattern is indeed testable.

Database and HealthCare are two applications that apply the AnyEntity design pattern. The main idea that can be drawn from studying the above two applications is that there is no need to change the core design. IO are the only sections that can be pluggable for infinite number of applications without even changing the core design. Many such scenarios can be illustrated where it can be proved that AnyEntity pattern is testable.

7.2.13   Business Issues

7.2.13.1   Business Models

The stable AnyEntity design model is generic, and so it is extendible by plugging in specific IOs that are specific to the application domain, thereby resulting in numerous stable and adaptable business models.

> A business model is a conceptual tool that contains a big set of elements and their relationships and allows expressing the business logic of a specific firm. It is a description of the value a company offers to one or several segments of customers and of the architecture of the firm and its network of partners for creating, marketing, and delivering this value and relationship capital, to generate profitable and sustainable revenue streams. [9,10]

Usually, a business model describes the business of any company. There are various components of a business model:

1. Infrastructure
2. Offering
3. Customers
4. Finances

The infrastructure consists of some important capabilities, a network of alliances, and different rules for configuration. The capabilities are required to manage the business model of any company. The alliance established by the companies ensures that there is high level of customer satisfaction and wise usage of finances. In addition, the rules of configuration should be enforced in such a manner that it is beneficial for both customers and businesses. The offering component usually describes the products and the perquisites that any business will provide.

The customers' components describe its valuable customers, the relationship between customers, and the channels through which the businesses reach their customers. The finances component consists of the cost infrastructure and the revenue generated by the company. The cost structure describes various strategies of finances followed by the companies to earn revenue.

7.2.13.2   Enduring Business Theme

The EBT of this pattern is *existence*. It is a generic goal and it remains similar for a variety of the applications in varied fields. By using this common goal for designing pattern, stability, and extensibility is achieved.

EBTs are the classes that present the enduring and basic knowledge of the underlying industry or business. They are extremely stable and form the nucleus of the stable model. BOs are the classes that map the EBTs of the system into more concrete objects. BOs are tangible and internally stable, but they are externally adaptable through hooks. IOs are the classes that map the BOs of the system into physical objects. For instance, the BO "Agreement" can be mapped in real life as a physical "Contract," which is an IO.

The goal of stable analysis patterns was to develop models that capture the core knowledge of the problem and present it in terms of the EBTs and the BOs of that problem. Consequently, the resultant pattern will inherit various stability features and hence it can be reused to analyze the same problem whenever it appears.

The AnyEntity stable analysis pattern has

- *Existence* is an EBT.
- AnyDomain, AnyParty, AnyState, AnyEntity, AnyType, and AnyProperty are BOs.

Existence of AnyEntity is an inherent property of any object we experience and any idea we form, and thus it is an enduring theme. In business entities, such as databases, programmers, and applications are all entities and a common theme for all of them is *existence*. Therefore, existence is an EBT.

[References: Referred the template provided by Dr. Fayad]

7.2.13.3 Business Rules

The business rules for AnyEntity pattern as derived from Figure 7.1 are

- An entity must be abstracted correctly.
- Entity must have an existence.
- Entity may be either external or internal to the system.
- Entity must be a part of the domain so chosen by AnyParty.

Whereas according to Meta model depicted in Figure 7.6, the business rules are

- There must be a reason for a person to exist.
- A person carries out activities in order to achieve the reason.

Thus, as seen from the contrasting business rules, it is obvious that stability model provides generic rules that can be applied across any applications and business domains. On the other hand, the rules from Meta model are specific to a particular application.

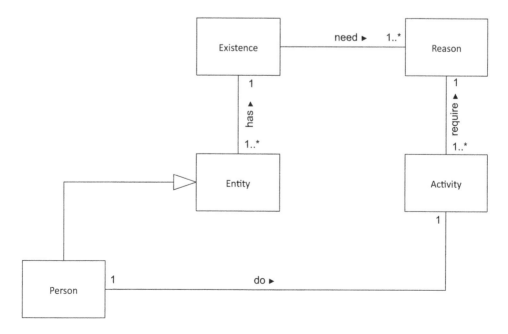

FIGURE 7.6 Meta model for AnyEntity.

7.2.13.4   Business Model

The stable AnyEntity design model is generic, and so it is extendible by plugging in specific IOs that are specific to the application domain, thereby resulting in numerous stable and adaptable business models.

7.2.13.5   Business Integration

Because AnyEntity pattern provides hooks, integration of the AnyEntity design pattern to existing business model is quite easy. Integrating AnyEntity pattern to a particular application domain will require the plugging of application specific IOs to the hooks.

7.2.13.6   Business Transformation

Use of AnyEntity design pattern by a company in the development of any application strengthens its business transformation process because AnyEntity pattern quickens the development process by providing the framework for the application. Also, since AnyEntity pattern is tested and verified, chances of generating bugs are minimal.

7.2.14   KNOWN USAGE

The AnyEntity design pattern can be used in any application that uses entity as a theme. Since a person is an entity, entity is used in almost all applications that involve human interaction.

7.2.15   RELATED PATTERN AND MEASURABILITY

A person is a type of entity and entity has existence that requires a reason. A person carries out activities for some reasons that invariably require activity. Based on this information, the Meta model in Figure 7.6 can be modeled.

7.2.15.1   Traditional Model (Meta Model) versus Stable Model (Pattern)

Traditional modeling (Meta model) generally results in an unstable model which is not adaptable, scalable, and extensible. On the other hand, analysis/design pattern created by using the SSM is adaptable to changing requirements and it can be scaled easily with minimum effort. In Traditional modeling, since we only design, as much is needed for a specific application in question, and not thinking of its applicability in other domain, the resulting application is only very specific. While stability in mind, we use general enduring concepts and hence the resulting pattern can be used for building numerous applications. In short, the resulting pattern can serve as building block for diverse application domains. With Traditional modeling approach, the resulting design is tangible enough to be comprehended while with stability model approach, the pattern is intangible. Designing in Traditional modeling arises from the problem statement, while stability model path needs experience to derive solution by intuition and by thinking out of the box. However, implementing successful pattern requires expertise, meticulous work, and detailed understanding of the problem domain.

7.3   SUMMARY

This chapter proposed a stable design pattern for Entity. The pattern has been developed in such a way that it represents an *Entity* using software stability principles. The important themes of entity have been extracted from different domains and they were represented as EBTs and BOs. This is demonstrated in the detailed model of the pattern. Thus, by identifying the EBTs and BOs of the system, we have created a model, which is applicable for entities across multiple domains.

The concept of entity is modeled solely based on the stability model by using the EBTs. This model can be used for different domains, while IOs can be extended according to the application. The model represents the core knowledge of the pattern in different applications and it is presented as EBTs and BOs. The model is explained with two applications that perform extremely well based on this model.

Though building a stable design pattern for AnyEntity, which can be reused and reapplied across diverse domain, is difficult and requires thorough understanding of the problem, it is worth the effort and time. AnyEntity pattern results in a reusable, extensible, and stable pattern.

The correct identification of EBT and BO for AnyEntity is the most challenging task and it requires some technical and practical experience. Once EBT and BOs are correctly identified, next challenge is to determine the relationship between EBT and BOs, so that AnyEntity pattern can hold true in any context of usage for partition. Once this is achieved, depending on the application, the IOs are attached to the hooks so provided by BOs. Therefore, by using AnyEntity pattern as a basis, an infinite number of applications can be designed by simply plugging the application specific IOs to the pattern. This results in a reduced cost, effort, and stable solution. Hence, AnyEntity design pattern is very useful.

7.4 OPEN AND RESEARCH ISSUES

An entity is a human and it is live and active. Taking its meaning to be true, it is an excellent candidate to create a stable design pattern. AnyEntity has an existence and its existence denotes semitangible and externally stable character. Creating a stable development pattern on an entity is challenging task as this chapter demonstrated in its earlier section. However, some experience and technical know-how will help find out the real business theme of the word, *Entity*. In addition, a developer will also be able to find out its essential BOs and IOs that eventually help in designing a stable development pattern that is applicable in various domains of usage.

However, a few numbers of research issues are still pending that can prevent a developer from creating a stable pattern. For example, a challenge remains to seek out the real meaning of an entity. An animate character is an easier entity that can be represented in a stable pattern while an inanimate one could be more challenging and tedious to apply to the notion of stable analysis pattern. Although, AnyEntity pattern is easier to apply to various contexts of life, a developer must ensure that a few numbers of rules are followed before creating a stable pattern. Database and healthcare are the two areas where one can apply the principles of stable pattern. However, it might be challenging to apply stable pattern principles to many other areas of human activities, as humans and their thinking are extremely dynamic and changing. This may pose some serious problems to a developer, as all aspects of humans should be considered before creating a pattern.

7.5 REVIEW QUESTIONS

1. (T/F) AnyEntity is always a human. List some common entities that exist in the world.
2. (T/F) AnyEntity cannot be inanimate. Does an entity signify human action?
3. What is meant by the term "existing"? Can the term "existing" be used in any other context than what has already been mentioned in this chapter?
4. List the synonyms of "entity." Can these terms be used interchangeably with respect to the AnyEntity pattern? Why an entity, as mentioned in a standard English dictionary, a fit candidate to design a stable pattern?
5. What are the requirements for an "entity"? Describe each of them?
6. Draw and describe the class diagram for the stable AnyEntity pattern?
7. List differences between the AnyEntity pattern described here and the traditional methods.

8. List some design and implementation issues faced when implementing the AnyEntity pattern. Explain each issue.
9. Give some applications where the AnyEntity pattern is used.
10. What lessons have you learnt by studying the AnyEntity pattern.
11. The AnyEntity design pattern can be applied and extended to any domain (True or False).
12. List some of the domains in which the AnyEntity design pattern can be applied.
13. List some domains that you believe the AnyEntity pattern would not apply. Explain your reasons.
14. List three challenges in formulating the AnyEntity design pattern.
15. List three different constraints in the AnyEntity pattern.
16. Is the AnyEntity pattern incomplete without the use of other patterns? Explain briefly.
17. (T /F) The AnyEntity pattern describes a system.
18. Present the sequence diagram for applicability of the AnyEntity stable analysis pattern in the e-commerce domain.
19. What do you think are the implementation issues for the AnyDomain BO when used in the AnyEntity design pattern?
20. What do you think are the implementation issues for the AnyParty BO when used with the AnyEntity design pattern?
21. Describe how the developed AnyEntity design pattern would be stable over time.
22. List some of the testing patterns that can be applied for testing the AnyEntity design pattern.
23. List three test cases to test the class members of the AnyEntity pattern.
24. Name the different types of entities that the AnyEntity pattern may represent.
25. Is AnyEntity pattern repeatable in real-life scenarios?
26. What is the EBT for this term?
27. Name some of the BOs and IOs.

7.6 EXERCISES

1. Illustrate with a class diagram the applicability of the AnyEntity design pattern to a major league sports association club members. Additionally, use the AnyParty pattern to represent the ownership group of a sports club.
2. Highlight and describe with a class diagram of this pattern by applying its principles to a college debate team members. In addition, use this pattern to represent the ownership group of the debate club.
3. Consider the example of a scout team. Create a class diagram for this pattern and apply its principles to the team while designing the stable pattern. Link the local scout troupe to the top most hierarchy (in this case, the world scout association) and represent the global ownership.
4. Rotary International is a well-known nongovernmental association with millions of members worldwide and it is controlled by its headquarters in Chicago. Take the example of a local Rotary club and its members. Define their role by noting down the notion of an entity. Create a stable pattern by including the entire leadership from world headquarters, to the local club, create a class diagram, and find out all BOs and IOs. Represent the entire scenario by using the principles of stable design pattern approach.
5. Think of any scenarios that have not been described in this chapter. List them with a short description.
6. Try to create a use case and interaction diagram for each of the scenarios you thought of in the above question.

7.7   PROJECTS

Video Game Concept: A video game has a main character and a party of three members under his/her control. There are many characters similar to the main character in the game world; use the AnyUser pattern to model them. The game characters party members are creatures of a wide variety. Use the AnyEntity pattern to model these creatures.

Main Character (AnyUser) ---owns→ group (AnyParty) <-belongs to--- 1..3 Creatures (AnyEntity)

Come up with corresponding class diagram, use case, and sequence diagram as shown in the solutions in this chapter.

REFERENCES

1. Ralph Cudworth (1617– June 26, 1688), last modified on April 17, 2017, https://en.wikipedia.org/wiki/Ralph_Cudworth
2. M.E. Fayad and A. Altman. Introduction to software stability, *Communications of the ACM*, 44(9), 2001, 95–98.
3. M.E. Fayad. Accomplishing software stability, *Communications of the ACM*, 45(1), 2002, 95–98, 111–115.
4. M.E. Fayad. How to deal with software stability, *Communications of the ACM*, 45(4), 2002, 109–112.
5. P. Chen. The entity-relationship model—toward a unified view of data, *ACM Transactions on Database Systems*, 1(1), 1976, 9–36. doi: 10.1145/320434.320440.
6. A.P.G. Brown. Modelling a real-world system and designing a schema to represent it, in Douque and Nijssen (eds.), *Data Base Description*, North-Holland, 1975, ISBN 0-7204-2833-5.
7. P. Seynon-Davies. *Database Systems*, Palgrave, Basingstoke Houndmills, UK, 2004, ISBN 1403916012.
8. ER 2004. in *23rd International Conference on Conceptual Modeling*, Shanghai, China, November 8–12, 2004.
9. A. Osterwalder. The business model ontology—A proposition in a design science approach. PhD thesis, University of Lausanne, 2004.
10. A. Osterwalder, Y. Pigneur, and C.L. Tucci. Clarifying business models: Origins, present, and future of the concept. *Communications of the Association for Information Systems*, 2005.

8 AnyData Stable Design Pattern

> To write it, it took three months; to conceive it three minutes; to collect the data in it all my life.

F. Scott Fitzgerald [1]

AnyData is a very general concept with wide ranging number of applications in diverse contexts. The AnyData stable design pattern focuses on analyzing the general and most important concept of AnyData, which is acquiring Knowledge or Information. Since AnyData pattern is introduced based on the principles of stable design pattern, it is easier to apply this pattern for use in different application contexts. This is possible by hooking the unstable industrial objects to the stable business objects according to the application under study. AnyData pattern models the core knowledge of any data, making it easy to reuse the pattern and build upon it to model different kinds of scenarios utilizing data rather than analyzing the same problem each time from the scratch. This pattern can be used to model any kind of data for any domain and it can be reused for building a new model too. In this chapter, AnyData design pattern based on stability model is described and few scenarios where this pattern can be used are introduced.

8.1 INTRODUCTION

Data refers to individual facts, statistics, or items of information. Data is a plural of datum, which is originally a Latin noun meaning "something given." Data are always raw. They simply exist and have no significance beyond their existence (in and of itself). It can exist in any form, usable or not. It does not have meaning by itself. In computer parlance, a spreadsheet generally starts out by holding data. Data represents a fact or statement of events without relation to other things. Example: Facts regarding raining are the data, but they have no meaning without a specific context or consequence being stated.

Data may exist in various forms—as number or as text on pieces of paper, or as bits or bytes stored in electronic memory, as facts stored in a persons mind, as a graph (pie chart or bar chart) giving overall view and comparison, and as a drawing or animation giving pictorial view. This variety factor in representing data gives a greater flexibility in terms of representation of data. However, sometimes the term data is used to distinguish *binary* machine-readable information from textual human-readable information. For example, some *applications* make a distinction between *data files* (*files* that contain binary data) and *text files* (files that contain *ASCII* data).

AnyData can be used in various areas, each one of them having its own context and rationale behind using the term data. Notable areas, where data storage and analysis is required may include agriculture, education, public health, financing, real estate, and weather. In agriculture and weather, data are collected, processed, and then analyzed to obtain trends. Next, these trends are used to predict future scenarios. Whereas in financing, scenario data are stored mainly to keep track of the transactions for future references and in some cases for future predictions. In public health, data are maintained to keep track of patients, illness, etc. and is crucial. In an education scenario, data are used to calculate literacy level or educational facilities available. *Database Management Systems* (DBMS) are used to store data in most of the above stated scenarios. In science and computing, data in a broad sense are *numbers, characters, images,* or other outputs from devices to convert physical quantities into symbols. Such data are further *processed* by a human as an *input* into a *computer, stored* and processed, or transmitted (*output*) to another human or computer. Data processing commonly occurs by several stages. Mechanical computing devices are classified according to the means through which they represent data. An *analog computer* represents a data as a voltage, distance, position, or other physical quantity. A *digital computer* represents data as a sequence of symbols drawn from a fixed *alphabet*. A *computer program* is a collection of data, which can be interpreted as specific instructions. In DBMS, data files are the files that store the *database* information, whereas other files, such as *index* files and data dictionaries, store administrative information, known as *metadata*.

Since the function of AnyData depends on the context in which AnyData is being used, the traditional approaches to software design will not yield a stable and reusable model. However, by using SSM, AnyData can be represented in any context by using a single model. The SSM requires creation of knowledge map by identifying underlying EBT and BO [2–4]. By hooking IOs, that is specific to each application, the model can be applied to any application domain. The resulting AnyData pattern is stable, reusable, extendable, and adaptable. Thus, any number of applications can be built by using this common model. The AnyData design pattern tries to capture the core knowledge of AnyData that is common to all the application scenarios listed above to design a stable design pattern.

The main objective of this chapter is to design a stability model for AnyData by creating knowledge map of AnyData. This knowledge map or core knowledge will serve as a building block for modeling different applications that occur in diverse domains. The rest of this chapter presents AnyData, a stable design pattern based on the software stability model.

8.2 ANYDATA DESIGN PATTERN DOCUMENT

8.2.1 PATTERN NAME: ANYDATA STABLE DESIGN PATTERN

Data are essential for deriving information and accumulating knowledge. The objective of AnyData is to make available complex data of system or section to gain knowledge by organizing that data as information. Organizing data helps in ease of achieving information by ensuring manageability. The main purpose of organizing data may be for eliciting quick references in business-related companies such as financial and technical. The data should be maintained in secured places and they should help us retrieve sensitive data even after many years. The goal of pattern is to generalize the idea of data, so that it can be used as a basis for organizing and managing information.

The AnyData design pattern represents knowledge in an understandable form about any aspect of an entity and thereby achieves a general pattern that can be used across any applications. This pattern is required to model the core knowledge of AnyData without tying the pattern to a specific application or domain; hence, the name AnyData is chosen. Again, since data is semitangible and externally stable, AnyData name is chosen and it is a BO.

8.2.2 KNOWN AS

The terms data, information, and knowledge are frequently used for overlapping concepts. As these three concepts are ambiguously defined in most subject matters, they sound very similar and they are used interchangeably in common context. Nevertheless, they differ slightly in their usage.

Information is a collection of facts from which conclusions may be drawn. Knowledge on the other hand is an organized body of information, or the comprehension and understanding consequent on having acquired and organized a body of facts. For example, "statistical data" is information, while knowing number of aquatic organisms breeding in summer and in a particular ocean based on analyzing statistical data of marine animals is knowledge. Data are information that might or might not be correct, reliable, organized, and informative. For accumulating information and subsequently knowledge, processing and analyzing needs to be carried out [5,6].

8.2.3 CONTEXT

Data are generally used to provide information and knowledge about an entity as a whole or as a part to understand its characteristics. The objective of data can differ from one application to the other. Data can also be used to accomplish the factor of simplicity by breaking down complex data into simple manageable parts and absorbing the needed information. This approach is generally used for deriving trends and patterns. Data needs to be organized and filtered for future use for storing information. However, if data are lost, misplaced, or corrupted, because of improper management, it can lead to serious issues. It is very crucial that data are maintained in its inherent form, when storing in data stores for future usage. Fraud/exploitation involves acquiring any data and corrupting the same in the process of transmission from one domain to other. Such data frauds can affect the entire system.

AnyData pattern can be applied to various applications that include agriculture, education, computing, public health, real estate, financing, weather, database management systems, and so on. The goal of acquiring data from all applications is to gather information on past and/or current and predicting future trends based on the past and current data. The advantages of storing data and organizing it include availability, manageability, and performance. AnyData pattern is applied to different domain to achieve different set of goals. The AnyData design pattern also deals with the issue of abstracting various kinds of data with different characteristics to gain knowledge.

As highlighted above, AnyData design pattern can be applied to numerous applications in diverse disciplines. In the later section, application scenarios that illustrate the applicability of AnyData pattern are shown. AnyData concept is illustrated in two distinct scenarios, like healthcare system and weather forecasting, to demonstrate how data can be used to gain knowledge from different applications.

Data are something that are meaningful, when all the distinct pieces of information are organized in a special and specific way. Data should be handled carefully; else, it may have disastrous consequences. For example, suppose a message is encrypted and send over to the destination and the receiver decrypts using a different key and the words in the message are jumbled, the entire message becomes useless and even harmful.

AnyData design patterns can be used in many domains. In any project, there should be some data that helps in reaching the goal post for a specific project. For example, when a company tries to launch a new product in the market, it will conduct a survey to study the usability of the product for the public use. For this survey, the company would identify prospective customers whom they will interview, the type of methodology that will be used for the interview, kind of questionnaire to gather the required information. Once data are collected, they are analyzed and a report or a statistics is created. Here, data becomes an integral part of this project. Data could be anything and in any form. In the abovementioned project, data are the opinions of the customers who will use a particular product. These data bits are collected either by conducting interviews, or simply by filling up a questionnaire (both Internet and paper), or through telephonic conversations. This data set is very important and sensitive to the company, as well to the customer, to improve quality of the product and to enhance the company's revenue and market share.

1. *Medical Data*: When a patient (AnyParty) visits a doctor (AnyParty) for a medical (AnyCriteria) checkup (AnyEvent), the doctor will record body temperature, blood pressure, heart rate, etc. (AnyData). This data provides information (Knowledge) to the doctor about the patient's health (AnyAspect). The data can relate to patient's heart, blood, etc. (AnyEntity). The data could be of many types (AnyType) like body temperature, heart rate, etc. and it can be represented as charts or as text in a report (AnyForm). Data can be divided into many domains based on the tests conducted like blood test data (AnyDomain). The data are then stored in a database (AnyCollection) for creating a detailed medical history.

2. *Web site Crash Matrix Data*: Most of the web sites have a facility to log/record (AnyEvent) crash matrix. Crash (AnyAspect) matrix sends the data (AnyData) on the time of failure (AnyCriteria) of the web site (AnyEntity) to the distant server. This will provide information (Knowledge) to engineers (AnyParty) and managers (AnyParty) on the time of web site crash and possible reasons for crash. The data can be of many types (AnyType) like crash count, crash time, etc. and it can be represented as charts or in a report form (AnyForm). The data can be divided into domains like time and count (AnyDomain). The data are stored in a database (AnyCollection) for future use and reference.

3. *Employment Data*: Companies and firms (AnyParty) maintain employee (AnyParty) records (AnyData). The data can be about contact information (AnyAspect) of the employee. This provides information (Knowledge) to managers and HR department to contact (AnyEvent) them. The data can be of many types (AnyType) like contact data, salary information, etc. (AnyCriteria) and can be represented as excel sheets (AnyEntity) or in a web form (AnyForm). The data can be divided into domains like financial information and personal information (AnyDomain). The data are then stored in a database (AnyCollection) for future use and references.

8.2.4 PROBLEM

The AnyData design pattern covers the concept of data collection in the context of diverse domains that have different functionalities, usage, and nature of data. The pattern focuses on gaining knowledge from data by analyzing, filtering, and storing the data. At the same time, the process used for filtering and storing should be flawless; otherwise, data loss may lead to serious issues. It is necessary to model a pattern to target generic domain that might contain different guidelines for storage and utilization of data for gaining information. In addition, the motive behind gaining information varies and differs widely. For example, one of the purposes may be to gather information, while other purposes may be making future predictions for using the same. Again, sensitive data needs to be protected by making it available only to authorized users. Thus, modeling a generic design pattern that can be applied to all these domains makes sense and rational. However, one should remember below listed points before modeling a AnyData pattern.

- The model should capture the core knowledge of the data problem.
- The model should be reusable to model the data problem in any application without losing data.
- The ultimate goal for AnyData must be established along with the other patterns that are used in the design pattern of AnyData.
- Identification of correct aspects of any entity, for which data are to be collected.

Since AnyData is used in different context and in different domains, building a generic model without losing data while storing is really challenging. By using SSM, this problem could be solved and a generic model might be modeled for different domains. This model is illustrated and described in the solution section.

1. *Within Context*: Since "data" is a very broad term, the pattern should be applicable specifically to the given domain.
2. *Reliability*: The data collected should be reliable; otherwise the goal of the design pattern will not be accurate and hence the pattern will not be stable.
3. *Wide Abstraction*: Existing data patterns are limited to data that are specific to a particular domain. As a result, it is quite hard to apply these patterns to data in other domains.
4. *Have several adequacies*, such as types, size, layout, representation, format, etc.
5. *Type Orientation*: It should orient and focus itself to the overall goal of the pattern.
6. *Knowledge*: The ultimate aim or goal of the design pattern should be to gather knowledge, as it creates stability, and hence becomes an enduring pattern.
7. *Formation or Tuned up*: Available data should be precise and they should help users form a specific usage scenario for the pattern.
8. *Existence*: A pattern that represents a foundation for modeling any data should verify the existence of the data too.
9. *Accuracy*: Accuracy of the data is one of the important requirements of any data pattern for complete understanding of the nature and scope of the problem.

8.2.4.1 Functional Requirements

1. *AnyData*: AnyData represents any information about AnyEntity. It depicts the characteristics of AnyAspect of the entity. AnyData has a unique ID, a type, and characteristics. It contains information, represents knowledge, and provides insights about AnyEntity.
2. *Knowledge*: This is the EBT for AnyData. The knowledge comes from the information provided by AnyData. The knowledge can be gained by reading the textual documentation or by seeing the pictorial representation of AnyData like charts or graphs. Knowledge has a subject and a type. It promotes learning and provides wisdom.
3. *AnyAspect*: AnyAspect represents a particular feature about AnyEntity. AnyData provides information about AnyAspect of AnyEntity and that information is used as knowledge. AnyAspect has a unique ID, a characteristic, a description, a name, and a type. AnyAspect relates to AnyEntity and categorizes its particular feature. It also provides details about that feature.
4. *AnyEntity*: AnyEntity represents any object or resource whose data are collected in order to get information. For example, when a patient goes for heart checkup, doctor gathers data about the patient's heart, and thus heart is AnyEntity in this case. AnyEntity has a unique ID, a name, a function, a role, and an attribute. It symbolizes an object, plays a role, and has a function to perform.
5. *AnyType*: AnyType represents the type of AnyData to be identified by AnyForm. For example, AnyData of a patient can be of many types like medical data, contact information data, etc. AnyType categorizes data into various categories and divides AnyData into many parts.
6. *AnyForm*: AnyForm illustrates the representation of AnyData in any form such as textual, pictorial, graphical, binary, etc. AnyForm has a unique ID, a name, a type, and a description. It helps in visualizing data. In addition, it illustrates data in a better way and displays it to AnyParty.
7. *AnyDomain*: AnyDomain represents the classification of AnyData based on certain qualities. AnyDomain has unique ID, a type, a description, and a title. It creates partitions in data and simplifies it. It also classifies and categorizes data into various domains.
8. *AnyCollection*: AnyCollection represents a group of objects belonging to a particular category or domain. AnyCollection has a unique ID, a type, criteria, and a title. It groups data, saves it, and creates history of it.
9. *AnyParty*: AnyParty represents a person or a company that collects data and is interested in getting information out of the data to convert it into knowledge. AnyParty has a name, a unique ID, email address, and contact information. AnyParty collects data, cleans it and stores it in a database.
10. *AnyEvent*: AnyEvent represents an event that leads to the discovery of data. The event could be a scientific research or hiring of a new employee. AnyEvent has a unique Id, a title, a description, date when it occurred, time of its occurrence, and a location associated with it. AnyEvent leads to the discovery of data; it provides information and knowledge about what data are needed.
11. *AnyCriteria*: AnyCriteria represents the rules and regulations that define the aspect. The data are collected based on these rules, so that they fit into a certain aspect. AnyCriteria has a unique ID, a description, and a title. It also provides rules, defines aspect, and helps in collecting data.

8.2.4.2 Nonfunctional Requirements

1. *Useful*: AnyData should be useful to AnyParty. The ultimate goal behind collecting AnyData is to get information and knowledge out of it. If the data are not useful, then they will not provide any information and knowledge to the party and the effort and purpose of collecting the data will not be fulfilled. Raw data in its original form is not practical and user should be able to generate only useful data from it.

2. *Consistent*: AnyData should be consistent. If at one place contact information means to contact a party, while at other it means how many times the party was contacted, then a lot of effort will be needed to clean up the data. Hence, it is very essential that the data should be consistent. Inconsistence data will require heavy sorting into comprehensible forms and it may result in spending a lot of time, effort, and money.

3. *Valid*: AnyData should be a valid data. If the data is not valid then it cannot be used to generate information or knowledge. In addition, the information that it will provide on the entity would be invalid and hence will not solve the purpose. Hence, being valid is a very important quality factor of AnyData. For example, weather forecasting and monitoring, weather experts need precise data on several parameters like temperature, pressure, directions of oceanic currents, seasonal information, and past weather models. However, any inclusion of invalid data in the modeling system would render it faulty and incorrect.

4. *Complete*: AnyData should be complete. If it is not complete and when there are missing pieces, then mining information would be very difficult and lots of effort will be required to find the missing parts to complete the data. Incomplete data will force a forensic investigator to search for missing details and it might turn out be solving a complex jigsaw puzzle.

5. *Precision or Accuracy*: AnyData should be accurate and precise. If the data contains lots of anomalies and errors and when it is incorrect, than gathered information on AnyEntity will be useless; hence, would need work to make it accurate.

6. *Available*: AnyData should be available at the time when it is needed. For example, if a manager wants to contact an employee, he or she should be able to get contact details quickly and in time; otherwise, the entire purpose of using that data will become a useless exercise.

7. *Timely*: AnyData should be available in a timely manner. If a party has to spend a lot of time to get the data, then the data might not be very useful. Thus, it is very important that the data should become available in a timely manner.

8. *Manageable*: AnyData should be manageable. If the data are in the form that cannot be managed or if the data are very voluminous that AnyParty cannot manage, then it is of no practical use and it will need additional effort to manage it in a proper form.

9. *Verifiable*: AnyData should be verifiable easily, so that its validity can be confirmed. If a party is not able to verify data, then it is very difficult to prove the available data for reliability and validity. Hence, it will be very difficult to trust the authenticity of the data.

10. *Accessible*: AnyData should be easily accessible and it helps users save time and money. If a party cannot access the data in time, then it is of no use. Thus, the data should be available easily and quickly and it should be accessible by AnyParty at any time.

11. *Conformity*: AnyData should confirm to specific rules, standards, and law. If it does not confirm to the rules, then the data might term as illegal or not useful. In other words, illegal data are not only dangerous, they could also lead to disastrous consequences.

12. *Integrity*: AnyData should have impeccable integrity. It should be whole and undivided data and should be well structured and defined. It should be trustable, authentic, and honest.

8.2.5 CHALLENGES

Challenge ID	0001
Challenge Title	Data Classified in Too Many Types
Scenario	A doctor classifying patient data in too many types
Description	Consider a scenario, when a doctor collects patient's data for identifying a disease. He suggests the patient to take 100s of tests. Each test result is classified as one type. It will be difficult for the doctor to identify the pattern of symptom from so many types.
Solution	Only the required number of type should be created and the data should be classified accordingly

Continued

Challenge ID	0002
Challenge Title	Too Much Data about an Entity
Scenario	A company wants to store employee information
Description	Consider a scenario, when a company wants to store information on its employees. If the company stores too much of data on them, like their parent's address and their profession, then it will be difficult to find the relevant information when needed urgently. Moreover, precious space will be wasted, which is not useful at all.
Solution	Only the relevant data should be stored in the database
Challenge ID	0003
Challenge Title	Data in Too Many Forms
Scenario	100s of charts representing population distribution
Description	Consider a scenario, when a person wants to represent population distribution data in the form of a chart. If he presents 100s of charts, then no one will understand the purpose and the main information might be lost. It is always better to represent data in only those forms that are necessarily required.
Solution	Use only required number of forms to represent data to avoid confusion.
Challenge ID	0004
Challenge Title	Data in Too Many Categories
Scenario	Medical data in too many categories
Description	Medical data are very critical. In emergencies, doctor needs to find the data very quickly. If the data are divided into 1000s of categories, then it will be difficult for doctor to remember all of them and to find the required data in short time.
Solution	Only the required number of category should be used. There should also be an upper limit on the number of categories that can be used.

8.2.6 Solution

The solution for AnyData pattern is demonstrated by using a class diagram and CRC cards that includes the role of each classes and its operation.

Figure 8.1 below shows the class diagram of AnyData pattern.

8.2.6.1 Pattern Structure and Participants

The pattern structure and participants of AnyData pattern are presented below.

8.2.6.1.1 Classes

AnyData: This class represents any information characterized by AnyEntity. It depicts the characteristics of AnyAspect of AnyEntity.

8.2.6.1.2 Patterns

Knowledge: This class represents the EBT; the concept of knowledge comes from AnyData. The knowledge can be gained by reading the textual documentation or by seeing the pictorial representation.

AnyAspect: This class represents a particular feature about AnyEntity.

AnyEntity: This class represents any object or resource, whose data are collected in order to gain information.

AnyType: This class represents the type of AnyData to be identified by AnyForm.

AnyForm: This class illustrates the representation of AnyData in any form, such as textual, pictorial, graphical, binary, etc.

AnyCategory: This class represents the classification of AnyData based on certain category.

AnyCollection: This class represents a group of objects belonging to a particular category or domain.

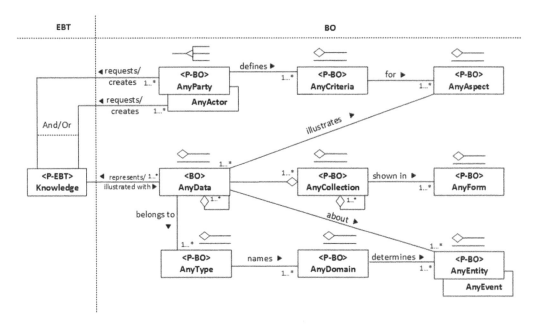

FIGURE 8.1   AnyData stable design pattern.

8.2.6.2   Class Diagram Description

The class diagram provides visual illustration of all the classes in the model along with their relationships with other classes. Description of the class diagram is given below.

1. AnyData (BO) represents Knowledge (EBT).
2. Knowledge (EBT) is obtained on AnyAspect (BO) of AnyEntity (BO).
3. AnyEntity (BO) is any object that is characterized by AnyData.
4. AnyData can be represented in AnyForm (BO) like textual or graphical form.
5. Again AnyData (BO) also belongs to AnyType (BO).
6. AnyData (BO) can be a part of some other AnyData (BO) to provide more description and details.
7. Grouping of related data results in forming of AnyCollection (BO). Thus, AnyData (BO) becomes a part of AnyCollection (BO).
8. Classifying similar data as well as AnyCollection (BO) together, results in forming AnyCategory (BO). Thus, AnyData (BO) and AnyCollection (BO) are part of AnyCategory (BO).

8.2.6.3   CRC Cards

	Collaboration	
Responsibility	**Client**	**Server**
Knowledge (Knowledge) (EBT)		
Represents the state or fact of knowing or awareness	• AnyData • AnyAspect	• impartFamiliarity() • spreadAwarness() • promoteLearning()

Attributes: Subject, source, aspect, type

Continued

	Collaboration	
Responsibility	**Client**	**Server**

AnyData (AnyData) (BO)

Represents any information about any entity that can be used for future purposes	• Knowledge • AnyEntity • AnyForm • AnyType • AnyCategory • AnyCollection • AnyData	• representCharacteristic() • containInformation() • passKnowledge()

Attributes: type, characteristic, data, entityName, form, categoryName

AnyAspect (AnyAspect) (BO)

Represents features of any entity	• AnyEntity • Knowledge	• characterize() • describe()

Attributes: noOfAspect, characteristic

AnyEntity (AnyEntity) (BO)

Shows any object or resource whose data are collected in order to gain information	• AnyAspect • AnyData	• representData() • symbolizeObject()

Attributes: attribute, function, role

AnyForm (AnyForm) (BO)

Presenting AnyData in various forms like textual, graphical, etc.	• AnyData • AnyType	• dispay() • illustrateData()

Attributes: type

AnyType (AnyType) (BO)

Represents different types of AnyData to represent data	• AnyData • AnyForm	• exhibitDataType() • prospectBy() • identify()

Attributes: text, numeric, alphanumeric

AnyCollection (AnyCollection) (BO)

Represents a group of objects, belonging to a particular category or domain	• AnyData • AnyCategory • AnyCollection	• storeDataByCategory() • group() • actAsCollection()

Attributes: type, criteria

AnyCategory (AnyCategory) (BO)

Represents the classification of AnyData based on certain category	• AnyCollection • AnyData	• classify() • categorize()

Attributes: type, classification

8.2.7   Consequences

The AnyData design pattern is generic enough to serve as a building block for applications in diverse domain.

Using the AnyData design pattern for AnyEntity will require that the AnyData have been represented by correct type and adequate form. In addition, the distribution of sensitive data must be protected by use of authorization and authentication methods, so that misuse of knowledge can be avoided. However, this does not mean that the pattern is incomplete, as this is the nature of patterns—they need to be used with other components.

One significant advantage with AnyData design pattern is that the pattern has been derived with stability in mind. It will capture the enduring knowledge of business and its capabilities, and it will stand the test of time. Nevertheless, the negative aspect with it is that it might result in loss of data or incorrect absorption of knowledge when data storage and filtration is not done properly. The AnyData pattern has the following benefits.

8.2.7.1   Reusability

Data can be reused for more than one purpose. For example, data can be collected to monitor ongoing process, as well as to predict future outputs of process by analyzing the current values. Hence, AnyData pattern can be used for all these cases and hence it aids reusability.

8.2.7.2   Flexibility

The AnyData pattern can be flexible to use, because related data from any database and any form and type can be used, thereby reducing the effort of gathering data from scratch.

AnyData design pattern is created with stability in mind; hence, it will be enduring. Since AnyData pattern works mainly on achieving the goal and until the goal of the project remains the same, the design will not change. In AnyData pattern, the goal is considered as enduring and hence it is identified as EBT.

One of the less important drawbacks is that the designer should be aware of the modeling heuristics for ex, no stars, no dangling, no tree, etc. However, once the designer gets accustomed to it, it becomes quite easy. One of the main concerns of the AnyData pattern is that one has to be very careful in selecting the sources of the data because when the data collected are malicious, then the entire design will go wrong.

Some of the benefits of the AnyData design pattern include.

8.2.7.3   Extensibility

One of the advantages of using AnyData pattern is that if a developer wants to extend his/her application with few more classes, only the IOs will change without touching EBTs and BOs. Therefore, there would be very little impact on the entire design.

8.2.7.4   Adaptability

Another advantage of using AnyData pattern is that it can adapt to any type domain, since it is designed for reuse in mind.

8.2.8   Applicability with Illustrated Examples

The AnyData pattern is developed in such a way that it captures the core functionality of all the applications that use data. This pattern is expected to play a role in many different applications, where data of any sort are required. In this section, the applicability of the AnyData pattern in two distinct scenarios is demonstrated by using a use case description and behavior model like sequence diagram. AnyData pattern can be used solely to model a problem with wide range of applications, thereby demonstrating its reusability.

The generated domain-independent pattern can be used to model data of any entity that includes wide range of product, application, etc., making this pattern a general pattern within some specific application. Because of its general applicability, this pattern is valuable and worth documenting.

8.2.8.1 Application: AnyData in Healthcare to Prescribe Medicine

Today, in order to prescribe medicines to patient, a thorough knowledge of his/her past conditions is needed. This approach helps in prescribing medicines quickly and without any complications. Using a patient's medical history, information regarding patient's allergy to a particular drug, his/her current medication, frequency of occurrences of medical problems, etc. can be identified, which can assist in precise prescription of medical process and medicines for patient.

8.2.8.2 Application: Collection of Data for Weather Forecast

Stable weather information describes daily observation of atmosphere, such as daily temperatures, moisture levels, showers, and expected hurricanes of a specific location. Collecting weather information based on the weather conditions and publishing through media to the people helps spreading knowledge to masses.

8.2.9 MODELING ISSUES

8.2.9.1 Abstraction

A stable design pattern is a pattern that designs a system which lasts forever. A stable design pattern will have three parts, an EBT, a BO, and an IOs.

EBTs are the classes that represent the main goal of the underlying business. As the word Enduring Business Theme says, EBTs are very stable and enduring. BOs are the classes that help EBTs to interact with the more specific classes of the system. BOs are concrete and are connected to EBTs via hooks. IOs are the classes that are very specific to the domain under design. For example, "AnyData" is a BO and corresponding IO from a survey domain will be "PopulationData."

The traditional model of designing is very specific to the domains and is hence not stable over time. If something is altered, the design will no longer be stable and has to be redesigned again. With stability design pattern, that issue is being resolved. Since the goal, which is identified as EBT is not being changed over time; the entire stable design pattern will remain the same.

The EBT of AnyData design pattern is

- Knowledge

BOs of AnyData design pattern are

- AnyData
- AnyCollection
- AnyEntity
- AnyEvent
- AnyForm
- AnyType
- AnyAspect
- AnyCategory

8.2.9.2   Simplicity versus Complexity

While modeling a system using stable design pattern, one needs to know the modeling heuristics; for example, there should not be tree structure, star topology, sequence, dangling classes, etc. in the model developed. Traditional model never restricts the modeler how the model should look like. In that way, it is a bit complex for a modeler at a beginner level. Nevertheless, the stable model provides flexibility to the developer who wants to extend it to a more detail level. Identifying the EBTs, BOs, and IOs could be a bit complex. However, once the goal of the system is identified, it is easy to determine the EBT.

8.2.10   Design and Implementation Issues

AnyData design pattern is designed, so that it can be adapted and reused in diverse domains. This section describes some of the issues faced in designing and implementing a stable AnyData pattern.

To implement AnyData analysis pattern, one needs to identify Enduring (EBTs) and BOs. The EBT for AnyData pattern is Knowledge. Identifying EBTs is the hardest and challenging part in designing a pattern because EBTs are conceptual and one needs intuition to identify them.

Another challenge is identifying correct supporting BOs and then connecting them correctly to EBTs in the design pattern. An important issue in implementing AnyData pattern is to ensure that there is no hook up between EBTs and IOs.

Designing generalized pattern that is applicable to all possible business domains is bit time consuming, as it requires identification and finding interrelatedness of all possible business objects for the corresponding enduring business theme. The pattern designed is accurate only when the designer has a thorough knowledge of the problem domain and has accurately identified EBTs, BOs, and IOs to accommodate all concepts of stability model.

EBTs, BOs, and IOs together should make a complete story in order to be the system to be stable. If at some point, the relation between the EBTs, BOs, and IOs does not add up, then that design will not be successful.

Identifying a goal of any project is extremely important and rather intricate. EBT represents the goal of the project. Therefore, unless a person has a deeper understanding of the project and its goal, it is quite impossible to create a stable design. Hence, the level of expertise of the designer is extremely important.

Another challenge in using AnyData design pattern (rather any design pattern) is the integration with any other design pattern because it needs more skill level and knowledge on the integration and a better understanding of all the design patterns used.

While using two stable design patterns together, one has to make sure that the business rules does not lose its integrity.

8.2.10.1   Initialization Pattern

8.2.10.1.1   Layered Initialization

When multiple implementations of an abstraction are needed, a class is defined to commonly encapsulate the logic and subclasses to encapsulate different specific logics.

When we need multiple implementations of an abstraction, we usually define a class to encapsulate common logic and subclasses to encapsulate different specialized logic. When common logic is needed to make a decision on which specialized subclass to create, this wont. This problem is solved by using the Layered Initialization pattern by encapsulating the common and specialized logic in separate classes.

We can use Layered Initialization pattern in the cases like when we want a specific subclass to be created. Also, this pattern may be used when the information that which subclass to be created cannot be decided at compile time. Often Layered Initialization pattern is being used in combination with Delegation and Composite patterns to hide the complexity from the user.

8.2.10.2   Implementation Example

```
Consider the DataQuery example:
public class DataQuery {
      private DataQueryFactoryIF factory;
      public setFactory(DataQueryFactoryIF factory) {
      …
      }
public DataQuery(String query) {
      …
      String dbName = null;
      DataQueryImplIF dataQuery = DataQueryImplIF)factory.
      createDataQueryImpl(dbName);
      …
      }
}
public interface DataQueryFactoryIF {
      }
class MyDataFactory implements DataQueryFactoryIF {
      private static Hashtable hT = new Hashtable();
      static {
            hT.put("INVENTORY",dataQuery.OracleQuery.class);
            hT.put("INVENTORY",dataQuery.OracleQuery.class);
            …
            }
      public DataQueryFactoryIF createDataQueryImpl(String dbName) {
            Class clazz = (Class)hT.get(dbName);
            try {
            return (DataQueryFactoryIF)clazz.newInstance();
            } catch (Exception e) {
            return null;
            }
      }
      }
```

Constraints of the Layered Initialization Pattern are

- Any service class uses exactly one ServiceImplIF interface.
- Service object uses exactly one ServiceImplFactoryIF interface to access a ServiceImplFactory object [6].

Layered Initialization is very specific to certain applications and it cannot be applied to all the domains. Whereas, AnyData design pattern can be applied to any domain; hence, it is stable.

Some issues in implementation of AnyData analysis pattern are described next. Knowledge (EBT) is not inherited and it has associative relationship with AnyData (BO) and AnyAspect (BO). AnyParty (BO) will inherit various parties involved in application context during the implementation phase.

8.2.11   Related Pattern and Measurability

An existing data pattern is shown below. It has some inherent problems like zero reusability because it is designed for a specific purpose. The data access object manages the connection with the data source to obtain and store data.

Class Diagram (Figure 8.2)
Sequence Diagram (Figure 8.3)

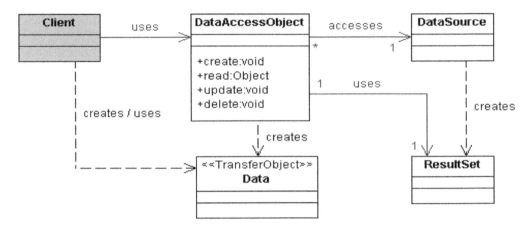

FIGURE 8.2   Class diagram for data pattern by use of traditional approach.

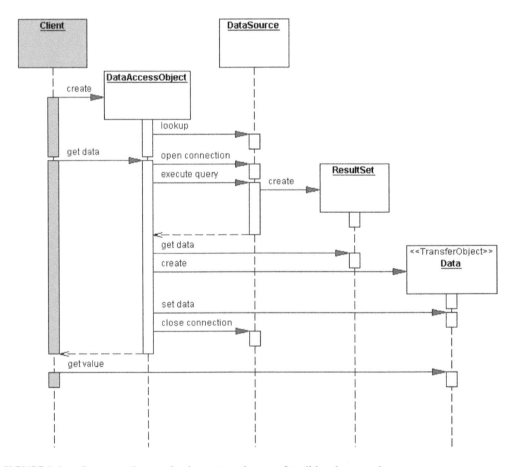

FIGURE 8.3   Sequence diagram for data pattern by use of traditional approach.

8.2.11.1   Traditional Model (Meta Model) versus Stable Model (Pattern)

Traditional modeling (Meta model) generally results in unstable model which is not adaptable, scalable, and extensible. On the other hand, analysis/design pattern created by using SSM is adaptable to changing requirements and it can be scaled easily with minimum effort.

In traditional modeling, we only design, as much is needed for a specific application in question, not thinking of its applicability in other domain; hence, the resulting application is very specific. However, with stability factors in mind, we usually use generally enduring concepts; hence, the resulting pattern can be used for building numerous applications. In short, the resulting pattern can serve as a building block for diverse application domains.

With traditional modeling, the resulting design is tangible enough to be comprehended; while with a stability model, the resulting pattern is intangible. Designing in traditional modeling comes from the problem statement. With stability approach, we need enough experience, as one would derive solution by intuition and thinking out of the box. However, implementing successful pattern requires expertise, meticulous work, and detailed understanding of the problem domain.

8.2.12 Business Issues

8.2.12.1 Business Rules

In any business operation, one of the main issues that we come across is scope creep. Therefore, by identifying the ideal scope for the project will reduce the design issues by a significant amount. Once we identify the scope of the project, identifying EBTs, BOs, and IOs should not be a difficult task.

Business rules describe the operations and constraints that apply to an organization in achieving goals. For example, from Figure 8.2, AnyData "shown in" AnyForm is a business rule. The business rules that interconnect the IOs only will be changed, as the domain changes because the EBTs and BOs remains the same for AnyData pattern, and AnyData design pattern is stable here.

"**Business rules** or **business rulesets** describe the operations, definitions and constraints that apply to an organization in achieving its goals" [7–9].

Some of the business rules that make AnyData design pattern meaningful are

1. AnyEntity should be characterized by AnyData
2. AnyData should belong to AnyType
3. AnyData should be shown in AnyForm
4. AnyForm should belong to AnyType
5. AnyAspect should be an aspect of AnyEntity
6. Knowledge should be about AnyAspect
7. AnyData should represent Knowledge
8. Knowledge should be illustrated with AnyData

Whereas the business rules in the traditional model are

1. Client should use DataAccessObject to access DataSource to create ResultSet
2. Client should use DataAccessObject which should use ResultSet

8.2.12.2 Business Model

AnyData design pattern is a reusable and hence it can be applied to any domain by attaching the domain-specific IOs. IOs must be related to the pattern domain. Thus, AnyData provides a stable and adaptable model.

8.2.12.3 Business Integration

Because AnyData patterns provide hooks, the integration of the AnyData design pattern to existing business model will be quite easy. Integrating AnyData pattern to a particular application domain will require the plugging of application-specific IOs to the hooks.

8.2.12.4 Business Transformation

Use of AnyData design pattern by a company while developing any application strengthens its business transformation process. This is because AnyData pattern quickens the development process by providing the framework for the application. Again, AnyData provides flexibility in storing data in any form and type suitable for the application. Hence, it is beneficial for companies using this pattern. Also, since AnyData pattern is tested and verified, chances of getting bugs are minimal.

While transforming the business, a design created using AnyData pattern will be more stable than the traditional model. As long as the business goal remains the same, AnyData pattern will work because it has been created to work for any domain. EBTs and BOs remain same and the IOs can be hooked to BOs without any hassles.

8.2.13 KNOWN USAGE

The AnyData design pattern can be applied in below listed applications. However, the need for data is different in each case and so is the mechanism used to collect the required data; but the core concept remains the same.

- *Media*: It collects data about current affairs, political issues, etc. to spread information and knowledge to the public.
- *Software Industry*: Data about application are collected, stored, and analyzed for various purposes.
- *Database*: It acts as a store of data. It can maintain data and help retrieving the same as and when needed. It also provides a platform for manipulating data.
- *Health*: Information about patient, doctor, patient's health, etc. is maintained for use.
- *Finance*: Data regarding various stocks and mutual funds are collected to predict future market trends.
- *Weather Information Collection*: Data about weather conditions such humidity, temperature, rainfall, etc. are collected and analyzed to forecast weather conditions.

8.3 SUMMARY

AnyData is modeled by using SSM by identifying the corresponding enduring business theme and BOs. This model can be used for different domains and IOs can be extended according to the application. The model represents the core knowledge of the pattern in different applications and is presented as EBTs and BOs. The model is explained with two applications that perform well based on this model.

Though building a stable design pattern for AnyData that can be reused and reapplied across diverse domain is difficult and requires thorough understanding of the problem, it is worth the effort and time. Modeling AnyData pattern by using SSM will result in a reusable, extensible, and stable pattern.

The correct identification of EBT and BOs for AnyData is most challenging task and requires some experience. Once EBT and BOs are correctly identified, next challenge is to determine the relationship between EBT and BOs, so that AnyData pattern can hold true in any context of usage for data. Once this is done and depending on the application, the IOs are attached to the hooks provided by BOs. Thus, using AnyData pattern as a basis, infinite number of applications can be built by just plugging in the application-specific IOs to the pattern. This results in reduced cost, effort, and stable solution. Hence, AnyData design pattern is very useful.

8.4 OPEN AND RESEARCH ISSUES

1. Many data applications have business rules and or constraints that dictate how the data are handled. How will implementing these business rules fit into the AnyData design pattern?
2. Is the AnyData pattern database agnostic? Does the underlying implementation/technology/method of a database matter to the AnyData pattern?

8.5 REVIEW QUESTIONS

1. (T/F) AnyData design pattern is domain specific.
2. (T/F) AnyData is never a human.
3. (T/F) AnyData cannot be inanimate.
4. (T/F) The AnyData pattern describes a system.
5. (T/F) The AnyData design pattern can be applied and extended to any domain.
6. Define AnyData design pattern.
7. List two major applications of the AnyData design pattern.
8. Why does the AnyData pattern have "Any" as a prefix?
9. List any challenges in formulating the AnyData pattern.
10. List any constraints that crop up during the formulation of the AnyData pattern.
11. Draw the design pattern solution for the AnyData design pattern.
12. List the participants (EBTs and BOs) in the AnyData design pattern.
13. Illustrate the detailed model with a diagram for the AnyData design pattern.
14. Draw the CRC cards for two of the BOs in the AnyData design pattern.
15. List four main benefits and usage notes for the AnyData design pattern.
16. List the use cases in classifying patient data applying the AnyData design pattern.
17. List the use cases in classifying laser printer in an inventory applying the AnyData design pattern.
18. List some of the patterns related to the AnyDesign pattern.
19. List two design issues for the "Knowledge" EBT during the process of linking from the analysis phase to the design phase.
20. List two important implementation issues for the AnyType BO during the process of linking from the design phase to the implementation phase.
21. List two scenarios that would not fit within the context of the AnyData design pattern.
22. List three business issues in the context of the AnyData design pattern.
23. Briefly explain how software stability concepts have been incorporated in the AnyData design pattern.
24. List a couple of research issues relevant to the AnyData design pattern.
25. How will the AnyData design pattern be stable over time?
26. Explain briefly how the AnyDesign pattern provides a high level of extensibility.
27. What is meant by the term "data"? Can the term "data" be used in any other context other than what has already been mentioned in this chapter?
28. List the synonyms of "data." Can these terms be used interchangeably with respect to the AnyData pattern?
29. What are the requirements for a "data"? Describe each of them.
30. Draw and describe the class diagram for the stable AnyData pattern.
31. List main differences between the AnyData pattern described here and the traditional methods.
32. List some design and implementation issues faced when implementing the AnyData pattern. Explain each issue.
33. Give some applications where the AnyData pattern is used.
34. What lessons have you learnt by studying the AnyData pattern?
35. List some of the domains in which the AnyData design pattern can be applied.
36. List some domains that you believe the AnyData pattern would not apply. Explain your reasons.
37. Is the AnyData pattern incomplete without the use of other patterns? Explain briefly.
38. Present the sequence diagram for applicability of the AnyData stable analysis pattern in the e-commerce domain.
39. What do you think are the implementation issues for the AnyCollection BO when used in the AnyData design pattern?

40. What do you think are the implementation issues for the AnyEntity BO when used with the AnyData design pattern?
41. Describe how the developed AnyData design pattern would be stable over time.
42. List some of the testing patterns that can be applied for testing the AnyData design pattern.
43. List three test cases to test the class members of the AnyData pattern.
44. What is the EBT for AnyData. Justify it with the help of an example.
45. What is the main aim of AnyData (SDP)?
46. "Since AnyData pattern is introduced based on the stable design pattern, it makes it easier to employ this pattern in many different applications." Explain how.
47. Define and describe AnyData.
48. How is the word data derived?
49. Explain how data are required in the field of agriculture and weather.
50. Why is it essential to organize data?
51. Explain the difference between data, information, and knowledge.
52. List some of the uses of AnyData Design Patterns.
53. List some of the properties of AnyData SDP.
54. Identify five contexts from day-to-day life where AnyData SDP can be applied.
55. Write three scenarios for AnyData SDP that is not mentioned in this chapter.
56. List the function requirements of AnyData SDP and describe them.
57. List the nonfunction requirements of AnyData SDP and describe them.
58. Identify three challenges related to AnyData SDP.
59. Identify a few constraints of AnyData SDP.
60. Draw a class diagram of AnyData SDP. Explain the participants and justify their existence.
61. Write down CRC cards for the class diagram of AnyData SDP.
62. What are the consequences of AnyData SDP?
63. What are the advantages of AnyData SDP?
64. Draw a class diagram for any application of AnyData SDP. Describe it. Write a use case for it and its description. Draw a sequence diagram for the use case and describe it.
65. Draw traditional model for AnyData.
66. Compare the tradition model of AnyData with the stable model. Which one do you like more and why.
67. List a few design and implementation issues with AnyData SDP.
68. List a few business rules of AnyData SDP.
69. List some business models where AnyData SDP can be applied. Explain them.
70. What did you learn from this chapter?
71. List a few tips and heuristics that you have learnt.

8.6 EXERCISES

Explain with an example the applicability of the AnyData design pattern in classifying students enrolled in a University. Draw the class diagram by including all the EBTs and BOs. The data of concern are student records, transcripts, grades, and metadata.

1. Illustrate with a class diagram, the applicability of the AnyData design pattern to a major league sports team's talent scout. Pick a sport that you are familiar with, list and detail the statistics that are relevant for measuring a player's talent.
   a. Draw a class diagram of this scenario.
   b. Draw a sequence diagram for this scenario.
2. Think of any scenarios that have not been described in this chapter. List them with a short description.

3. Try to create a use case and interaction diagram for each of the scenarios you thought of in the above question.
4. Illustrate with a class diagram and describe the applicability of the AnyData SDP in the following scenario: A company wants to store the employee data in their database.
5. Draw with a class diagram and describe the applicability of AnyData SDP in a scenario where a college keeps a record of student and professors data.
6. Identify a few challenges and constraints of AnyData SDP that are not listed in this chapter.
7. Identify a few advantages of stable model over traditional model that are not discussed in this chapter.

8.7 PROJECTS

1. Hate Crime Data Collection Project.
   a. Explain the above scenario.
   b. Draw a class diagram for it.
   c. Document a detailed and significant use case.
   d. Create a sequence diagram of the created use case of c.
   http://www.osce.org/odihr/datacollectionguide?download=true.
2. Football (soccer) management is a popular genre of video games. This genre of video games is strongly data driven with algorithms that determine the skill ceiling of a player. However, this is all derived from real-world sports statistics. Apply the AnyData pattern and other patterns you have learned in this book to develop a managerial video game simulation of a sport you are familiar. Common roles in the game include, game day lineups, substitutions during a game, scouting talent in other leagues/teams, and developing young players. Assume the algorithms that drive the game are implemented for you; only model the data access/management side of the game. Create class diagrams, CRC cards, and sequence diagrams.
   a. Draw a class diagram for it.
   b. Document a detailed and significant use case.
   c. Create a sequence diagram of the created use case of c.
3. Government wants to store the population data.
   a. Explain the above scenario.
   b. Draw a class diagram for it.
   c. Document a detailed and significant use case.
   d. Create a sequence diagram of the created use case of c.
4. Mining expert wants to store the data to mine and generate patterns.
   a. Explain the above scenario.
   b. Draw a class diagram for it.
   c. Document a detailed and significant use case.
   d. Create a sequence diagram of the created use case of c.
5. An E-commerce company stores order history for all customers.
   a. Explain the above scenario.
   b. Draw a class diagram for it.
   c. Document a detailed and significant use case.
   d. Create a sequence diagram of the created use case of c.
6. *1.0 Application*: AnyData in healthcare to prescribe medicine.
   a. Explain the above scenario.
   b. Draw a class diagram for it.
   c. Document a detailed and significant use case.
   d. Create a sequence diagram of the created use case of c.

7. Application (2): *Collection of data for weather forecast.*
    a. Explain the above scenario.
    b. Draw a class diagram for it.
    c. Document a detailed and significant use case.
    d. Create a sequence diagram of the created use case of c.

REFERENCES

1. F. Scott Fitzgerald (September 24, 1896–December 21, 1940), last modified on June 21, 2017, https://en.wikipedia.org/wiki/F._Scott_Fitzgerald
2. M.E. Fayad and A. Altman. Introduction to software stability, *Communications of the ACM*, 44(9), 2001, 95–98.
3. M.E. Fayad. Accomplishing software stability, *Communications of the ACM*, 45(1), 2002, 111–115.
4. M.E. Fayad. How to deal with software stability, *Communications of the ACM*, 45(4), 2002, 109–112.
5. P. Checkland and S. Holwell. *Information, Systems, and Information Systems: Making Sense of the Field*. John Wiley & Sons, Chichester, West Sussex, December 1997, 278 pp.
6. T. Bevis. *Java Design Pattern Essentials*, 2nd Edition, Ability FIRST, Essex, UK, October 2012.
7. B. Von Halle. *Business Rules Applied*. Wiley, New York, NY, 2001, 592 pp., ISBN 0-471-41293-7.
8. T. Morgan. *Business Rules and Information Systems: Aligning IT with Business Goals*. Addison-Wesley, Boston, MA, 2002, ISBN 0-201-74391-4.
9. R.G. Ross. *Principles of Business Rule Approach*, First Edition, Addison-Wesley Professional, Boston, MA, February 15, 2003, 400 pp., ISBN 0-201-78893-4.

9 AnyEvidence Stable Design Pattern

Facts are stubborn things; and whatever may be our wishes, our inclinations, or the dictates of our passions, they cannot alter the state of facts and evidence.

John Adams [1]

Evidence is a very important concept that deals with basic issue to find proof for a given proposition. The main idea behind this pattern is to develop a stable model, which is easily applicable to all the applicable and possible domains. This pattern provides an ultimate solution to AnyEvidence, so that we need not try to find solution again; right from the scratch, every time we need evidence in an applicable domain. The major applications of this pattern are judicial law and major areas of sciences and scientific research. This pattern could be used as a stand-alone system or it can be used as a part of other pattern.

The *AnyEvidence* pattern is a very important notion that deals with the core in depth to find a proof for any proposition. The plan behind this pattern is to develop a stable model, which helps in reusability of the pattern under any application and in any area of domain without having to develop the model again from scratch.

9.1 INTRODUCTION

Evidence is a lexicon, which denotes something that plays a key role in our daily life. We just cannot imagine a domain without proposition and AnyProposition cannot be proved without evidence. So, everyone needs a stable model for AnyEvidence, to avoid developing a model again from start and afresh. Hereby, we represent the pattern by using EBTs and BOs [2–4].

There is just one ultimate goal for any pattern. The ultimate goal (EBT) of *AnyEvidence* is knowledge. *AnyEvidence* can be a presentation of knowledge. The evidence pattern is a stability model that consists of one EBT and several BOs. It helps us in solving any kind of problem that investigates AnySource such as research, observation, empirical study, etc. The evidence design pattern is a solid base for any proof or proposition. The evidence can be in AnyForm such as testimony, journal, analysis result, etc.

Evidence, in a general sense, refers to anything that is used to demonstrate the truth of an assertion. It indicates specialized meanings and comprehension, when applied to diverse fields such as policy, scientific research, and criminal investigation. In scientific research, evidence is accumulated through observation of phenomena. In addition, this evidence is compared with the existing theory to either accept or reject it.

In data mining, evidence can be used to identify a particular pattern that exists in the data. For example, evidence of proof of renting a car or hotel booking made along with air ticket will enhance knowledge about buying pattern of a particular traveler and this important information can be very useful for the travel industry. In turn, this could become be extremely beneficial to the system to integrate car rental and hotel deals with that of air ticket deals.

In the case of a criminal investigation, available evidence is used to find out the person who is responsible for the purported criminal act and the evidence is also used to know other important details about crime.

In the case of software security, available evidence is used to investigate a breach in security, determine the type of breach and the time at which the breach occurred, in order to determine the eventual cause for the breach.

In all of the domain examples highlighted above, the core concept of evidence remains the same, which creates the need for creating a generalized model to solve this problem. With the *AnyEvidence* stable model proposed here, we can achieve the factor of reusability while developing an application.

In this chapter, we will develop both a stable model and a traditional model for AnyEvidence pattern and contrast/compare them with each other, so that we will be achieving our goal of accepting this model for its application in AnyApplication domain. We will also illustrate two examples of these patterns to demonstrate their reusability in different domains.

9.2 ANYEVIDENCE DESIGN PATTERN DOCUMENT

9.2.1 PATTERN NAME: ANYEVIDENCE STABLE DESIGN PATTERN

The name of this pattern is *AnyEvidence* stable design pattern. Evidence is a much generalized concept. There are alternate names to evidence like proof and indication. However, they represent a specific domain, while *AnyEvidence* is in a generalized form and it presents the true essence of the problem modeled in this chapter. *AnyEvidence* has *Any* prefix, because it is not an enduring concept. It is an intuitive way to achieve final goal knowledge.

9.2.2 KNOWN AS

AnyEvidence is also popularly known as the proof. Nevertheless, proof is not the right or apt name to be used for the problem. Proof does not capture true essence of the modeled pattern. Hence, evidence is the most appropriate term for this stability pattern.

Similarly, other names such as clue, sign, symptom, lead, guess, assertion, indicator, and substantiation, which are synonyms of evidence, cannot replace the all-embracing term for modeling this stability pattern.

Clue: According to Merriam-Webster online dictionary, a clue is something that assists a person find something, understand something, or solve a mystery or puzzle [5]. It is also an understanding or knowledge about something. Both clue and evidence denote different meanings. A clue is a sort of tool that helps someone to find or understand something. However, it does not act as an evidence to help someone to prove the finding or understanding.

Sign: A sign, according to Merriam-Webster online dictionary, is something (an action or event) which shows that something else exists, is true, or will happen in future [5]. The terms "sign" and "evidence" are different, as the former does not lead to the furnishing of evidence. Rather, it simply implies that something exists and that it might happen in the future.

Symptom: A symptom is also something that signifies that existence of something else. A symptom could be temporary event and it may cease to exist after sometime. On the other hand, evidence is permanent that eventually proves that something actually occurred or something definitely exists.

Lead: Lead means to bring some conditions or conclusions and this is entirely different from the word "evidence." Some events or conditions may eventually lead to the creation of "evidence." In other words, a lead may act as a tool to generate "evidence."

Guess: Guess is purely speculative and it leads to formation of an opinion or it might furnish some answers about something when we do not know much about it. In other words, mere speculation may never lead to the formation of "evidence."

Assertion: Assertion is the deliberate action of stating clearly to make others aware of a specific situation. The person who makes an assertion is firm and steadfast. However, an assertion may not be just enough to say that it is an "evidence" for something.

Indicator: An indicator is a sign that demonstrates an existing condition or it may show the existence of something. An indicator can never act as an "evidence" because it is purely speculative and a belief that something exists.

Substantiation: A pointer demonstrates the state or condition of something that exists. However, to substantiate something, one needs a solid piece of evidence that is competitive and definite. Although, substantiation is almost similar to the word evidence, it is not "evidence" per se.

9.2.3 CONTEXT

The pattern of AnyEvidence can recur in many of the concepts like crime investigation, in any case of jurisdiction, any research on any domain, data mining, and software testing. To prove this, let us consider an example in real time. Consider an e-banking site, through which we will transfer some money from our account to other's account. Soon after the successful transaction, we will receive an e-mail regarding the records of transaction as evidence of sending money. We may also receive an SMS regarding the transaction as an evidence to prove it.

In the domain of jurisdiction, AnyEvidence plays a very crucial role. Any claims made by any party must be supported by appropriate evidence to make it acceptable for all parties concerned. Any judgment is given only based on the presentation of AnyEvidence. *Digital evidence* or *electronic evidence* is relevant information stored or transmitted in *digital* or electronic form that a party to a *court case* may use at *trial* (Sidebar 12.1).

On the other hand, in scientific research areas, a scientist will collect evidence to support a theory or hypothesis. If the experiment is simple and straightforward, AnyEvidence is collected in the form of a series of readings. If experiments relate to an exploration, then the evidence is collected in the form of fossils or research samples. For example, to prove existence of particular species in the past, fossilized examples of those species are collected from the area.

Another important applicability of AnyEvidence pattern is in the domain of data mining. Data mining, when used to find out target opportunities, a consumer's purchase decisions and ideas play an important role as an evidence to identify buying patterns and habits.

Another example of use of AnyEvidence pattern is when an individual wants to open a bank account. The bank will ask for documented evidence to establish user identity. Another example is receiving a detailed SMS message and e-mail alerts to confirm successful online transaction; both of them are evidences that support the claims for any successful transaction.

The following two examples provide the practical representation of AnyEvidence: job application and identification of disease.

1. *Job Application*: When a person (AnyParty) applies for a job in a company (AnyParty), he or she has to provide a resume (AnyEntity) at the company's career web site. The resume provides knowledge (EBT) about the person's accomplishments (AnyEvidence) that can be proved by providing degree certificates (AnyForm). The certificates also contain the issued date (AnyPeriod). The university or professor (AnySource) can also provide a recommendation letter (AnyProposition) that proves and supports the achievements of that person.
2. *Identification of Disease*: A doctor (AnyParty) identifies (knowledge) a disease (AnyEntity) that a patient (AnyParty) is suffering from, by conducting blood test (AnyForm) and presents it to patient in the form of a test report (AnyEvidence). The tests are valid only for a particular time (AnyPeriod). After that time, the tests are conducted again. The diagnostic lab (AnySource) also supports the doctor's identification of the disease (AnyProposition) based on the tests.

9.2.4 PROBLEM

Following aspects are the driving forces for modeling AnyEvidence stable design pattern.

1. *Wide occurrence*: Evidence is used in many different domains. That creates a need for a generalized model.

2. *Lack of stable model*: There is no stable model for this problem is applicable to all the applicable domains, by just adding domain-specific elements. Stable pattern makes it easy to achieve the factor of reuse.
3. Availability
4. *Structural Adequacy*: This is a rating of the member force and the strength of the pattern to model the pattern. AnyEvidence, because of its applicability in various domains and hence supporting reusability, has a high rating of structural adequacy, when compared to using any other terms.
5. *Admissibility*: The AnyEvidence design pattern is flexible to accommodate IOs from various domains in order to extend for different applications. Hence, the AnyEvidence design pattern is admissible.
6. Categorization
7. *Impactability*: The AnyEvidence design pattern is stable enough not to have any impacts on the EBT and IOs, when extended to be applicable in various domains.
8. *Knowledge*: Knowledge is chosen as the pattern EBT, because knowledge is an enduring concept and it has the ability to be all encompassing for applications in various domains.
9. *Initiation of Source by Party*: AnyParty IO achieves knowledge by AnySource that is derived from AnyEvidence.

9.2.4.1 Functional Requirements

1. *AnyEvidence*: Represents the evidence itself. The ultimate goal of *AnyEvidence* is knowledge. *AnyEvidence* may be in any form such as blood test, x-ray, document, audio, video, etc. AnyEvidence has a unique ID and a status. In addition, AnyParty collects it, and thus it has a party associated with it. It guides AnyParty in discovering knowledge and is used to support the discovery.
2. *AnyParty*: Represents the evidence seeker. It models all the parties that deal with *AnyEvidence*. Party can be a person, organization, a political party, country, software program, or a legal group with a specific orientation. AnyParty starts using AnySource like observation, experimentation, etc. to find the *AnyEvidence* for any entity. AnyParty has a name, a type, and an authority level. In addition, any party plays a specific role. Any party also initiates a process, explores possibilities, collects evidence, and achieves knowledge or results.
3. *AnySource*: Represents the source through which the evidence is discovered. For instance, CIA can investigate the activities of a terrorist plot. The source mentioned here is a person, who is involved as an undercover agent investigating the terrorist activity, and he or she provides photos, addresses, and transactions in the form of evidence. AnySource has a unique ID, a title, and a type. It develops and builds upon evidence; it satisfies the discovery and proves the knowledge.
4. *AnyForm*: Represents the form of the evidence. This means that it is a physical form of the evidence and not the form in a document. The form of *AnyEvidence* defines the form of the evidence, which is very essential element of *AnyEvidence*. AnyForm has a unique ID, a type, and has a time associated to it. It also provides the proof of discovery and is dated to a particular period.
5. *AnyProposition*: Represents the initialization of evidence. This class presents a core element for the evidence to be proved. The proposition must support a form of evidence. AnyProposition has a unique ID, a status, a title, and a description. It supports the evidence and used to acquire evidence in some cases. AnyProposition is involved in a particular discovery.
6. *AnyEntity*: Represents the basic object on which the source is initiated by the party to determine the evidence in order to prove the proposition. AnyEntity can be any basic object such as software project, crime scene, chemical, etc. AnyEntity has a unique ID, a title, and a type. It contains knowledge, supply knowledge, and involves party with it.

7. *AnyPeriod*: Represents the period within which the evidence comes out. *AnyEvidence* has a time factor associated with it. The form also supports the period, because it should come out within that period. This period can be in the metrics of time only. AnyPeriod has a date and a time associated with it. It also has a unique ID and a unit in which the date and time is specified. This AnyPeriod provides a date; it also records time and denotes a period range.

8. *Knowledge*: Represents the ultimate goal of *AnyEvidence*. Knowledge is an enduring concept. *AnyEvidence* is to be presented with the help of knowledge. This class presents all the characteristics and attributes of evidence, depending up on the entity. Knowledge can have a type. It has a status and it needs to be validated by some party. Knowledge is presented as any evidence. It also provides wisdom and involves AnyParty with it.

9.2.4.2 Nonfunctional Requirements

1. *Relevance*: AnyEvidence must be relevant to the knowledge. If the evidence does not prove the knowledge that a person is interested in, but proves any other knowledge, then it will not be relevant for that person, and hence it will be of no use. Thus, being relevant is one of the important quality factors of AnyEvidence.

2. *Achievable*: AnyEvidence must be achievable, so that it can be achieved and presented in time to prove the discovery or the knowledge. If the evidence exists but cannot be achieved, AnyParty cannot use it as a proof, and the evidence will loose its significance.

3. *Justifiable*: AnyEvidence should be justifiable, so that it can be used to prove the theory or discovery of knowledge. If the evidence cannot justify the knowledge, it would be termed as any evidence. Thus, being justifiable is one of the most important quality factors of AnyEvidence.

4. *Timeliness*: Any Evidence should be presented on time. There is always a time factor associated with any evidence. For example, if a person is blamed for a crime although, he or she should present the evidence in defense to the court in a timely manner and before the court gives a final verdict. Otherwise, the evidence will be of no use.

9.2.5 CHALLENGES AND CONSTRAINTS

9.2.5.1 Challenges

Challenge ID	0001
Challenge Title	Too Many Propositions
Scenario	A doctor proposing or making too many proposition to a patient for conducting test.
Description	Consider a scenario, when a patient is not well and when he or she consults a doctor. If the doctor proposes a patient to undergo 100s of laboratory tests before identifying the disease, the patient will lose faith in the doctor. Moreover, this will consume a lot of time and the disease might spin out of control.
Solution	Only the required number of tests should be proposed in this scenario or in general, only the most relevant testing propositions should be made.
Challenge ID	0002
Challenge Title	Knowledge Presented as Too Many Evidences
Scenario	A person convicted for a crime presents evidence of innocence based on the knowledge.
Description	If a person presents 100s of evidences based on a particular knowledge, the most important evidence might be overlooked. Moreover, the judge might feel that the person is trying to fake the evidence.
Solution	Only the required number of instances of evidence should be presented from the existing set of knowledge. Moreover, the evidence that is most relevant for the situation should be used.

Continued

Challenge ID	0003
Challenge Title	Too Many Evidences Discovered about a Particular Entity.
Scenario	Person convicted of a murder, presenting evidence that he or she never possessed the murder weapon.
Description	If a person presents 100s of instances of evidences based on one particular knowledge, the most important evidence might get overlooked. Moreover, the judge might feel that the person is trying to fake the evidence.
Solution	Only the required number of evidence should be presented from a set of knowledge. Moreover, the evidence that is most relevant in the situation should be used.
Challenge ID	0004
Challenge Title	Too Many Parties Achieve Knowledge Through a Source.
Scenario	Too many students studying from one book.
Description	In this scenario, a book acts as an evidence of a discovery and is a source of knowledge. If 100s of students study from the same book, this book might be destroyed and might not be useful to anyone in the future. However, if multiple copies of books are circulated, so that there is one book for each available for one small group of students, then each student will get a fair chance to read this book and the condition of this book can be maintained too.
Solution	There should be an upper limit set on the number of parties that can use a source to extract knowledge out of it.

9.2.5.2   Constraints

1. Knowledge has one to many AnyParty involved.
2. One-to-many Entity gives one knowledge.
3. Knowledge is presented as one-to-many evidences.
4. One evidence takes one form.
5. One-to-many evidences discovered about one entity.
6. One evidence comes out in one period.
7. One-to-many party gets evidence through one source.
8. One form is dated to one period.
9. One source proves for one-to-many propositions.
10. One proposition supports one form.
11. One or more entity provides knowledge.
12. One or more parties achieve knowledge.
13. Knowledge can be presented as one or more evidence.
14. One or more evidence is discovered by an entity.
15. Every evidence takes a form.
16. Every evidence has a time period associated with it.
17. One or more party can act as a source.
18. A source provides proof for one or more propositions.
19. Each proposition supports a form of evidence.
20. Each form of evidence is dated for a particular time period.

9.2.6   Solution

9.2.6.1   "AnyEvidence" Stable Model

Figure 9.1 shows AnyEvidence stable design pattern. AnyParty use AnySource to support AnyProposition. AnyProposition needs AnyEvidence for AnyEntity. AnyEvidence can be of AnyForm and it is attached to AnyPeriod. This way AnyEntity, gives knowledge to AnyParty (Figure 9.1).

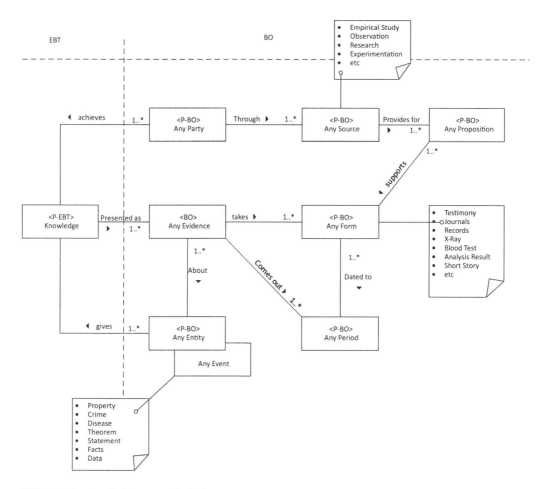

FIGURE 9.1 AnyEvidence stable design pattern.

9.2.6.2 Class Description

9.2.6.2.1 Participants

> *Knowledge*: Represents the ultimate goal of *AnyEvidence*. Knowledge is an enduring concept. *AnyEvidence* is to be presented with the help of knowledge. This class presents all the characteristics and attributes for evidence depending up on the entity.

9.2.6.2.2 Patterns

> *AnyEvidence*: Represents the evidence itself. The ultimate goal of *AnyEvidence* is knowledge. *AnyEvidence* may be in any form like blood test, x-ray, document, audio, video, etc.
>
> *AnyParty*: Represents the evidence seeker. It models all the parties that deal with *AnyEvidence*. Party can be a person, organization, political party, country, software program, or a legal group with specific orientation. AnyParty starts using AnySource like observation, experimentation, etc. to find the *AnyEvidence* for any entity.
>
> *AnySource*: Represents the source through which the evidence is discovered. For example, CIA can investigate the activities of a terrorist plot. The evidence gathered here is by a person, who works as an underagent on the terrorist activity and evidence will be in the form of photos, sketches, addresses, and transactions.

AnyForm: Represents the form of the evidence. This means that it is a physical form of the evidence and not the form in a document. The Form of *AnyEvidence* defines the form of the evidence, which is a very essential element of *AnyEvidence*.

AnyProposition: Represents the initialization of evidence. Thus, this class presents a core element for the evidence to be proved. The proposition must support a form. It is important that some propositions may not support all evidences. Thus, this class might fail in few cases.

AnyEntity: Represents the basic object on which the source is initiated by the Party to determine the evidence to prove the proposition. AnyEntity can be any basic objects such as software project, crime scene, chemical, etc.

AnyPeriod: Represents the period within which the evidence comes out. *AnyEvidence* has a period associated with it. The form also supports the time because it should emerge within that period. This period can be in metrics of time only.

9.2.6.3 CRC Cards

	Collaboration	
Responsibility	**Client**	**Server**
Knowledge (Knowledge) <EBT>		
Presents the knowledge to the	AnyEvidence	present()
evidence	AnyParty	supply()
		involve()
Attributes: knowledgetype, status, validatedBy		
AnyParty (AnyParty) <BO>		
Initiate the source to find details of the	Knowledge	initiate()
Evidence	AnySource	achieve()
		collect()
		explore()
Attributes: partyname, partytype, authorityLevel, role		
AnySource (AnySource) <BO>		
Investigates, depending on the type of	AnyEvidence	prove()
domain initialized by Any Party, to	AnyParty	satisfy()
satisfy a Proposition		develop()
Attributes: sourceId, sourceTitle, sourceType		
AnyEvidence (AnyEvidence) <BO>		
To discover about any entity in any	AnyEntity	Take()
form	AnyPeriod	come()
	AnyForm	discover()
		support()
		guide()
Attributes: evidenceId, evidenceStatus, collectedBy		
AnyPeriod (AnyPeriod) <BO>		
Denote the time within which the	AnyEvidence	provideDate()
Evidence has come out.	AnyForm	recordTime()
		denote()
Attributes: periodId, periodUnit, perioudFor		

Continued

Responsibility	Collaboration	
	Client	Server
AnyProposition (AnyProposition) <BO>		
Support the form of the evidence and	AnyEvidence	support()
are proved by the source.	AnySource	provide()
		acquire()
Attributes: propositionId, propositionStatus		
AnyEntity (AnyEntity) <BO>		
Contains all the elements to be	Knowledge	contain()
discovered about Knowledge.	AnyEvidence	supply()
		involve()
Attributes: id, title, entityType		

9.2.7 Consequences

Use of this pattern can greatly simplify any investigative work.

This pattern can very easily achieve reuse. Although, it will require that new elements based on situation, that is, if evidence needs to be stored for further reference, it will need AnyLog When *AnyEvidence* needs to be displayed, and then this model will need AnyMedia.

9.2.8 Applicability with Illustrated Examples

9.2.8.1 Case Study: A Hospital Scenario

Mr Grant is a patient, who has been suffering from bouts of headache for a long time. He consults a doctor and asks him about the problem that he is facing. The doctor starts investigation by asking about other symptoms that might be troubling Grant, records them and eventually prescribes him an MRI test. Mr Grant visits the Diagnostic Department and undergoes a scan as required by the doctor. The Diagnostic Department furnishes him a scan report. Mr Grant shows it to the doctor who analyzes the report to identify the problem. Eventually, the doctor concludes that the patient has a tumor in the brain for the last 6 months.

9.2.8.2 Scenario Use Case

1. Use Case # 1
2. **Use Case Name:** Check Health Status.
   Actors and Roles:

Actors	Roles
AnyParty	Doctor
	Patient

Classes, Corresponding Attributes, and Interfaces: Table 9.1

3. **Use Case Description:**
   a. The patient (AnyParty) consults a doctor (AnyParty) about a recurring headache problem and seeks medical solution.
   b. The doctor starts an investigation on the HumanBrain, of the patient (AnyParty).
   c. The patient takes an MRIScan and collects a prepared report as AnyEvidence.
   d. The patient submits the report to doctor (AnyParty), who analyses it and finds out AnySource of medical complaint.
   e. The doctor (AnyParty) confirms that patient (AnyParty) has BrainTumor and prescribes medication and surgery.

f. The MRIScan report acts as AnyEvidence to confirm the presence of BrainTumour in AnyParty (patient).
4. **AnyEvidence in a Hospital Example** (Figure 9.2).
5. **Sequence Diagram** (Figure 9.3).

TABLE 9.1
Check Health Status

Class	Type	Attributes	Operations
Knowledge	EBT	knowledgetype status description id	present() supply() involve()
AnyParty	BO	partyname email id contactNumber	initiate() achieve() collect()
AnySource	BO	sourceId sourceTitle sourceType	prove() satisfy() develop()
AnyEvidence	BO	evidenceId evidenceStatus collectedBy	take() come() discover()
AnyPeriod	BO	periodId periodUnit perioudFor	provideDate() recordTime() denote()
AnyProposition	BO	propositioned propositionStatus description	support() provide() acquire()
AnyEntity	BO	id title type category	contain() supply() involve()

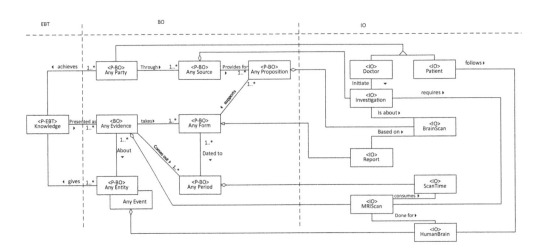

FIGURE 9.2   AnyEvidence in the hospital scenario.

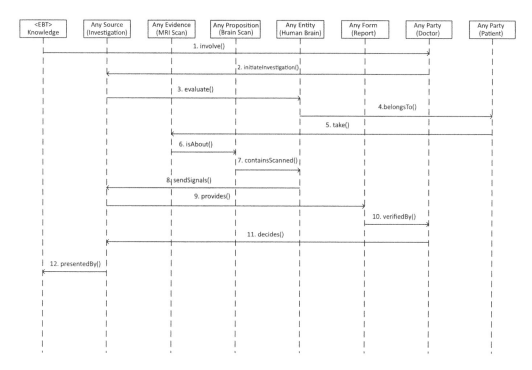

FIGURE 9.3 AnyEvidence sequence diagram of check health status.

9.2.9 Modeling Issues

9.2.9.1 Abstraction (for AnyEvidence BO)

Abstraction is a modeling issue during the process of determining the ultimate EBT for AnyEvidence BO. It is necessary to capture all the scenarios in various domains in order to find out the EBT for AnyEvidence BO. The following method describes the process of finding out the ultimate EBT, which is the knowledge as suggested in the stability pattern for AnyType.

Information, wisdom, lore and erudition are terms that are similar in meaning to the term knowledge. Nevertheless, none of them are as domain nonspecific as knowledge, and hence knowledge is the ultimate EBT for AnyEvidence. This is abstraction and it is a very important concept to understand modeling as it helps in the process of concluding the ultimate EBT for the AnyEvidence BO.

As illustrated in the stability pattern for AnyEvidence, AnyParty, AnyProposition, AnyPeriod, AnyForm, AnySource, AnyEntity, and AnyParty are the other supporting BOs for modeling AnyEvidence.

9.2.9.2 Use of a Static Model (Figure 9.4)

Consider the above-furnished case, where various modes of evidence are depicted, as inheriting from AnyEvidence BO. In this case, the model is static and it restricts the number of child classes to 'n' that is not appropriate, when modeling a real-time situation. Hence, it is always better to use aggregation, which makes the "whole" (AnyType) container and makes it more flexible for the applicability.

9.2.10 Implementation Issues

The ultimate goal of AnyEvidence is knowledge and AnyEvidence is means to achieve this goal. So, it is concluded that AnyEvidence is a BO.

Hook for AnyParty, AnyEntity, and AnyForm should represent generalization. On the other hand, AnyPeriod, AnyProposition, AnySource signify aggregation.

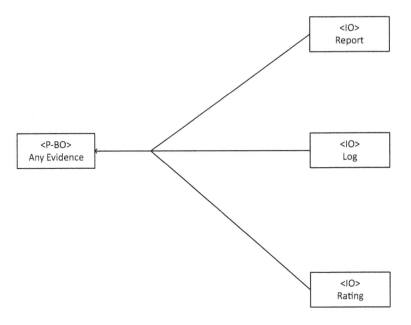

FIGURE 9.4   Use of a static model: use aggregation instead of inheritance.

9.2.11   ENDURING BUSINESS THEMES

Knowledge: The knowledge class completely depends on the domain of the evidence. AnyEvidence must be presented with the knowledge that encapsulates it. Therefore, the design issues of EBT completely depend on the BOs. Since the EBT pertains to AnyEvidence, it cannot be mentioned as a single class. Hence, a hook cannot be added here.

9.2.12   BUSINESS OBJECTS

- *AnyEvidence*: The IOs for AnyEvidence cannot be inherited always. Mostly, the IOs like report or journal do not share interfaces and properties, so that we can decide on only the part of the class. The design issue, hence, becomes complicated as it progresses.
- *AnyParty*: The IOs for AnyParty will always inherit from it. Most of the times, the IOs such as doctor, project team, and country consists of common persons who are involved in them. Therefore, they share most of the interfaces and properties. Hence, AnyParty can be generalized with IOs.
- *AnySource*: The IOs for AnySource cannot be inherited always. Usually, the IOs such as observation, research, and inspection do not share much of the interfaces and properties, so we can decide IOs only as the part of the class. The design issue between BOs and IOs can always be hooked as a part to add stability (i.e., aggregation).
- *AnyForm*: The IOs for AnyForm will always inherit the class. Generally, the IOs such as report and testimony have a definite shape and material to be used to make them. Hence, they have many common properties involved in them that can be inherited. Therefore, they share most of the interfaces and properties. Eventually, AnyParty can be generalized with IOs. A hook can be easily added to the code of the BO.
- *AnyProposition*: The IOs for AnyProposition cannot be inherited always. Usually, the IOs such as a software project plan do not share similar interfaces. Therefore, as they do not share much of the interfaces and properties, we can decide IOs only as part of the class. This design issue between BOs and IOs is that they can always be hooked as a part to add stability (i.e., aggregation).

- *AnyEntity*: The IOs for AnyEntity will always inherit the class. Under normal circumstances, the IOs such as human brain and software project are instances of object that share general properties. As they have common properties involved in them, they can be readily inherited. Therefore, they share most of the interfaces and properties. AnyEntity can be generalized with IOs. A hook can be easily added to the code of the BO.
- *AnyPeriod*: The IOs for AnyPeriod cannot be inherited always except only in some of the rarest cases. Most of the times, the IO as schedule can only be part of the class and possess different interfaces. Therefore, as they do not share much of the interfaces and properties, we can decide IOs as only be part of the class. The design issue between BOs and IOs is that they can always be hooked as a part to add stability as an aggregation to maintain stability.

9.2.13 Business Issues

Business Rules: The business rules for the pattern are general in application and they involve most of the domains, while the EBT is applicable to most of the applications. Nevertheless, the BOs can only be a part of the application.

Business Standards: The business standard for this application is very useful in most of the applications, as it is an extremely stable model and it describes almost all BOs related to the IOs that have a standard in a domain.

Business Patterns: The business patterns for AnyEvidence pattern could be described as one the most useful methods to find any evidence for any source. Thus, the BOs of the pattern describe most of the patterns here.

9.2.14 Known Usage

This pattern can be used to make an intelligent system to take decision and add knowledge base based on the evidence for a healthcare system or for conducting a criminal investigative work.

*Related Pattern and Measurability (**Figure 9.5**):* the comparison table between the two software development models is given in Table 9.2.

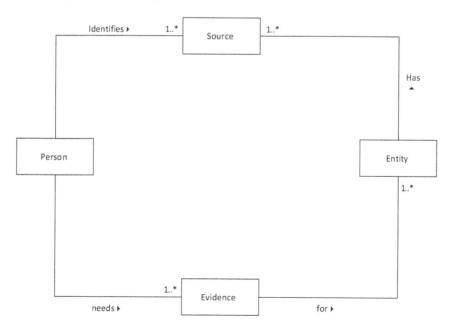

FIGURE 9.5 Traditional model of evidence.

TABLE 9.2

Traditional Model vs Stability Model of Evidence

	Traditional Model	Stability Model
Reusability	Traditional model cannot be reused because it is specific to one problem and does not analyze the system a whole. It is not designed to accommodate different problem. It deals only with tangible objects	Stability model can be reused repeatedly by just changing IOs. The EBTs and BOs remain stable and require no modification. The stability model is modeled after analyzing other similar problems, so it provides excellent reusability
Understanding	Traditional model deals with tangible object, so it does not provide solutions various problems	In stability model, other similar problems are analyzed to identify EBTs and BOs, so we have better understanding of the problem. For instance, the term "Pleasure" analyzes the pleasure gained from good and bad activities, so it can be reused for many problems
Maintenance	Maintenance in traditional model is difficult and every expensive	Maintenance in stability model is easy and less expensive because it has the core knowledge for all related problems
Analysis	Analysis is limited in Meta model and it allows limited changes	Analysis plays a vital role in stability model, so the stability pattern remains stable for long time
Iteration	Traditional model requires less thinking and iteration	Stability model requires more thinking and iteration to come out with the right model. During each round of iteration, few BOs are added or deleted and the model gets better with every instance of iteration
Upgradeability	Traditional model provides very limited upgradeability. Most of the times, it requires the entire system to be remodeled	Since the stability model is built to accommodate other similar system, it is easy to upgrade and extend the system with minimal cost
Cost	Development cost is less. However, the maintenance cost exceeds the development cost. At time, it might even be 2000 times that of the development cost	Initial development cost is high, but after development, the remolding expense for various applications is very less

9.3 TIPS AND HEURISTICS

9.3.1 FOR EVERY PATTERN, THERE SHOULD BE ONLY ONE EBT

In the stability model, we understood that every pattern should have one ultimate goal and that should be the EBT. We have selected the EBT, which is not only the goal of one application but also for other similar applications. Therefore, the pattern remains stable and can be reused repeatedly by changing its IOs.

9.3.2 NEED ITERATION TO BUILD STABLE PATTERN

Building the stable pattern requires multiple iterations. During each round of iteration, BOs will change either by adding or by changing to get the right pattern.

9.3.3 THERE SHOULD NOT BE ANY DEPENDENCY BETWEEN THE CLASSES

Dependency between the classes should be avoided in a class diagram. In dependency, if one class fails, all the classes that depend on class will also fail. We have carefully examined our class diagram and avoided dependency between the classes.

9.3.4 DESIGN HEURISTICS

1. The AnyEvidence design pattern mainly serves to improve knowledge of AnyEntity depending on the data domain objects that are represented by AnyParty.
2. An in-depth understanding of the pattern and a great deal of intuition and experience are required to determine the appropriate EBT for the AnyEvidence pattern. This is because, EBTs are the ones that remain constant and make up the core of the software system.
3. The knowledge EBT cannot talk directly to the IOs (which are domain and scenario specific). The BOs, namely AnyParty, AnySource, AnyEvidence, AnyPeriod, AnyForm, AnyProposition, and AnyEntity, help in providing coherence between the EBT and the IOs.
4. Identifying the most pertinent BOs always requires time and effort.
5. Reusability is introduced with BOs, as they provide the hooks to which the IOs (belonging to various domains) can be attached.
6. Use of the correct BOs also makes the pattern more adaptable to changes within the IOs.
7. A proper illustration and description for the class diagram and the sequence diagram help understand the problem in a minute detail.
8. The design pattern should be sweeping in order for it to be extended for further use in multiple domains.
9. The AnyEvidence design pattern requires careful and calibrated work, as it is a stable model.

9.4 SUMMARY

The presented model shows and describes all the essential elements for knowledge enhancement through extensive evidence collection. The purpose of this chapter was to identify and represent the core concept of any evidence. This chapter has also exhibited two examples of application of this model, which shows how easy it is to apply this model in other domains. Stability model helps in developing a robust, reusable, upgradeable system when compared with the Meta model. Thus, this model can be easily extended to model any other application that requires using AnyEvidence.

9.5 OPEN AND RESEARCH ISSUES

The term evidence has different connotations and meanings and under different context of usage. Often, finding out the most appropriate meanings for the term "evidence" is extremely challenging. In fact, this leads to the detection of the EBT for the pattern; in this chapter, knowledge was found to be the EBT, while other terms were ruled out and they assigned the functionalities of BOs. A notable research issue in the field of creating stable patterns is choosing the right type of EBT out of many research terms. In fact, most of them are not as domain specific as one specific term that eventually turns out to the EBT for the pattern. In other words, the issues of abstraction and modeling an EBT are the two important open issues that needs due consideration from pattern developers.

9.6 REVIEW QUESTIONS

1. Define the AnyEvidence stability design pattern.
2. Why do you consider AnyEvidence as the most appropriate BO in the pattern?
3. List a few application domains for the AnyEvidence design pattern.
4. List four challenges while formulating the AnyEvidence design pattern.
5. AnyEvidence as a stable design pattern is domain and application specific—say True or False.
6. List any two constraints while formulating the AnyEvidence design pattern.

7. List the participants in the AnyEvidence design pattern.
8. Illustrate the class diagram for the AnyEvidence design pattern.
9. List one tradeoff in using this pattern.
10. Document the CRC card for the AnyEvidence BO.
11. The AnyEvidence pattern involves the use of other patterns too. Briefly explain.
12. List two main advantages of using the AnyEvidence stable design pattern.
13. List two applications of the AnyEvidence design pattern.
14. Briefly describe why knowledge is the most appropriate EBT for use in the AnyEvidence design pattern.
15. Why do you think that the AnyEvidence design pattern would be stable over time?
16. Can you list some of the research issues related to the AnyEvidence design pattern?
17. Why the AnyEvidence design pattern provides extensibility over a period.
18. List two modeling issues that can crop up, when modeling the stable design pattern for AnyEvidence.
19. List one scenario that will not be covered by AnyEvidence design pattern.
20. List three test cases to test the participants of the AnyEvidence design pattern.
21. List some of the related design patterns used in formulating the AnyEvidence stable design pattern.
22. List some of the implementation issues for AnyEvidence BO during the process of linking from the design phase to the implementation phase.
23. List some of the constraints in developing the AnyEvidence design pattern.
24. Briefly describe the core problem tackled by the AnyEvidence design pattern.
25. Briefly explain how software stability concepts have been used to implement the AnyEvidence design pattern.

9.7 EXERCISES

1. With the aid of a class diagram, explain how the AnyEvidence design pattern is used in plant hybridization domain in order to cultivate plants with a particular dominating character. Draw CRC cards for all the EBTs and BOs involved in the problem.
2. Illustrate with a class diagram and describe the applicability of the AnyEvidence SDP in following scenario: A marketing company of a juice is trying to tempt people to buy the product by proving that it contains lots of vitamin D.
3. Draw with a class diagram and describe the applicability of AnyEvidence SDP in a scenario, where a person is charged with a crime, but he is trying to prove himself innocent.
4. Identify a few challenges and constraints of AnyEvidence SDP that are not listed in this chapter.
5. Identify a few advantages of stable model over traditional model that are not discussed in this chapter.

9.8 PROJECTS

Using the scenario of plant hybridization as described in Section K.8, explain how AnyEvidence may be chosen as the stability pattern for modeling the scenario. Illustrate with the corresponding class diagram, CRC cards, use cases, use case diagram, and the sequence diagram.

1. Attracting people to buy a juice by proving that it has healing properties.
   a. Explain the above scenario.
   b. Draw a class diagram for it.
   c. Document a detailed and significant use case.
   d. Create a sequence diagram of the created use case of c.

2. Student needs to provide evidence that he did not cheat in exam.
   a. Explain the above scenario.
   b. Draw a class diagram for it.
   c. Document a detailed and significant use case.
   d. Create a sequence diagram of the created use case of c.
3. A student provides evidence to prove his research.
   a. Explain the above scenario.
   b. Draw a class diagram for it.
   c. Document a detailed and significant use case.
   d. Create a sequence diagram of the created use case of c.
4. A person presenting evidence of his innocence to court.
   a. Explain the above scenario.
   b. Draw a class diagram for it.
   c. Document a detailed and significant use case.
   d. Create a sequence diagram of the created use case of c.

REFERENCES

1. John Adams (October 30 [O.S. October 19], 1735–July 4, 1826), last modified on June 24, 2017, https://en.wikipedia.org/wiki/John_Adams
2. H. Hamza and M.E. Fayad. A pattern language for building stable analysis patterns, in *9th Conference on Pattern Language of Programs (PLoP 02)*, Illinois, September 2002.
2. M.E. Fayad and A. Altman. An introduction to software stability, *Communications of the ACM*, 44(9), 2001, 95–98.
3. H. Hamza. A foundation for building stable analysis patterns, MS thesis, Department of Computer Science, University of Nebraska-Lincoln, Lincoln, NE, 2002.
4. Merriam-Webster Dictionary. Encyclopædia Britannica Online. 2015. https://www.merriam-webster.com/, retrieved June 24, 2015.

SIDEBAR 9.1   E-EVIDENCE OR DIGITAL EVIDENCE

Digital evidence or *electronic evidence* is relevant information stored or transmitted in digital or electronic form that a party to a court case may use at trial [1]. Before accepting digital evidence, a court will determine whether the evidence is relevant, whether it is authentic, if it is hearsay and whether a copy is acceptable or the original is required [1]. Courts are allowing presentation of digital evidence in various formats like e-mails, digital photographs, ATM transaction logs, word processing documents, files saved from accounting programs, spreadsheets, Internet browser histories, databases, the contents of computer memory, computer backups, computer printouts, digital video or audio files [2], and any other social media records and dialog, digital chatting, medical records.

Although federal courts in the United States have allowed presentation of digital evidence during trials, concerns are raised over the lack of established norms and regulations, especially related to the authenticity and tampering of the presented records [3,4]. At times, digital evidence records are easy to modify, tamper, and duplicate, which eventually could lead to falsification of evidence. Often, digital evidence is not admitted by a court of law because of the absence of authorization and nonissue of warrant for seize and investigate [1,5]. In addition, courts are largely concerned with the reliability of presented digital evidence [4]. As courts became more familiar with digital documents, they backed away from the higher standard and have since held that "computer data compilations … should be treated as any other record." US v. Vela, 673 F.2d 86, 90 (5th Cir. 1982) [1,6].

References

1. E. Casey. *Digital Evidence and Computer Crime*, Second Edition. Elsevier, Amsterdam, Netherlands, 2004, ISBN 0-12-163104-4.
2. Various. E. Casey, ed. *Handbook of Digital Forensics and Investigation*. Academic Press, Cambridge, MA, p. 567, 2009, ISBN 0-12-374267-6, retrieved September 2, 2010.
3. A. Richard. 2012. The Advanced Data Acquisition Model (ADAM): A process model for digital forensic practice (PDF).
4. D.J. Ryan and G. Shpantzer. Legal Aspects of Digital Forensics (PDF), retrieved August 31, 2010.
5. State v. Schroeder, 613 NW 2d 911—Wis: Court of Appeals 2000. 2000.
6. US v. Bonallo. Court of Appeals, 9th Circuit. 1988, retrieved September 1, 2010.

10 AnyPrecision Stable Design Pattern

> It is the mark of an instructed mind to rest satisfied with the degree of precision which the nature of the subject admits and not to seek exactness when only an approximation of the truth is possible.
>
> **Aristotle [1]**

AnyPrecision pattern models and structures the core knowledge of precision in a domain-independent way, which is usable in any type of application processes, where a fair degree of precision is always considered important and critical for the measured data. This pattern can be used as the core logic, for precision is any application, rather than implementation of it as a part of the application. To achieve this cherished goal, we will use the concept of "Software Stability Model" (SSM) [2] to identify and isolate the core knowledge of precision from the application specific knowledge. Several scenarios need deep exploration and probing to show the applicability and reusability of this pattern.

10.1 INTRODUCTION

This chapter introduces the notion of precision in a simple and comprehensible manner by using the concepts of SSM. This chapter also develops a stable design pattern of the same and discusses the applicability of pattern in a number of applications from different domains. The usage and utilization of precision may vary from very simple to very complex, and it may also rely on the usage in many domains for many different purposes. Here are some samples of usages, utilization, and purposes of precision:

The use of precision is deeply rooted in our civilization: In other words, civilizations depend on a correlation process, where many things work together in combination to achieve a single goal. To do that, each action for individual entity has to be very precise and measured; otherwise, it will never work in a group. For example, to assemble a machine, all individual parts must be very precise and up to date; otherwise, it will not fit.

The degree of precision varies based on the need: For example, while building a house, bricks can have precision in millimeters, whereas, in an electronic chip, the dimension of the transmitter has to be very precise in micron levels. Precision is the exactness of data; if the Measurement is consistent in respect of actual findings, then it is very precise.

In a mathematical term, it is representing by the digits soon after the decimal point: Precision is very important and critical in Measurement and design process as it dictates the exactness of any information or any mechanism. For example, if the temperature of boiling water is measured a couple of times and if the Measurement gives the values of: 210.1°F, 210.2°F, 210.6°F, 210.8°F, 210.9°F, and 210.5°F, then it could be said that the distribution is displaying, that the Measurement is quite accurate and precise, because all the measured values are leaning toward the high range.

From engineering to commerce, and from medicine, to day-to-day life: Precision plays a very important role. If the computer clock is not precise and accurate, then one cannot bid at right time on eBay. The reliability of the result depends on the level of precision. If the precision level is high, then one could always expect better performances.

Sometimes accuracy and precision are interchangeable: However, accuracy mostly tells us regarding the reliability of the results; whereas, precision is always related with the perfection of the method that is used to obtain the result.

Precision can also be used to represent the abstract things like expressing ideas and answering questions: There is no doubt that precision is very important and critical in every field or aspects of life, and it could be associated with various domains; therefore, we need a model that could accommodate various situations without making much changes in design.

Although, precision sometimes is described as a property of a measuring instrument: Its usage and importance is far beyond than just a simple property of the instrument. We cannot simply judge the effect of precision based on the instrument itself without taking into account the context where it is used and what is expected precision from the involved parties in a specific given application. We have considered all these aspects of precision while using this pattern.

The degree of precision varies widely based on the testing capability of the system and the expected result. Different domains have their own limitations or pitfalls and testing requirements; for example, when we are measuring our height, we can use any ruler that can measure in inches; however, if we have to measure the dimension of a microprocessor chip, it has to be very accurate in the micron level. Any pattern in precision should enforce the required precision level based on the requirements of the domains.

This chapter intends to isolate the core knowledge of precision concept from the application specific logic and present it later as a stable analysis pattern to provide the reusable core for other applications sharing the same core concept. In order to achieve this goal, the AnyPrecision pattern is designed based on the SSM [2–5]. The SSM provides a stable and *reusable* core for multiple applications that are sharing the same core knowledge [2]. Software stability describes the knowledge as design pattern named AnyPrecision. The pattern is documented by using Fayad's pattern documentation template in Chapter 4 [6]. This chapter is not a scientific study of what precision is and is now about the "why" of precision. It is about building a reusable pattern called AnyPrecision based on the foundations of SSM, discuss the various issues related to the usage of the pattern, be it business related or implementation related, and then to demonstrate methods to use the pattern in any application scenario, by providing two case studies that demonstrate application of it.

10.2 PATTERN DOCUMENTATION

10.2.1 Name: AnyPrecision Stable Design Pattern

Precision is an accurate science or knowledge, which specifies the degree of exactness; it is unrelated to any specific application or domain. As the concept is generic in nature, the AnyPrecision described in this chapter, is also very stable and generic and not tied to any application specific logic whatsoever. The patterns describe the core logic behind precision and how different entities interact with each other, to achieve the precision in the context where the pattern will be applied. Therefore, the name AnyPrecision is chosen as the name of the pattern, where the word "Any," always signifies any domain or context, where the pattern can be applied.

10.2.2 Known As

Precision can also be known as exactness. People often misunderstand accuracy with precision. However, these two terms have different meanings; accuracy is the degree of conformity with a standard. Accuracy is measured in the quality of a result, whereas precision relates to the quality of the operation by which the result is obtained.

10.2.3 Context

Precision is mainly associated with the Measurement, but sometimes it can also relate to abstract things like expressing ideas precisely or providing precise answers to a question. In the context of Measurement, it can relate to engineering precision, which describe how precisely different objects are put together to build a structure, like how to design different component to build a bridge. In data mining, precision describes the accuracy of the data and the accuracy of the analysis.

Precision is used everywhere and it exists in almost all fields, from a laser guided missile system to agriculture and even in our daily life; it plays a very important and critical role. During day-to-day communication, if the words are precise and up to the point, then people can understand dialogues easily and effectively. However, every application needs precision, but there is no easy way to express that logic into a pattern, which are easily measurable and monitored. In this chapter, we have developed a stable design pattern AnyPrecison, which is usable in any application and in any given domain, where the aspect precision is too important and needed to be monitored and measured.

Precision can describe the limitation of instruments in Measurement; for example, different instruments have specific jobs to do and perform. Depending on the need and requirement of the jobs, it is possible to build the instrument by keeping a predefined precision level. It is quite important to match that level with the requirement of the application where the instrument is in use. That is why precision is so important and critical in the context of any Measurement because it provides the most important metric in any Measurement.

Precision can also be used to find the usability of any mechanism; for example, in text mining, precision is used to verify if the algorithm used to return search result is actually returning the cases for which the user is looking for. In this case, precision is in use to measure user satisfaction, which is different from measuring a length, but the underlying usage of precision is identical.

1. *Computer clock and bidding system*: In the field of e-commerce, if the computer clock is not precise (AnyMetric) and accurate (AnyModel), then a bidder (AnyParty) cannot bid (AnyData) at right time (AnyCriteria) on eBay (AnyParty). Thus, a person wants to make sure that his or her computer clock is working precisely (AnyPrecision) and efficiently (Measurement) before bidding.

2. *Precision in communication*: During day-to-day communication, if the words are precise (AnyPrecision) and up to the point (AnyMetric), then people (AnyParty) can understand dialogues (AnyData) easily (AnyCriteria) and effectively (AnyModel). This would make communication more efficient (Measurement).

10.2.4   PROBLEM

Too Mathematical: Most of the present models, for example, decimal model and percentage model deal only with the aspect of mathematical precision; for example, in a floating point number, how many digits should be considered after the decimal sign. Precision plays a major role in mathematics and statistics, but it can be usable in other fields of study too. For example, precision is very important in text mining. The quality of any text-mining algorithm depends on how precisely it can narrow down to the result of a search. The existing model as discussed in the beginning cannot be applied under such cases.

Major requirements with the present precision model are described below.

10.2.4.1   Functional Requirements

Functional Requirements can be classified as

1. *Internal Requirements*: As the name suggests, these requirements are internal to the goal, meaning the nontangible things that are needed for an accurate Measurement. The internal requirement changes or shifts as the method of Measurement changes and so are the precision of the Measurement. These requirements are not directly visible to others but are tightly intervened to the EBT. Some of them are
   a. *Method of Measurement*: There are various means of measuring the same thing. The accuracy of the Measurement also depends on the method adopted for measuring, as the tools or mechanism used for measuring can produce varied results. Hence, the selection of appropriate method for measuring is the first step.

    b. *Accuracy of device/tool used*: Precision in Measurement is largely dependent on the error range of the device used. Moreover, the condition and age of the device can also affect the precision in Measurement.

    c. *Experience of the party/actor*: The person or party doing or carrying out the Measurement also affects its precision. The major role is played by the knowledge and experience of the person. A beginner tends to make more mistakes, when compared to an experienced person.

2. *External Requirements*: These represent all the BOs that are important for Measurement. Based on the goal to achieve, the BOs change their definition and scope of application. The BOs related to Measurement are described below.

3. *Subject*: The need for Measurement can arise in any area and Measurement acquires different forms for different subjects. For example, the ways things are measured in mathematics are quite different from the techniques used for measuring while painting or cooking. As a result, a generic model of precision is necessary to overcome all these differences. The main goal of this chapter is to create such a model of precision, so that it can be applied in any subject and is useful also.

4. *Data*: Here, data means input data and output data. As precision has to be applicable in any subject, so it will be drawn out from the Measurement of different types of data. Therefore, precision model should be able to deal with any kind of data and draw out correct statistics from it.

5. *Model/Type*: Different type of data are measured and presented in different formats and thus different models are required for presentation. Thus, precision data should be expressible in any kind of model and should be convertible among various models.

6. *Measurement*: This concept uses different techniques and instruments. Every instrument has some precision range; while measuring data, any techniques should clearly define the precision range that should include the Measurement and instrument precision. The proposed precision model should cover this point.

7. *Party/Actor*: Measurement carried out by or for different parties or actors by using same technique and same instrument sometimes differs in their results. The proposed precision model should overcome this drawback.

8. *Indicators*: Precision model should be able to tell its users (parties/actors) the inferences of the Measurement carried out like how accurate is the Measurement and technique used by it.

9. *Criteria*: While measuring, conditions of accuracy are defined by party/actor based on techniques and instruments used. The model developed in this chapter should be stable enough not to change as the party/actor defines more and more conditions. The idea is to develop a flexible stable design pattern that remains unaffected by the constraints imposed while measuring.

10.2.4.2 Nonfunctional Requirements

1. *Extensibility*: Any pattern on precision should provide ways and means by which they can be extended to provide the functionality that is needed for the specific domain.

2. *Accuracy*: The accuracy of Measurement depends on the specific requirement of the application. The pattern should provide capability to indicate the required precision level for a specific domain.

3. *Preciseness and Exactness*: In other words, precision provides exactness in Measurement. The degree of exactness varies based on the mechanism used in the process; the pattern should provide ways to monitor the exactness provided by the mechanism used in the Measurement process.

4. *Consistency*: The pattern should also provide an effective way to measure the consistency in the results, by validating precision level needed for an application, with the precision provided by the mechanism used in the Measurement process.

10.2.4.3 Properties

Properties of the goal refer to operations and attributes that are associated with it. Each of the operations and attributes defined for EBT has some constraints attached to it. Hence, it becomes very important to determine all the operations, attributes, and their constraints that can affect the pattern of AnyPrecision.

1. Operations of AnyPrecision and the constraints associated with them are as follows:
   a. tellsConsistency():
      i. Consistency of AnyPrecison depends on the method of Measurement and the tools used.
      ii. It also depends on the skills of the person who is doing the Measurement.
   This operation specifies that the precision of Measurement should not vary with changes in methods or tools used for Measurement. For example, either a ruler or Vernier Caliper can measure a small length or they should produce the same consistency in result.
   b. specifiesAccuracy():
      i. Precision tells us about the accuracy of result. Hence, it depends heavily on the manner in which precision is determined.
      ii. Accuracy also relies on the error range of the instrument.
   This operation also depends heavily on the error of the measuring instrument. Precision should change and modify, when Measurement is carried out through different instruments of same type, but having different error range definition.
   c. communicatesExactness():
      i. This operation depends on the way in which one interprets precision.
      ii. The exactness also depends on the quantum of error allowed and is acceptable.
   This operation has a direct relation with the units used for defining precision. The smaller the units to specify precision, the more exact will be the Measurement. For example, the precision specified in cm is less accurate than the one that is defined in mm.
   d. validatesPrecision():
      i. Depends on party or actor involved.
      ii. Rules defined.
      iii. Mechanism used.
   By this operation, the correctness of the process of Measurement is defined. The smaller the value of precision, the more accurate will be the process.
   e. indicatesUnit():
      i. It depends on area of Measurement.
      ii. It also depends on the error range of instrument.
      iii. Type of Measurements.
   This operation depends on the method/tool used for Measurement. For example, the error range of the tool is in cm unit, and then the precision unit has to be in same unit or smaller unit, say cm or mm.
   f. runsPrecision():
      i. Number of rules defined.
      ii. Factors involved in determining precision.
   This operation is necessary in order to certify the exactness of Measurement. Every Measurement should have some precision defined.
2. Attributes of AnyPrecision and constraints associated with them:
   a. Unit:
      i. Criteria based on which Measurement units are dependent.
      ii. How much accuracy of result is needed?
      iii. Error range of the instrument.

    b. Value:
        i. Error ranges of the instrument.
        ii. Acceptable possible error.
    c. Range:
        i. Method of Measurement.
        ii. Units of Measurement.
    d. Data:
        i. Mechanism used.
        ii. Instruments used.
    e. Accuracy:
        i. Mechanism and tools used.
        ii. Units of precision.
    f. measurementTechnique:
        i. The type of Measurement to be carried out.
        ii. Acceptable range of error.
        iii. Units of Measurement.
    g. Model:
        i. Units used.
        ii. Type of Measurement.

10.2.4.4 Constraints

1. The first and foremost constraint in requirement is understanding the context of the pattern. The pattern adopts different forms depending what needs to be measured.
2. In the requirement, the rules defined for pattern should also be laid down in a proper manner.
3. The metric to be used for Measurement should be decided beforehand as it greatly influences the interpretation of precision.
4. The instruments used also affect the aspect of precision to a great deal. The error range of all the instruments should also be recorded properly and a detailed study should be done, as what effect they can have on the precision of Measurement.

10.2.4.5 Example

To give a broader view of BOs and in what manner they relate to the goal of Measurement, some examples are provided:

1. *AnyParty/AnyActor*: Either a single person or a group of persons can carry out the Measurement. For example, a group of scientist experimenting something by the navigation crew, etc.
2. *AnyMetric*: Measurement can be carried out through various ways. For example, by using some measuring device or through mathematical equations, etc. This BO decides the unit of Measurement.
3. *AnyModel*: This BO represents the model of units to be used in Measurement.
4. *AnyCriteria*: These can be the rules defined for measuring technique used or the rules of the instrument.
5. *AnyPrecision*: This BO represents the accuracy of the result and is usually specified in a range with a positive or negative value.
6. *AnyData*: This represents the input to the pattern. It can be anything a cloth, a land, software, etc.

Given the above set of issues, the problem now is how we can design and develop a model that is based on the core knowledge of precision, which can be reused by different applications in different domains and can be extended as and when needed.

10.2.5   Challenges and Constraints

AnyPrecision pattern should resolve the following weaknesses, constrains, and challenges:

10.2.5.1   Challenges

Challenge ID	0001
Challenge Title	Too many Criteria Imposed
Scenario	Communication
Description	To make communication precise and more efficient; if a criterion is imposed to speak less and slowly, then it is easy to follow it; but if there are 1000 s of criteria imposed, then the person would keep on thinking about them and would actually not be able to express clearly what he or she is trying to say.
Solution	Only the required number of criteria should be imposed by actor/party
Challenge ID	0002
Challenge Title	Many Parties Requesting Measurement
Scenario	Toy car working with the help of a remote controller
Description	It is expected that the remote controller works precisely with car. But if there are 1000 s of people playing with their car in the same ground, then the remotes may not work as efficiently due to the interference of other connections.
Solution	There should be a threshold on the number of actors/parties requesting Measurement

10.2.5.2   Constraints

1. One or more AnyActor/AnyParty should request Measurement.
2. AnyActor/AnyParty should specify zero or more AnyCriteria.
3. AnyCriteria influences AnyMetric.
4. Measurement is taken by running one or more AnyMertic.
5. AnyPrecision is a part of AnyMetric.
6. AnyMetric produces AnyModel.
7. Measurement is done to measure AnyData.
8. AnyModel is related to one or more AnyData.

10.2.6   Solution

Very few attempts have been made to design pattern that are based solely on precision; most of the existing design on precision are application specific and they can only work in a particular set of data and can handle only specific problems. In this pattern, we have tried to overcome that problem by separating the application specific logic from the core logic. We have also tried to implement the core logic based on EBT and BO, which is stable and generic, therefore, can be used in any application in any domain (Figure 10.1).

10.2.6.1   Class Diagram Description

1. AnyParty/AnyActor requests Measurement
2. AnyParty/AnyActor imposes AnyCriteria
3. AnyCriteria influences AnyMetric
4. Measurement runs AnyMetric
5. AnyPrecision is part of AnyMetric
6. AnyPrecision produces AnyModel
7. AnyMetirc produces AnyModel
8. AnyModel related to AnyData
9. Measurement produces AnyData

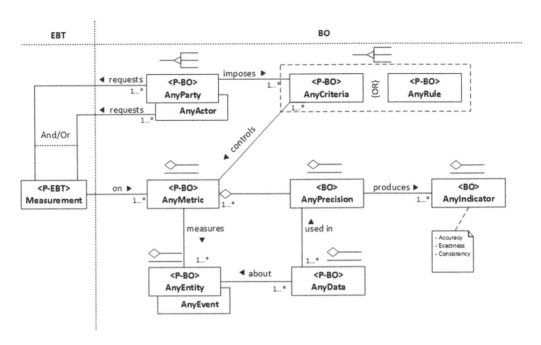

FIGURE 10.1   Class diagram of AnyPrecision pattern.

10.2.6.2   Pattern Structure and Participants

The participants of the interaction pattern are as follows.

10.2.6.2.1   Classes

AnyPrecision: This class describes the core knowledge of the stable design pattern, and
provides a suitable way to define the precision needed for any application based on the
mechanism used. It also provides a capability to monitor the output of any algorithm or
mechanism used in Measurement.

10.2.6.2.2   Patterns

Measurement: This represents the EBT, which is used by AnyPrecision, as the goal of the
pattern. Any actors can use Measurement to run several different mechanisms based on
the context.

AnyModel: This represents the model produced by the metric; it can represent accuracy,
exactness, or consistency.

AnyMetric: AnyMetric represents the method or procedure used to measure certain properties
of any entity based on the defined criteria. AnyMetric can range from mathematical algorithm, such as binary tree to any mechanical procedure, such as using a ruler to measure
the length of a table.

AnyData: This represents the input data to the Measurement process.

AnyParty/AnyActor: This represents the actors those are involved in the Measurement process. The actor can also be a verifier of the output who can help to determine the degree of
precision in the result data.

AnyCriteria: This pattern helps to define a criterion that is associated with properties or
mechanisms. For example, different tools have a limitation, while measuring any property of an object, such as a simple ruler that cannot be used to measure length smaller than
millimeter.

10.2.6.3   CRC Cards

Responsibility	Collaboration	
	Client	Server
Measurement (Measurement Handler)		
Defines the process by registering entities to be measured by using different mechanism	AnyParty AnyMetric AnyData	specifyMechanism() selectMetric() doMeasurement() measureEntity() givesResult()
Attributes: name, type, description, metric, precision, data, methodUsed, doneBy, rulesFollowed		
AnyMetric (Mechanism Descriptor)		
Represents the generic interface for different mechanism used to measure the entities	Measurement AnyCriteria AnyModel AnyPrecision	runMechansism() addProperties() registerConstraints() namesUnitForMeasurement() describes()
Attributes: name, constraints, procedure, properties, precision, parametersUsed, standard		
AnyPrecision (Precision Descriptor)		
Represents the precision that is defined by the Measurement process.	AnyModel AnyCriteria	runPrecision() validatePrecision() indicateUnits() specifiesAccuracy() tellsConsistency() communicatesExactness()
Attributes: units, value, range, data, measurementTechnique, model, accuracy		
AnyData (Data descriptor)		
Describes the object, which is getting measured	Measurement AnyModel	measures() properyToMeasured() hasModel() belongsTo() formCollection()
Attributes: id, name, property, model, application, format, belongTo, domain, context		
AnyModel (Precision indicator)		
Represents the degree of precision in the output	AnyPrecision AnyMetric AnyData	indicatePrecison() hasIndicator() formBase() showsCorrectness() pointoutToOriginal()
Attributes: type, value, precision, units, standard, proposedBy, popularity		

10.2.7   CONSEQUENCES

The benefits of using AnyPrecision pattern are listed as below.

10.2.7.1   Benefits

Scalable: The AnyPrecision pattern is a design pattern that is based on stability design model. The core knowledge of precision has been defined based on EBTs and BOs, whereas IOs can be added based on the application logic. The pattern is not dependent on any specific business logic; therefore, you can use it in any application as the core model.

Stable: It is possible to achieve stability through the EBT and BO. By definition, EBT and BO do not change over time, nor they change based on application requirements, and they are the core knowledge behind the precision concept.

Support different type of object: One can apply AnyPrecision pattern to different entities. For example, it can be a simple table to a spacecraft depending on the application.

Support different properties: The properties of the participating entity can be different based on application logic. In the case of Precision Navigation, velocity is used as the property, which requires a precision Measurement.

Flexible: AnyPrecision pattern can be applied in different domains, as we have discussed before. It can be used as a Precision Navigation tool for a submarine or can describe a process to measure a table using a ruler.

Adaptable: It is possible to apply the same pattern in different domains by using IOs. The patterns do not dictate the IOs, rather it specifies how the core knowledge can be laid out and application logic hooked into it very easily.

10.2.7.2 Limitations

No Implementation Logic: AnyPrecision is a conceptual and thoughtful model; yet, all the components used in this pattern display clear, concise names, and functionalities, which helps to visualize the pattern quite easily, but the pattern does not provide any actual implementation of the pattern of a specific context. In one way, this is a perceived limitation, but on the other hand, the pattern provides us with a common core, which is usable in different applications as the core knowledge is untied with the specific application logic.

No Monitoring of Result: As the quality of precision can only be judged on a specific context, it has not been considered as the core knowledge of precision pattern. This limitation is easy to overcome by attaching other stable analysis pattern, like Monitoring or Recoding with the AnyPrecision pattern.

10.2.8 Applicability with Illustrated Examples

We, hereby, present two case studies to show the application of the AnyPrecision pattern in different domains.

Use of precision Measurement in navigation and use of precision to verify the output of a text mining application: For the sake of simplicity and flexibility, we will just focus on the usage on AnyPrecision pattern in the problem space; therefore, the design does not include other classes or component involved in the specific application.

10.2.8.1 *Case Studies 1*: Precision Navigation

Precision Navigation is vital and important for providing accurate coordinate information for any moving object, ranging from missile to satellite. With the assistance of new and modern technology, we can precisely locate a moving object or can predict the trajectory of movement, which later helps us to send space missions to Mars. To provide such a high degree of precision, the application needs to be developed by keeping in mind the required precision, all the while considering the limitation of instruments that requires proper design and Monitoring of each component used in the system. You may also need to monitor any Measurement to enforce the demanded precision. To model such application, one needs the deep understanding of precision and in what manner certain degree of precision can be enforced and monitored. We have described one such Precision Navigation application based on the AnyPrecision pattern (Figure 10.2).

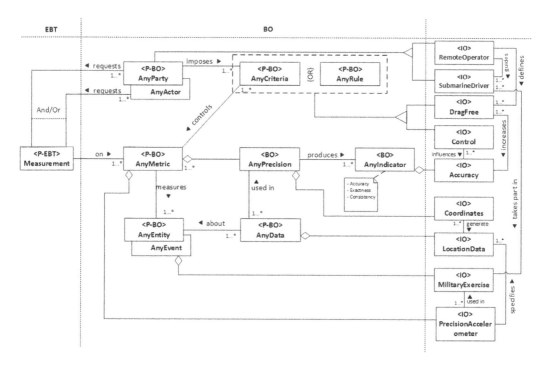

FIGURE 10.2 Class diagram of a Precision Navigation System for a submarine.

10.2.8.1.1 Class Diagram Description

1. Measurement is done by AnyParty/AnyActor.
2. RemoteOperator and SubmarineDriver belong to AnyParty/AnyActor.
3. SubmarineDriver drives submarine.
4. Submarine has AnyMetric(PrecisionAccelerometer).
5. AnyMetric(PrecisionAccelerometer) is influenced by AnyCriteria(DragFreeControlCriteria).
6. AnyParty imposes AnyCriteria(DragFreeControlCriteria).
7. AnyParty (RemoteOperator) is inherited by Satellite.
8. Satellite calculates AnyData(Location).
9. Location is dependent on Velocity.
10. Velocity is controlled by SubmarineDriver.
11. Velocity and Location are part of submarine.
12. Submarine has AnyMetric (PrecisionAccelerometer).
13. AnyPrecision is part of AnyMetric.
14. AnyMetric produces AnyModel.
15. Measurement uses AnyMetric.
16. AnyModel is produced by AnyPrecision.
17. AnyModel is related to AnyData.
18. AnyData is measured by Measurement.

10.2.8.1.2 Use Case Description

To provide clarity, a use case is presented along with the sequence diagram, which shows how precision can be used to guide a submarine.

Use Case Id: 1.0
Use Case Title: Guide submarine

Actors	Roles
AnyActor	SubmarineDriver
AnyActor	RemoteOperator

Class	Type	Attributes	Interfaces
Measurement	EBT	title, description, metric, methodUsed, doneBy, rulesFollowed	measures()
AnyParty/AnyActor	BO	name, responsibility, role, age	initiates() doesMeasurement()
AnyCriteria	BO	name, description, type, value, rule	influences() imposes()
AnyModel	BO	type, value, precision, units, standard	indicatesPrecision()
AnyPrecision	BO	name, description, units, range, accuracy	specifiesAccuracy() createsModel()
AnyMetric	BO	parametersUsed, name, constraint, procedure	produces() namesUnitForMeasurement()
AnyData	BO	Id, property, type, subject	formsCollection() relatedTo()
SubmarineDriver	IO	name, age, experience, birthDate, skills	drives() controls
RemoteOperator	IO	Constitution, type, locatedAt, size	locates() alerts() tracks()
Satellite	IO	rotationSpeed, height, type, name	calculates() constitutesRemoteOperator() tellsWay()
Submarine	IO	type, size, usage, made	operates() glidesUnderWater()
Velocity	IO	units, ofWhom, measuredBy	tellsSpeed() formulates()
PrecisionAccelerometer	IO	name, description, precision, mechanism	runs() confinesSubmarineOperation()
Location	IO	position, time, precision	relatedTo() dependsOnVelocity()
DragFreeControlCriteria	IO	name, description, numberOfRules	restricts()

Enduring Business Themes (EBT): Measurement
Business Objects (BO): AnyParty/AnyActor, AnyCriteria, AnyModel, AnyPrecision, AnyData, AnyMetric
Industrial Objects (IO): PrecisionAccelerometer, Location, DragFreeControlCriteria, RemoteOperator, SubmarineDriver, Satellite, Velocity, submarine

10.2.8.1.2.1 Description of the Use Case
1. Measurement is done by AnyParty/AnyActor, which can be any SubmarineDriver<<IO>> or RemoteOperator<<IO>>
   How is Measurement done? What are the methods that are used by AnyParty/ AnyActor?
2. SubmarineDriver drives submarine and is guided by PrecisionAccelerometer that is present in the submarine.

PrecisionAccelerometer is a part of AnyMetric.

How does PrecisionAccelerometer guide SubmarineDriver? Is accelerometer working correctly? How does accelerometer form a part of AnyMetric? How to read precision accelerometer?

3. AnyMetric is influenced by AnyCriteria(DragFreeControlCriteria) and these criteria are defined by AnyParty/AnyAtor(RemoteOperator and SubmarineDriver).

How does criteria influences AnyMetric and in what way and to what extent? How are these criteria decided by AnyParty/AnyActor?

4. RemoteOperator is inherited by Satellite that calculates the location of submarine.

How does Satellite calculate the location of submarine? What is the purpose? How accurate is the location?

5. Location is a type of AnyData and thus needs to be determined precisely. Hence, it forms a part of AnyPrecision.

Why location needs to be determined precisely? What are the factors that affect precision of data?

6. Location is dependent on Velocity of Submarine, which is controlled by SubmarineDriver

How velocity affects location? How does submarine driver control velocity?

7. AnyData(Location) is related to AnyModel and AnyModel is created by AnyPrecision, which specifies accuracy.

How does precision specifies accuracy and in what way? How are data related to AnyModel?

8. AnyPrecision is a part of AnyMetric and AnyMetric produces AnyModel.

How do metrics produce model? What kind of model? What is the difference between model and metric?

9. AnyData is measured by Measurement.

How are data measured by Measurement? How are units decided?

10.2.8.1.3   Sequence Diagram (as Shown in Figure 10.3)

10.2.8.1.3.1   Sequence Diagram Description

1. Measurement is done by AnyParty/AnyActor(SubmarineDriver, RemoteOperator).
2. AnyActor(SubmarineDriver) drives submarine.
3. Submarine has AnyMetric(PrecisionAccelerometer).
4. AnyMetric is influenced by AnyCriteria(DragFreeControlCriteria).
5. AnyCriteria is imposed by AnyParty/AnyActor.

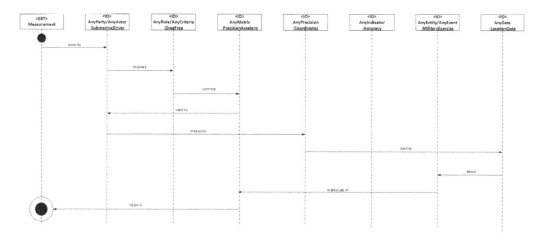

FIGURE 10.3   Sequence diagram of a Precision Navigation System for a submarine.

6. AnyParty/AnyActor(RemoteOperator) consists of Satellite.
7. Satellite calculates AnyData(Location).
8. AnyData(Location) is dependent on Velocity.
9. Velocity is a part of submarine.
10. Submarine has AnyMetric(precisionAccelerometer).
11. AnyMetric comprises of AnyPrecision.
12. AnyPrecision produces AnyModel.
13. AnyModelis related to AnyData.
14. AnyData is measured by Measurement.

10.2.8.2   *Case Studies 2*: Text Mining

The term precision is commonly used in the domain of text mining, as the Measurement of system performance. Precision denotes the ratio of spots correctly identified by the user, as on topic out of all the spots reported as being on topic. It is also very important to measure the precision and monitor the performance of the text mining application. In our case study, we have shown that how any text mining application can use AnyPrecision pattern to monitor and evaluate the performance of its algorithm. For the sake of simplicity, we have only focused on the area of precision in text mining application, while we are omitting other details deliberately on some purpose (Figure 10.4).

10.2.8.2.1   *Class Diagram Description*

1. Measurement is done by AnyParty/AnyActor (Disambiguator).
2. Disambiguator works on web site.
3. Web site is mined by query that inherits from AnyCriteria.
4. AnyCriteria influences AnyMetric.
5. Query form Cloud that are of two types, TextCloud and TagCloud.
6. Cloud has some Relevance in respect of web site and Relevance denotes AnyPrecision.
7. AnyPrecision is part of AnyMetric.
8. AnyMetric produces AnyModel(SpaceModel).

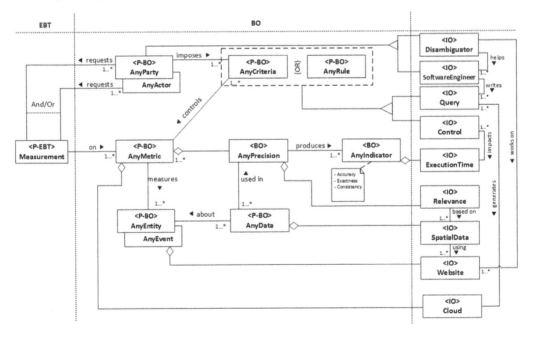

FIGURE 10.4   Class diagram of a text mining application.

  9. SpaceModel is possessed by Cloud.
 10. TextCloud and TagCloud are part of AnyData.
 11. AnyData relates to AnyModel.
 12. AnyModel is produced by AnyMetric.
 13. AnyData is measured by Measurement.

10.2.8.2.2 Use Case Description

To provide clarity, a use case is presented along with the sequence diagram, which shows how precision can be used measure the usefulness of a Search Algorithm used in a text mining application.

Use Case Id: 1.0
Use Case Title: Calculate Precision in a Text Search algorithm

Actors	Roles
AnyActor	Website
AnyActor	Disambiguator

Class	Type	Attributes	Interfaces
Measurement	EBT	title, description, metric, methodUsed, doneBy, rulesFollowed	measures()
AnyParty/AnyActor	BO	name, responsibility, role, age	initiates() doesMeasurement()
AnyCriteria	BO	name, description, type, value, rule	influences() imposes()
AnyModel	BO	type, value, precision, units, standard	indicatesPrecision()
AnyPrecision	BO	name, description, units, range, accuracy	specifiesAccuracy() createsModel()
AnyMetric	BO	parametersUsed, name, constraint, procedure	produces() namesUnitForMeasurement()
AnyData	BO	Id, property, type, subject	formsCollection() relatedTo()
Disambiguator	IO	type, application, description, language	works() derives()
Website	IO	type, graphics, language, data, history	informs() represents()
Query	IO	type, language, workDone, script	mines() generates()
Cloud	IO	description, numberOfWords, links, type	summarizes() represents
TextCloud	IO	numberOrWords, frequencyOfWords, presentation, type	tellsAboutWebsite()
TagClouds	IO	word, link, presentation, type	linksToOtherSite()
Relevance	IO	Importance, description, ofWhat	showsImportance()
SpaceModel	IO	description, type, notation	denotesCloud()

Enduring Business Themes (EBT): Measurement
Business Objects (BO): AnyParty, AnyCriteria, AnyModel, AnyPrecision, AnyData, AnyMetric
Industrial Objects (IO): Disambiguator, Website, Query, Cloud, TextCloud, TagCloud, Relevance, SpaceModel

10.2.8.2.2.1 Description of the Use Case

  1. Measurement is done by AnyParty/AnyActor. Dismabiguator (type of software) inherits from AnyActor and works on mining of web site.

How is Measurement done? What kind of software is disambiguator? How is mining of web site done? How does disambiguator helps in mining?

2. Web site is mined using some query that imposes restrictions on mining and hence forms a part of AnyCriteris that are imposed by AnyParty/AnyActor.

   How are query formed? How they help in mining? On what basis are the criteria defined? How are they imposed?

3. Query generates Cloud that summarizes the whole web site. There are two type of Cloud: TextCloud and TagCloud.

   What is Cloud? How are they generated? What do they represent? What are TextCloud and TagCloud? How the Cloud summarizes the complete web site?

4. Cloud represents the web site and thus has some Relevance in context of web site. Thus, Relevance forms a part of AnyPrecision.

   How does Cloud represents the whole web site? In what format? How much is Cloud relevant?

5. AnyPrecision creates AnyModel like SpaceModel that denotes Cloud.

   How does precision create model? What is space model? How the Cloud is denoted by space model?

6. AnyModel is produced by AnyMetric that is influenced by AnyCriteria.

   How does criteria influences metric?

7. TextCloud and TagCloud form a part of AnyData.

   What kind of data?

8. AnyData relates to AnyModel.

   How are data related to model? What is model?

9. AnyData is measured by Measurement.

   How Measurement is done? How are data measured and in what units? How are units decided for data?

10.2.8.2.3   Sequence Diagram (as Shown in Figure 10.5)

10.2.8.2.3.1   Sequence Diagram Description

1. Measurement is done by AnyParty/AnyActor(Disambiguator).
2. AnyParty/AnyActor(Disambiguator) works on mining of web site.
3. Web site is mined by AnyCriteria(Query).
4. AnyCriteria(Query) generates Cloud.
5. Cloud has AnyPrecision(Relevance).

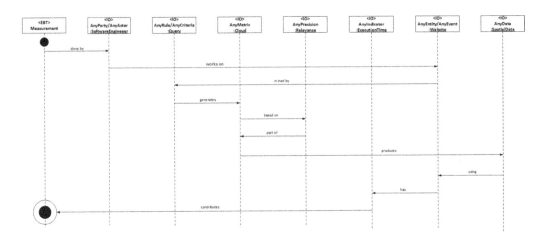

FIGURE 10.5   Sequence diagram of a Precision Navigation System for a submarine.

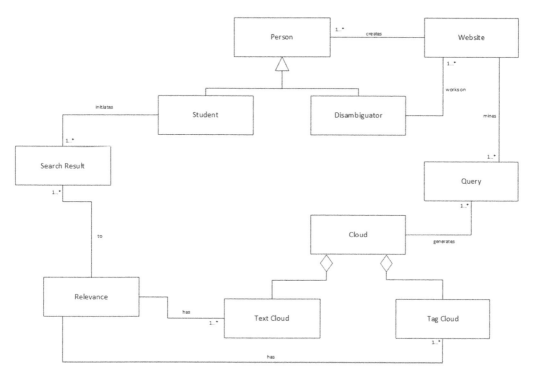

FIGURE 10.6   Traditional model for AnyPrecision.

  6. AnyPrecision(Relevance) is a part of AnyMetric.
  7. AnyMetric produces AnyModel(SpaceModel).
  8. SpaceModel is possessed by Cloud.
  9. Cloud consists of AnyData(TextCloud and TagCloud).
  10. AnyData is measured by Measurement.

10.2.9   RELATED PATTERNS AND MEASURABILITY (FIGURE 10.6)

10.2.9.1   Traditional Model versus Stable Model

- IOs form the *basis* of Traditional model, which are physical objects and are unstable. On the other hand, Stable Model relies on three concepts–EBT, BO, and IO. The EBTs represent elements that remains stable internally and externally. The BOs are objects that are internally adaptable, but externally stable, and IOs are the external interfaces of the system.
- Traditional model is hard to *reuse* if the requirement changes midway. Any changes in the requirements may cause complete reengineering of project with new requirements. The stable model is highly flexible and is reusable in wide domains and applications.
- Traditional model requires very high *maintenance cost* in terms of time, labor, and money. The system built by using Traditional model is neither extendable nor is it adaptable. Furthermore, the Stable Model is easily maintainable and extendable.
- Division of classes in stability model in categories of EBT, BO, and IO makes it simpler and easier to *understand* and apply as compared to a Traditional model. This is also one of the reasons for the scalability of stable models as EBT, BOs, and the relationship between them remains stable. IOs are the ones that change or modify with any given application. It is very easy to hook IOs of any application with the BOs and create a new model with relative ease that too with least effort.

- It is very easy to *identify, detect, challenges, and constraints* of an application with stability model as compared to Traditional model.
- Traditional models are very much susceptible to *dangling,* while dangling is never a problem in stability model, because of multiple inheritance and aggregation.
- It is very difficult to define correct *multiplicity constraint* in relationships in a Traditional model, while in stability model the multiplicity constraints are very much obvious.
- The *interdependency* among the classes in Traditional model is very high, such that a small change in a single class affects the whole model. While in stability model, the effect of change does not propagate through whole of the model and hence stability model is more stable than Traditional model.
- Limitations of Traditional model make it unsuitable for *large projects* having numerous goals, while stability model, because of its stable nature, can be applied to large projects very easily and their maintenance is very easy as well.
- One of the major differences between the two models is the *cost.* Cost involved in developing project by using Traditional model is very high, as compared to a stability model. This is because, while developing the user requirements keep on changing and a slight change in the requirement results in complete reworking of Traditional model, while this is not the case with stability model.

10.2.9.2 Measurability

10.2.9.2.1 Quantitative Measure

Factors on which quantitative measure can be applied are as follows:

1. *Quantity Aspect of EBTs, BOs, and IOs*: The more the number of patterns, the more it will result in lines of code while developing the system. In addition, as lines of code increase, error propagation rate will also increase, and it will be difficult to maintain accuracy in the pattern development. Quantitative aspects shows that EBTs, BOs, and IOs should be selected in such a way that it should cover all the necessary patterns required in modeling, and yet it should be developed in a manageable number of lines of code, which will result in lesser error propagation.
2. *Number of Classes*: The second aspect of quantitative metrics is that the stability model has fewer numbers of classes with the focus on explicit as well as implicit factors, when compared to a Traditional model. Stability model relies on the concept of EBTs, BOs, and pluggable IOs. As a result, the base pattern remains stable and has the capability of representing a large number of applications, by just hooking the appropriate IOs with the base pattern. This reduces the number of classes required to represent an application by a drastic amount.
3. *Cost Estimation*: Determining and developing Estimation or Measurement Metrics is far easier and less time consuming when compared to that of a Traditional model's because we know the base pattern of a stable model very well in advance.
4. *Coupling Among Classes*: Coupling represents how tightly the classes bind together and depend on each other. In Traditional model, coupling among classes is very high. As a result, even a small change or modification to any class in Traditional model ripples through and affects the entire Traditional model. While in a stability model, change in one class does not affect the whole model and remain restricted to that particular BO.
5. *Constraints*: They represent the multiplicity of the class and are very easy to define in stability model as compared to Traditional model.

10.2.9.2.2 Qualitative Measure

Qualitative measure of a pattern relates to usability, stability, scalability, and maintainability of the pattern. If you can use pattern for a number of applications without any significant changes in the system, then the pattern will have a number of qualitative qualities. Moreover, the pattern should be reusable in a wide variety of applications. Besides these features, the maintenance cost of the pattern should also be very low. Stable model approach to develop patterns supports all these features, while Traditional models do not. Patterns developed by using Traditional model are quite specific to many applications and thus cannot be used repeatedly. Moreover, they are meant for only one specific application. For a new application with the same base, an entirely new pattern has to be developed which incurs a lot of cost and resources. On the other hand, stability model is opposite to a Traditional model. One single pattern only supports a variety of different application with the same goal.

10.2.10 Modeling Issues, Criteria, and Constraints

10.2.10.1 Abstraction

One of the most important features of a stable pattern is its abstraction behavior toward an application. The SSM relies on this fact. The basis of SSM is EBTs and BOs, which encapsulates the pattern behavior from the application and thus, results in reusability of the pattern to a wide range of applications. Hence, it becomes necessary to debate on the selection of EBT and BOs for the pattern, so that they can support a wide range of applications without any major changes. Considering the case of Precision design pattern:

- Precision cannot be an EBT. Precision cannot be an ultimate goal of an application too. Hence, in this pattern, precision is taken as BO and AnyPrecision design pattern is developed.
- The next step is to find the ultimate goal or EBT for this design pattern. There can be number of goals associated with precision like accuracy, Measurement, and exactness. We chose Measurement as the ultimate goal and EBT of this pattern because accuracy and exactness are just other terms for precision; hence, it becomes difficult to relate to classes with different names, but same meaning in the pattern. Measurement on the other hand is a process that involves and requires precision and hence, becomes EBT of this pattern.
- Now, the next class or BO to consider is AnyMechanism or AnyMethod through which Measurement is performed. Instead of this BO, AnyMetric was chosen, because Measurement is largely dependent not only on the mechanism through which it is done, but also on the model of units in which it will be interpreted.
- Another BO that becomes necessary for this pattern is AnyModel, that represents the system of units based on the thing that is being measured.
- AnyData in this pattern represents something that needs to be measured and thus, acts as an input to this pattern. Now, Measurement has to be done by some person, so, AnyActor came in place.
- AnyActor is not the only one doing Measurement, as Measurement can be done by a group of people together also. Hence, AnyParty was also chosen.
- AnyActor/AnyParty depending on the type of Measurement they are doing, defines the rules or criteria of Measurement, which also sets the rules for AnyPrecision. Another BO, which becomes important for this pattern is AnyCriteria.

In the end, using all these BOs, a pattern was developed and this pattern is complete by itself and is able to support all the applications from any domain.

10.2.11 Design and Implementation Issues

In the design phase, the BOs and EBTs so decided are taken and a pattern is formulated by using them. This phase involves the tedious task of deciding on the attributes and operations for each EBT and BO. Once the attributes and operations are finalized, the constraints associated with each one of them are listed in order. Then, the relationship among BOs and EBTs are defined and a stable pattern is designed. The challenges and constraints associated with the pattern as a whole are also taken into consideration. After the design phase is over, the next phase is implementation phase. In this phase, the pattern is applied to any desired application. For this, the IOs are first defined based on the context of the application and then hooks are created between pattern and IOs of the application. Thus, in the implementation phase, the pattern is developed for the application, by simply hooking the IOs of the application to the BOs of the pattern. One way of developing the hooks is via **interface.**

Interface is a function that would list all operations of BOs in a combination required to connect BOs to IOs. Thus, BOs will connect to IOs via interface. It will also increase functionality. All the links, which are used to connect to IOs, will be included in interface.

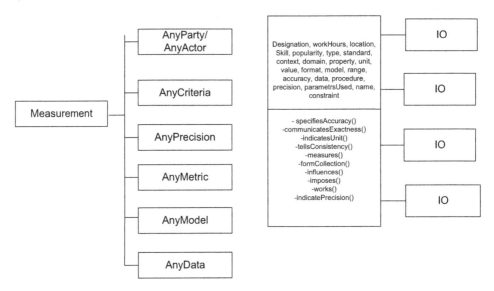

10.2.12 TESTABILITY

In general, patterns designed by using the stability models are more easily testable, when compared to the Traditional model. This is because the EBTs and BOs rarely change and they are applicable to other applications without any major changes.

In this project, the AnyPrecision design pattern is tested, by applying the pattern to two different applications, without any changes to the core pattern. This is achievable by plugging in the necessary IOs to the core dignity pattern.

In the same way, other application's IOs from any context can be plugged to BOs. The above pattern so developed will be testable, only when it is possible to apply it to any other scenario/application and produces the correct output.

Some scenarios in which the above pattern may not give correct output and will fail are

1. This pattern fails when the metric defined for the data is not appropriate. Then, the precision units will not be correct and hence the accuracy of the result cannot be verified.
2. If the party or actor is unable to define proper rules/criteria for the Measurement, they will not create effective influence on metric. Therefore, this is also one of the cases, where the above-developed pattern will fail.
3. The pattern may also fail, when an inexperienced person tries to take Measurement and is not aware of skills to use the metric and model for determining precision.

10.2.13 FORMALIZATION

AnyPrecision stable design pattern is formalized or finalized by describing in many-sorted, first-order languages. These languages consist of many sets of sorts and types. Every sort has a universe set, a set of functions and relations, and can have subsorts as well. Thus, universe of a sort is the union of universe of its subsorts.

XML is used for describing formalization because it is a simple text based language for representation. XML notations can be easily converted into any languages like Java, .net, or C. Syntax of dignity pattern in XML terms has been described below. Only a part of schema has been described because full description of schema is not possible here and will go out of scope of the document.

Note: According to mathematical logic conventions, constants over a sort have been defined as constant: Sort name.

```
<Pattern>
<Title>
      "measurement"
</Title>
<sort>
      <Title>
            "Measurement"
      </Title>
      <Sort>
          <Title>
                "Precision"
          </Title>
          <Sort>
              <Title>
                    "unit"
                  </Title>
              <Type>
                    String
              </Type>
              <Universe>
                      {degree,radian,minute, …}
              </Universe>
                </Sort>
          <Sort>
              <Title>
                    "value"
              </Title>
              <Type>
                      Float
              </Type>
              <Universe>
                      {+0.5,-.01,…}
              </Universe>
              </Sort>
          <Sort>
              <Title>
                    "accuracy"
              </Title>
              <Type>
                            String
              </Type>
              <Universe>
```

```
                            {...}
                        </Universe>
                </Sort>
                <Sort>
                        <Title>
                            "model"
                        </Title>
                        <Type>
                            Model: System
                        </Type>
                </Sort>
                <Sort>
                        <Title>
                            "data"
                        </Title>
                        <Type>
                            Input: Data
                        </Type>
                    </Sort>
        ....
            <Function>
                    <Title>
                        "specifiesAccuracy"
                    </Title>
                    <Type>
                        Constant: AccuracyConstant
                    </Type>
                    <Description>
                      Accuracy constant defined according to the application
                    </Description>
                </Function>
                <Function>
                        <Title>
                            "validatePrecision"
                        </Title>
                        <Type>
                                Boolean: True or False
                        </Type>
                </Function>
                <Function>
                        <Title>
                            "indicatesUnit"
                        </Title>
                        <Type>
                            String
                        </Type>
                    </Function>
        </Sort>
...
</Pattern>
```

10.2.14 BUSINESS ISSUES

10.2.14.1 Business Rules

- There should be some data ready for Measurement.
- Data given as an input to the pattern should be in a measurable state.

- There should be some party or actor for measuring data.
- The party or actor should define some rules/criteria, which directly influences the Measurement process.
- Measurement is based on some metric that must be defined properly.
- All the rules influencing the Measurement metric should be defined properly.
- The metric should be able to produce model for Measurement.
- The data and model should match with each other, for example. To measure any liquid appropriate system of units should be used.
- The range of precision should be allowable range and its units should match with the defined metric system.
- Any exceptions with the method of Measurement like the instrument error ranges and other factors should also be defined very clearly beforehand.
- The Measurement should be based on AnyCriteria that are decided on by AnyParty/ AnyActor.

Business rules control and manage the behavior of the system. They also impose constraints on the system and tell the system "what" it should do. Business rules are atomic in nature and thus you cannot break them down in to smaller pieces without losing valuable information. You may have to define them prior to defining requirements of the system.

Elements of business rules are

1. *Business Items*: This element corresponds to different classes that are forming the pattern. Stable pattern consists of classes at various levels: EBT, BO, and IO, and they have different functions and responsibilities at each level. Some of the business rules defined for business items are
   a. Each class should be capable of at least one function.
   b. Classes should be able to work independently.
   c. IO classes should interact with BO classes only.
   d. EBT classes should interact with BO classes only.
   e. Classes should be able to reflect the specificity of application.
   f. BO classes, when combined, should be able to represent a pattern depicting the meaning of EBT/s involved.
2. *Properties*: Properties in business language corresponds to attributes and operations of classes in stable language. Business rules related to properties are
   a. The operations defined for the class should be unique and generic, such that they can be used to represent any application.
   b. The class should be able to carry out the responsibility assigned to it.
   c. The attributes of the class must cover all the distinct aspects of the class.
   d. The operations defined for the class should be such that the class is able to perform them independently, as well as in cooperation with other classes.
3. *Relationships*: It presents the interdependency among classes and in what manner one class relates with to each other. Business rules defined at this level are
   a. One relation can connect only two classes.
   b. Every class should relate to another class through some relation. No classes in a pattern can standalone.
   c. Relation can be simple relation connecting two classes, that is, association or it can be "a kind of" or "a part of" relation.
   d. Every associative relation has some multiplicity. The default is one to one.
   e. Every association relation has some name that represents the type of connection between two classes.

4. *Facts*: These represent business or common terms that can occur in form of EBT or BO. Some of them are
   a. Measurement: A system of measuring.
   b. UnitSystem: Represents the system of units that can used to describe results of Measurement.
   c. Instrument: Means of Measurement like barometer and measuring tape.
   d. Accuracy: Represents the correctness of measuring process.
   e. Error: Represents accuracy error.
   f. Range: Represents the range of accuracy error.
5. *Constraints*: These represent the restrictions imposed on pattern. Some of the constraints are
   a. One or more criteria can influence AnyMetric.
   b. Measurement should have at least one Metric.
   c. Measurement can be done by one or many actor/party.
   d. AnyModel should relate to one or more AnyData.

Based on the above elements, some generic business rules for the pattern are as below:

1. AnyMetric should have proper definition for Measurement.
2. The units of precision should match with that of the defined Metric.
3. All the criteria affecting the decision of AnyMetric should be laid down properly.
4. AnyMetric and AnyPrecision should produce the same model.
5. Measurement should have at least one input in the form of AnyData.
6. AnyData should be in a state that can be measured.
7. All the exceptions with the method of Measurement should be determined beforehand.
8. AnyCriteria should be validated for Measurement.

10.2.14.2 Known Usage

AnyPrecision pattern is very generic, and it can be used in many day-to-day applications.
A number of scenarios where this pattern can be used are

- In laboratories, for measuring anything; it could be a physical laboratory, chemical, or even biological.
- This pattern can also be used in measuring distances and locations of stars, galaxies, or meteors in universe.
- This pattern can also be used in simple daily life activities, like measuring a cloth or fruits, etc.
- It can also be used in web mining and in search engines to determine the Relevance of web site content.
- This pattern can also be used in medical sciences for developing medicines or for new hybrid species.

10.3 TIPS AND HEURISTICS

- Measurement is a very important and critical part of everyone's life and precision is an essential part of Measurement; without determining the accuracy of Measurement, you cannot validate it.
- The stable pattern of Measurement is general enough for use in any kind of Measurement, from measuring a simple thing to measuring the distances and location in universe.
- The EBTs and BOs form the core of the pattern and these encapsulate the details of the pattern and its working from the application.

- Any application is easy to plug with the pattern and only IOs need proper definition.
- The rules or criteria defined for Measurement change with application and they greatly influence the accuracy of Measurement. Hence, you should define them properly and clearly.
- The method of Measurement should comply with the Metric that are defined for it.
- The units of precision and Measurement should match one another.
- One should record all the exceptions of the Measurement method in a proper manner.

10.4 SUMMARY

This chapter plans to present AnyPrecision pattern that models the core knowledge of precision in a domain independent way. This chapter also shows that it can be used in any application, where, for the measured data, precision is very important. It details and explains all the problems, challenges, and solution of AnyPrecision model. We have also discussed the limitations of this chapter and its pattern. When combined with other patterns, it will reveal all its strengths. Besides this, we have also presented two separate case studies to show the application of the AnyPrecision pattern under various domains. The first case study discusses precision Measurement in the Navigation. The Precision Navigation mainly explains and deliberates, we can precisely locate a moving object or can predict the trajectory of movement by using this new technology.

For an easy understanding and comprehension, the class diagram of Precision Navigation has also been provided. The other case study for AnyPrecision, presented in this chapter, is the Text Mining. This case study shows how we can achieve the desired precision in any text-mining algorithm. It also explains the importance of measuring the precision and Monitoring the performance of the text mining application. The sequence diagram has also been added for better and easy understanding. Anyone can use this pattern with other patterns to develop some stable analysis patterns and hence it is applicable in any domain. To develop AnyPrecision model, we have used the concept of "SSM" [2]. Further, SSM identifies and isolates the core knowledge of precision from the application specific knowledge.

10.5 REVIEW QUESTIONS

1. "Use of precision is rooted in our civilization." Explain with the help of an example.
2. How does degree of precision varies based on the need?
3. What does precision means in mathematics?
4. How does accuracy differ from precision?
5. Explain any application of precision in commerce.
6. How can precision be used to represent abstract things?
7. Name a few synonym of precision.
8. Define AnyPrecision SDP.
9. What does precision mean?
10. Describe a few contexts, in which precision is used in day-to-day life.
11. Write 2–3 scenarios where AnyPrecision SDP is applicable.
12. Discuss the functional requirements for AnyPrecision SDP.
13. What are the types of functional requirements?
14. Describe internal and external requirements.
15. Discuss the nonfunctional requirements for AnyPrecision SDP.
16. Discuss the properties of AnyPrecision SDP.
17. Write four challenges of using AnyPrecision SDP.
18. Write a few constraints of AnyPrecision SDP.
19. Draw class diagram for AnyPrecision SDP.
20. Explain and justify the participants in AnyPrecision SDP.
21. Write CRC cards for AnyPrecision SDP.

22. What are the consequences AnyPrecision SDP in terms of following:
    a. Scalability
    b. Stability
    c. Type of object
    d. Properties
    e. Flexibility
    f. Adaptability
    g. Implementation of logic
    h. Monitoring of results
23. Explain the application of AnyPrecision SDP in a scenario of navigation. Draw a class diagram for it. Write any one use case for this application and draw a sequence diagram for the same. Write test cases for the same.
24. Explain the application of AnyPrecision SDP in a scenario of text mining. Draw a class diagram for it. Write any one use case for this application and draw a sequence diagram for the same. Write test cases for the same.
25. Draw Traditional model for AnyPrecision.
26. Describe the adequacies used to evaluate any model.
27. Compare traditional AnyPrecision model with AnyPrecision SDP. Which one do you like more?
28. Compare both the models quantitatively in terms of number of classes, cost estimation, coupling among classes and constraints.
29. Compare both the models qualitatively.
30. Discuss the process of abstraction of AnyPrecision SDP.
31. Discuss the design and implementation issues of AnyPrecision SDP.
32. Discuss the testability of AnyPrecision SDP.
33. Write some business rules for AnyPrecision SDP.
34. Discuss some of the known usages of AnyPrecision SDP.
35. What did you learn from this chapter?
36. Write your findings, tips, and heuristics.

10.6   EXERCISES

1. Describe a scenario of an application of AnyPrecision in mathematics.
   a. Draw a class diagram of the application.
   b. Document a detailed and significant use case.
   c. Create a sequence diagram of the created use case of b.
2. Describe a scenario of an application of AnyPrecision in communication.
   a. Draw a class diagram of the application.
   b. Document a detailed and significant use case.
   c. Create a sequence diagram of the created use case of b.
3. Describe a scenario of an application of AnyPrecision in the functionality of steering wheel of a car.
   a. Draw a class diagram of the application.
   b. Document a detailed and significant use case.
   c. Create a sequence diagram of the created use case of b.

10.7   PROJECTS

1. Model an application of AnyPrecision in weather forecasting.
2. You are a part of a team that is making a 3-D movie. Model an application of AnyPrecision SDP in it.

3. Model an application of AnyPrecision in a motion control video game.
4. Model an application of AnyPrecision in a webcam chat.
5. Model an application of AnyPrecision in precision drilling.
6. Model an application of AnyPrecision in precision agriculture.

REFERENCES

1. Aristotle (384–322 BC), last modified on June 23, 2017, https://en.wikipedia.org/wiki/Aristotle
2. M.E. Fayad and A. Altman. Introduction to software stability, *Communications of the ACM*, 44(9), 2001, 95–98.
3. M.E. Fayad. Accomplishing software stability, *Communications of the ACM*, 45(1), 2002, 95–98.
4. M.E. Fayad. How to deal with software stability, *Communications of the ACM*, 45(4), 2002, 109–112.
5. H. Hamza. A foundation for building stable analysis patterns, Master thesis, University of Nebraska-Lincoln, Lincoln, NE, 2002.
6. M.E. Fayad. *Stable Analysis Patterns for Software and Systems*, Auerbach Publications, Boca Raton, FL, Taylor & Francis Catalog #: K24627, May 2017, ISBN-13: 978-1-4987-0274-4.

11 AnyCorrectiveAction Stable Design Pattern

Waiting too long before taking corrective action could be dangerous and cause long-term damage.

David Frost [1]

The stable design pattern AnyCorrectiveAction is based on the term corrective action [2], which generally refers to any countermeasure taken by any party, who seeks Maintenance by the solution to any problem. Maintenance can be of different types: corrective, adaptive, and predictive. AnyParty decides upon taking any corrective action, based on any evidence and the reason, and also the criteria that govern it. It involves one or more countermeasures taken to prevent the recurrence of the problem or the reason for the corrective action. In the end, a log is produced, which serves as a basis for verification of the fact that the corrective action has been accomplished and it meets the ultimate goal of Maintenance.

This chapter intends to apply the SSM [3] approach toward modeling a system that wants to perform Maintenance by employing corrective action and at the same time create a stable pattern, which can form a part of such a system. This approach also intends to focus on applying corrective action in a holistic way, so as to prevent unseen circumstances which might cause potential impedance mismatch between process and workflows at a later stage.

The SSM ensures very high reusability. This unique quality guarantees that a design once created can be used to model any application, in any domain, thus making the task of designing more efficient and making the system largely domain independent [4]. The main goal of this chapter is to model the system on the basis of an EBT. At this juncture, the ultimate goal or the EBT is Maintenance and AnyCorrectiveAction is what acts as a workhorse to achieve the goal of Maintenance.

Here, we have first designed a model, which defines the relationship between the EBT with the help of BOs. Later, we have gone ahead with modeling an application scenario, which portrays the reusability advantages of the stability model. A comparison between the traditional model and the stability model is also included to describe how the latter overcomes numerous drawbacks of the former. Different models, such as use cases, CRC cards, class diagrams, and sequence diagrams, have been used to give a better insight into the working of the project.

11.1 INTRODUCTION

The traditional model of software modeling fails to consider issues like high reuse stability, scalability, and impedance mismatch, and systems as well as patterns designed on those lines, fail to be generic enough so that they can be applied in any scenario regardless of the domain of application. Any software designed on traditional lines lacks flexibility of being adapted to any other application. Any software modeling performed by using this technique is never flexible and it will not be adaptable to future changes. It is just a static view of the system, and it also lacks the ability to depict the dynamic nature of the system.

Also, in the software development industry, numerous projects have failed in the past and they continue to do so. Adding to this, the ones which do not fail have huge costs of maintenance associated with them. This is primarily because of the flaw in the analysis and design of those systems and the way they have been modeled. They have been modeled by keeping one static configuration or only one application in mind and they lack the ability to dynamically adapt to changes. It is like constructing a big car manufacturing plant that produces just one car out of the facility and then restructuring the whole plant again to produce the second car and so on, which clearly is an extremely inefficient way of doing anything.

So, now we are trying to model on the lines of making a system dynamic. Instead of concentrating on a particular application as such, we are now concentrating upon a stable design pattern of corrective action and the EBT that it aims at implementing. The EBTs are realized through their workhorses, known as BOs, which are then applied to one or more applications. We will call this as the stability modeling technique for software modeling. It gives us the ability to cater to various business themes and makes the system domain independent rather than sticking to just one particular application of it. We also hereby attempt to show how the same system, which was created earlier by using traditional model, can be modeled again by using the superior stability model, thus giving it a larger purview and stability over time.

11.2   PATTERN DOCUMENTATION

11.2.1   Pattern Name: AnyCorrectiveAction Stable Design Pattern

The stable design pattern AnyCorrectiveAction is based on the term corrective action, which generally refers to any countermeasure taken by any party, who seeks Maintenance by the solution to any problem. Maintenance can be of different types: corrective, adaptive, predictive. AnyParty decides upon taking AnyCorrectiveAction, based on AnyEvidence and the reason and also the criteria that govern it. It involves one or more countermeasures taken to prevent the recurrence of the problem or the reason for the corrective action. At the end of it, a log is produced which serves as a basis for verification of the fact that the corrective action has been accomplished and meets the ultimate goal of Maintenance. AnyCorrectiveAction is a BO to a workhorse to realize the EBT of Maintenance.

11.2.2   Known As

1. *Betterment*: The AnyCorrectiveAction pattern is sometimes also confused with the term betterment, that is, making something that is there and making it better, but being confused as corrective action, but it is not quite the same; since, betterment is about making something that is there, which is already in good shape and later improving upon it. This is not exactly what a corrective action does. Our pattern is employed where we want to rectify a problem that exists instead of making something which may be good, even better.

2. *Prevention*: Prevention is again, not about taking a corrective action. It is rather about taking preventive measures that help the system to prevent the occurrences of any new problems that would need a corrective action in the future. It is, hence, not the appropriate term for our pattern.

3. *Correction*: Correction, in some cases, can also mean the comments given for what should have been the right thing, instead what is currently there [5]. Here, we are not talking about it and since correction encapsulates all such possibilities, that is, the responses to any problem or way of doing things that currently exists, it is not accurate enough to be fit into the definition of the pattern in question. The term correction can also be the measure of changes made in anything to achieve the desired level/state; it might even be considered sometimes as a punishment for some wrong deeds [6]. Therefore, because the term "correction" can have multiple meanings in different context, it is not accurate enough and it cannot be used, as the term which would represent the pattern AnyCorrectiveAction.

11.2.3   Context

The stable design pattern, CorrectiveAction can be applied to any domain in which we have a board, which requires regular Maintenance and which has a requirement of doing so because of a problem that exists. For example, in military, there can be a corrective action taken; for example, initiating a court-martial on an officer, if the problem of smuggling of arms was the case and there was evidence that supported the act of smuggling. Generally applied in military domains, this pattern can now be applied across other domains; and by the means of stability model, it has now attained a completely domain-independent form. It can be applied to any organization, that is, a company that wants to raise working capital can make a corrective action of liquidating unused assets, which were bought earlier, but are not being used or this can be applied to any university, where the university takes a corrective action of counseling students for not cheating, as a corrective action to maintain academic honesty or to counter the problem of plagiarism.

1. *CorrectiveAction in Military*: In military (AnyParty), there can be a corrective action taken (AnyCorrectiveAction), initiating a court-martial (AnyMechanism) on an officer (AnyParty), if the problem (AnyCriteria) of smuggling (AnyReason) of military (AnyType) arms (AnyEntity) was the case and there was written (AnyLog) evidence (AnyEvidence) that supported the act of smuggling.

2. *CorrectiveAction in University*: A university (AnyParty) takes a corrective action (AnyCorrectiveAction) against students (AnyParty) by punishing (AnyMechanism) them for cheating (AnyCriteria), if an evidence (AnyEvidence) proving plagiarism is found, as a corrective action to maintain (Maintenance) academic (AnyType) honesty (AnyEntity) or to counter (AnyReason) the problem of plagiarism.

11.2.4 PROBLEM

The pattern AnyCorrectiveAction must incorporate the way corrective action is taken by AnyParty to overcome any recurring problem. There has to be a reason that prompts Maintenance of the system for which the corrective action is required. The corrective action must be approved by an authority or AnyParty, and it must fall under the periphery that is defined by the policies or AnyCriteria set for by the authority. Finally, AnyCorrectiveAction must result into AnyEvidence that shows that the corrective action was taken and should also be supported by AnyLog. The approving authority has to be displayed the evidence and log to prove that the corrective action was already taken.

11.2.4.1 Functional Requirements

1. *Maintenance*: Maintenance should restore back the right state of the system in the right manner. Maintenance would be achieved by AnyMechanism that is specifically employed for this; however, it is the job of Maintenance as such, to make sure that the system now acts in the correct way that it should have, after the corrective action was taken.

2. *AnyParty or AnyActor*: AnyParty or AnyActor specifies AnyCriteria that governs the mechanism being used and the measures being taken in order to perform the corrective action. The one or more parties and persons must have someone with them, who has the expertise related to the domain in which the corrective action has to be done, so that the required corrective action(s) can be aptly identified, prioritized, and instituted for action. Also, the party here is the authority, who establishes the fact that the corrective action was properly carried out by examining the logs and evidence generated. So, the expert(s) is(are) needed, so that it can be judged whether the corrective action was properly taken or not.

3. *Issuing Authority*: There has to be some governing body that seeks Maintenance to a problem and approves the corrective action. It cannot be done in an ad hoc manner by just anyone.

4. *Criteria*: AnyCorrectiveAction must be governed by the criteria set forth by the issuing authority and must be performed within the boundary of one or more criteria that have been set. It cannot go beyond the policies that drive the business; otherwise, the corrective action would result in other problems for which more corrective actions would be required.

    Moreover, AnyCriteria goes beyond just what AnyParty or AnyActor has defined with respect to themselves and their internal concerns; certain governing criteria can also arise from the law of the land and the culture of the society that the system should cater to. Such criteria have to be met without fail.

5. *Criteria and CorrectiveAction Definition Process*: Defining AnyCriteria is also one point of concern especially, when multiple AnyParty and AnyActor are involved in the formation of one or more AnyCriteria. There has to be an effective coordination mechanism and a formal process of doing this, because the definition of criteria has to be carried out responsibly and effectively. Each participating member must review and approve the decisions being taken, and for approach, a formal process needs to be instituted.

6. *AnyCorrectiveAction*: Some AnyCorrectiveAction mechanism has to be used in order to perform CorrectiveAction and do Maintenance, so such a mechanism must exist. AnyCorrectiveAction would be governed by the criteria defined by AnyParty or AnyActor seeking Maintenance. It is this CorrectiveAction that achieves Maintenance. AnyCorrectiveAction in this case can be a type of AnyMechanism.

7. *Type of CorrectiveAction Used*: The type of *CorrectiveAction* mechanism used to attempt Maintenance through AnyCorrectiveAction has to be specific to the problem, which has to be addressed and it should be within context. Choosing the right mechanism is essential to the corrective action's appropriateness.

8. *Proof*: The corrective action must result into an evidence or a proof that shows that the corrective action was indeed taken to rectify the situation. Otherwise, the purpose of corrective action is defeated and it cannot be made sure that whether the corrective action was taken in an appropriate manner or not.

9. *Collection of Data*: There should be provisions for capturing the related data, which results from the effects of the corrective action and must be captured in an AnyLog, which can display the chronological order of the various actions and effects of this corrective action.

10. *Reason*: There has to be a well-defined and identified reason for the corrective action to be performed and hence the Maintenance to be undertaken. Generally for employing corrective action, the problem at hand is a recurring one. Thus, the reason has to be a recurring problem that occurs within the system. It has to be something internal to the system and external influences cannot be considered as reasons that require any corrective action on the system. One or more AnyReason define the type of mechanism to be used to achieve Maintenance, depending upon what the reason is and the various factors and needs associated with them.

11. *Type*: Maintenance can be of different types: corrective, adaptive, predictive. Each one of these types will lead to special type of entities that needs to be included in the corrective action process. It is essential that such "type" of Maintenance is clearly identified or a subsystem is out in place that identifies the right kind of Maintenance that has to be done.

12. *AnyEntity*: AnyEntity is something that is used by AnyMechanism that is being employed, to perform corrective action and do Maintenance. So, AnyEntity being used by AnyMechanism has to be accessible and interfacable with the system, so that it can be used in performing the required tasks. Choosing the right entity is crucial here, since it may be governed by certain access regulations and there may be some other limitations to the use of it; therefore, to maintain the cost-effectiveness, the entity must be chosen correctly. The Entity must therefore make the relevant Meta data available upon request, so that the other parts of the system can properly decide if it can be used or not and if so, can properly interface with it to get the job done.

13. *AnyLog*: AnyLog must have at least three files: the log of the problem that exists, the corrective action process to be executed, and the execution log generated from the corrective action process. AnyLog must be able to convey the related data to the stakeholders and if required, the system should be able to interface with the medium that any of the stakeholders wants to have the log on.

11.2.4.2 Nonfunctional Requirements

1. *Urgency*: At any time, there may be a certain set of problems that needs to be addressed and not just one problem. In such a scenario, one problem may be more serious than the other and might require urgent attention. The system must also take into account such situations, where a bunch of corrective actions are required to be taken, but we need to prioritize them based on the urgency of the same. This may be decided upon by the issuing party as to which one to do first and which one the next.
2. *Cost-Effectiveness*: The corrective action being taken must be done in a cost-effective manner. The system must not take forever to perform the tasks that are to be performed as a part of the process and still the entire process should be limited within a set amount of budget, so as to keep a balance between performance and costs incurred. Any extremes experienced here have the tendency of stalling the system, if it takes too long to finish the job, or making it impossible for the management to keep the system in use, if it bears an enormous cost.
3. *Exactness*: The corrective action taken should exactly conform to its description and must act fully and exactly as per the boundaries specified for it. Any more or less than the expected action would render it useless and may even result in a situation where one or more corrective actions are required to counter the ill effects that it created in the processes of attempting the corrective action that was instituted. If the corrective action is not exact to its definition, more corrective actions would be required and that would result in increased cost, time, and effort.
4. *Reliability*: The reliability of any corrective action is also desirable, because, if one is not sure about the outcome of the corrective action being taken, it cannot be predictable, and hence would not be as effective in addressing the problem as it should be, thereby defeating the whole purpose of serving it only partially.

Also, the reliability and consistency of AnyLog and AnyEvidence are required, because, if they are not reliable enough, there will not be a proof for the fact that the corrective was taken and its effectiveness cannot be measured.

11.2.5 CHALLENGES AND CONSTRAINTS

11.2.5.1 Challenges

Challenge ID	0001
Challenge Title	Too Many Criteria Imposed
Scenario	Corrective action to maintain a software
Description	Software Maintenance costs a lot of money. The Maintenance procedure has to follow some criteria so as to make sure that the process is efficient and cost-effective and the resulting software is of high quality. If there are a few criteria, then these can be taken care of, but in case where are 1000s of criteria, the task of corrective action becomes very challenging.
Solution	Only the required number of criteria should be imposed by actor/party.
Challenge ID	0002
Challenge Title	Many Parties Requesting Maintenance
Scenario	Corrective action in university to avoid plagiarism.
Description	If there are on an average 10 students, who are caught cheating every year, then it is manageable for the professor to take corrective action against by asking them to repeat the assignment. But in cases where all the students are caught cheating, it becomes difficult for the professor to find out a way to correct them.
Solution	Policies must be made in a way to avoid Maintenance, as Maintenance always comes at the cost of resources like time and money.
Challenge ID	0003
Challenge Title	Too Many Mechanisms Used to Maintain

Continued

Scenario	Maintenance of a software
Description	If software can be maintained by using just one mechanism or one corrective action like replacing a procedure with another procedure, then it would be easy to do so. But, if 1000s of mechanisms were needed, then it would be truly challenging, as the order of these mechanisms must be known. Also, some of the mechanisms can be overlapping as well, thus making it more difficult.
Solution	Flowcharts and other diagrams can help in the scenario, when there are many mechanisms to be followed.
Challenge ID	0004
Challenge Title	Too Many Entities Required by CorrectiveAction
Scenario	Maintenance of a furniture
Description	If an old piece of furniture requires simple painting and polishing, then it will not be very difficult to perform it. However, in the case where it is totally broken and need lots of things such as cloth, foam, nails, paint, etc., it would be beneficial to buy a new one than to fix the old one. Thus, the task of corrective action becomes more and more challenging as the type of entities it needs increases.
Solution	Identify the cost of maintenance. If the cost of gathering the entities and maintaining is too much, then it is better to replace it.

11.2.5.2   Constraints

1. One or more AnyActor/AnyParty should request Maintenance.
2. AnyActor/AnyParty should specify zero or more AnyCriteria.
3. AnyCriteria influences AnyCorrectiveAction.
4. Maintenance is achieved with the help of AnyCorrectiveAction.
5. Maintenance is done for one or more AnyReason.
6. AnyReason defines one or more AnyType of AnyEntity used for AnyCorrectiveAction.
7. AnyCorrectiveAction uses one or more AnyEntity.
8. AnyCorrectiveAction is taken as a result of one or more AnyEvidence.
9. AnyCorrectiveAction produces one or more AnyLog.
10. AnyEntity is also on one or more AnyLog.

11.2.6   Solution

11.2.6.1   Pattern Structure

<P-EBT> - Pattern EBT

<P-BO> - related Pattern BO, which, along with AnyCorrectiveAction and other pattern BOs, realize the EBT of Maintenance

11.2.6.2   Class Diagram Description (as Shown in Figure 11.1)

1. AnyParty or AnyActor seeks Maintenance, which is achieved through AnyMechanism
2. AnyParty or AnyActor defines AnyCriteria which controls AnyMechanism
3. Maintenance has AnyReason, which names AnyType, which specifies AnyEntity used by AnyMechanism
4. AnyCorrectiveAction results into AnyEvidence and gives rise to AnyLog

11.2.6.3   Participants

11.2.6.3.1   Classes

AnyCorrectiveAction: Any thought that is conceived by any actor.

11.2.6.3.2   Patterns

Maintenance: Represents the enduring concept of Maintenance; it includes the operations and attributes of the same.

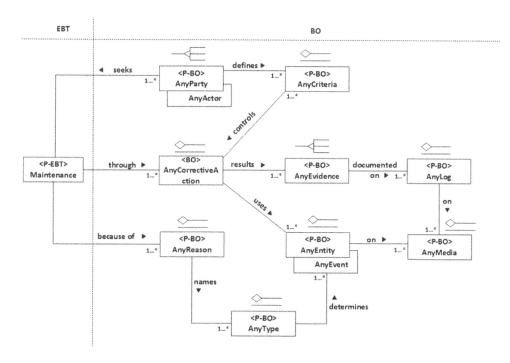

FIGURE 11.1 AnyCorrectiveAction stable design pattern class diagram.

AnyEntity:	Refers to any objects that are being used in the system.
AnyActor:	Person, animal, software, or hardware; something that acts on its own will.
AnyParty:	Any person, organization, country, or political party.
AnyEvidence:	Something that forms the basis for AnyCorrectiveAction.
AnyCriteria:	Refers to the basis that governs AnyMechanism through which Maintenance is achieved.
AnyType:	Is the type of AnyMechanism and it specifies AnyEntity that shall be used by AnyMechanism.
AnyCorrectiveAction:	The method employed to achieve Maintenance
AnyLog:	The record that is generated after AnyCorrectiveAction is taken.
AnyReason:	The problem that exists and which requires corrective action.

11.2.6.4 CRC Cards

	Collaboration	
Responsibility	**Clients**	**Server**
<P-EBT> Maintenance (Maintenance)		
To achieve consistency	AnyParty	maintain()
	AnyActor	renew()
	AnyCorrectiveAction	increaseLifeSpan()
	AnyReason	fix()
		adapt()
		perfact()

Attributes: type, duration, degree, frequency, phase, when?, who, reason(),

Continued

Responsibility	Collaboration	
	Clients	Server
<P-BO> AnyCorrectiveAction (AnyCorrectiveAction)		
To take countermeasure and produce evidence of it	Maintenance	takeCorrectiveMeasure()
	AnyCriteria	restore()
	AnyEvidence	produceLog()
	AnyEntity/AnyEvent	resultEvidence()
Attributes: duration, name, type, serialNumber, purpose		

The CRC cards are designed based on the guidelines as mentioned in References 6 and 7. They help us in defining a complete picture of how a common set of operations, with individual responsibility of each class and the related operations served by it, can be utilized to form the basis of an application scenario. Such an application that is built on top of this pattern, utilizing one or more operations from the generic operations that are offered by the BOs, has been shown in Section 11.5 (Applicability). The CRC cards name the class, responsibility, and its collaborations. The CRC cards also name a role for each class, which is useful for identifying the class responsibility.

Each class should have only one and unique responsibility. The collaboration consists of two parts: clients and server. Clients are classes that collaborate and have relationship with the named class. Server contains all the services that are provided by the named class to its own clients. It is worth to point out that in documenting CRC cards for stable patterns, we deal with any pattern that are included within the main pattern itself as a class. That is, each subpattern will be represented by a CRC card that documents its responsibility and collaborations as a black box. To avoid any confusion, and for simplicity, we do not care about how the subpattern handles its reasonability according to its internal structure, and all what we care about here whether this subpattern will perform the task as a black box, by leaving all other details to the second abstraction level of the pattern description.

11.2.7 CONSEQUENCES

The AnyCorrectiveAction stable design pattern presents us with many challenges that have to be tackled. In order to meet those challenges and to fulfill them, we would need to design the system taking into consideration certain capabilities that the system must have, and doing so might result in many consequences that the system will face. Here are a few of those:

1. To be able to support a large amount of media for storing AnyLog generated by the system, it would need to have the capabilities for interfacing with the different types of media that are available. This would result in the system being of a large size, a considerable part of which would be something, which deals with this interfacing than anything else.
2. Also, if the user is to define the way to handle the storage of this, then we need to have this support for different storage mechanisms. If these storage facilities are on the Internet and need an access mechanism, then the system needs to cater to that too, and things like access rights management, etc. also need to be implemented.
3. These features may also call for support for plug-in and an appropriate API would be needed, so that the consumer can make use of the available options within the system, as well as add his/her custom devices by integrating the necessary interfacing plug-in into the system.
4. Doing so, we may also have to consider many issues like security, because we would allow the system to be customized by the client to a great extent and hence whatever the plug-in does need to be monitored or approved before it can be added, or alternatively such things have to be "legally" taken care of and further investigation needs to be done in this area.
5. For the system to be cost-effective and to deal with the issues like critical corrective actions, especially in those cases, where more than one corrective actions are of a very high priority

and it is necessary to handle both of them as quickly as possible, then we need to empower the system with efficient scheduling capabilities, where the system can effectively decide which entity to use and at what time and for how long. The system needs to plan out for minimum access of the entities, if they are to be paid for on a *pro rata* basis and hence the system should be able to arrange the set of activities, on the fly, in a way which strikes a balance between cost and performance, while staying within the cost and time constraints.

11.2.8 APPLICABILITY

11.2.8.1 Application 1: Company Trying to Reduce Losses

This application is about a company, which is undergoing losses because some of its employees lack minimum attendance. So, in order to reduce these losses incurred because of low attendance, the company takes corrective action in that regard.

11.2.8.1.1 Class Diagram

Class Diagram Description (as shown in Figure 11.2)

1. AnyParty or AnyActor seeks Maintenance, which is achieved through AnyMechanism
2. AnyParty or AnyActor defines AnyCriteria, which controls AnyMechanism
3. Maintenance has AnyReason, which names AnyType, which in turn specifies AnyEntity used by AnyMechanism
4. AnyCorrectiveAction results into AnyEvidence and gives rise to AnyLog
5. Company has a policy, which issues one or more warning to employee
6. Employee undergoes counseling, which improves attendance
7. Attendance decreases loss
8. Balance sheet reflects the loss

11.2.8.1.2 Use Case

Title	Reduce Losses Due to Improper Attendance		
Id	1.1		
Actor	**Roles**		
AnyParty	Company, Employee		
Class	**Type**	**Attribute**	**Operation/Interface**
Maintenance	EBT	duration, degree	renew()
AnyReason	BO	validity, significance, stimulus	corroborate()
AnyParty	BO	partyName, motto, type, size	defineCriteria()
AnyMechanism	BO	name, effectiveness, strength	achieve()
AnyEntity	BO	description, name	helpAccomplishTask()
AnyCriteria	BO	clauseNumber, type, domain, name	control()

Continued

Class	Type	Attribute	Operation/Interface
AnyCorrectiveAction	BO	duration, name, type, serialNumber, purpose	produceLog() resultEvidence()
AnyEvidence	BO	timestamp, assertion, reliability	prove()
AnyLog	BO	name, size, media, logNumber, timestamp	record()
Loss	IO	cost, type	affectNegatively()
Company	IO	name location objective	seekMaintenance()
Policy	IO	text, clause, name	instritureWarning()
Counseling	IO	duration, reason, domain	motivateEmployee() improveMentality()
Attendance	IO	registerName, size	showPresense() keepTrack()
BalanceSheet	IO	name, day, month, year	elicitateFinance() storeRecord()
Employee	IO	id, name	undergoCounselling()
Warning	IO	type, instruction	inform() issueWarning()

Description

1. AnyParty seeks Maintenance
    1.1 Which has AnyReason
    1.2 AnyReason determines AnyType

TC: Maintenance is done by what? Who seeks Maintenance? Maintenance is based on what? Why is Maintenance required? What determines AnyType? How?

2. Maintenance
    2.1 Is achieved through AnyMechanism,
    2.2 A company seeks Maintenance because of the problem of Loss and
    2.3 Uses Warning to intimate to the employee about it as it is being created because of employee's low attendance

TC: What achieves Maintenance? How is Maintenance achieved? Why is Maintenance needed?

3. AnyType
    3.1 Specifies AnyEntity which
    3.2 Is used by AnyMechanism
    3.3 To achieve Maintenance

TC: AnyType does what? How? What specifies AnyEntity? Based on what? What does AnyMechanism use to achieve Maintenance?

Continued

4. AnyParty
    4.1 Defines AnyCriteria
    4.2 Company defines a policy which
    4.3 Governs the code of conduct of employees
    4.4 And describes the process of issuing warning to employees
TC: Who defines AnyCriteria? AnyParty does what? Based on?

5. AnyCriteria
    5.1 Controls AnyMechanism
    5.2 Issues Warning to the Employee
    5.3 Uses Warning to intimate to the Employee about low Attendance, since it is creating Loss
TC: What does AnyCriteria do? What controls AnyMechanism? Is this control needed? How? Who decides on AnyCriteria?

6. AnyCorrectiveAction
    6.1 Results into AnyEvidence
    6.2 And gives rise to AnyLog
    6.3 The employee undergoes counseling which
    6.4 Helps him/her get perspective and understand the expectations of the company along with undergoing introspection
    6.5 And Attendance is increased as a result
TC: What does AnyCorrectiveAction result? What else does it give rise to? What produces AnyEvidence? Or AnyLog? What do they do?

7. AnyLog is
    7.1 Related to AnyMechanism
    7.2 Which achieves Maintenance
TC: What is related to AnyMechanism? How? AnyLog is related to what? What achieves Maintenance?

8. AnyLog
    8.1 Helps prove that any AnyCorrectiveAction was taken
    8.2 Undergoing counseling helps the employee to
    8.3 Improve Attendance
    8.4 Which reduces Loss
    8.5 Which is shown by the balance sheet
TC: What gives a proof of AnyCorrectiveAction? What proves that AnyCorrectiveAction was taken? What else does this do? How?

Alternatives

1. Company seeks Maintenance to the problem of loss
2. Which is being generated due to lack of Attendance
3. Company institutes a policy for incentives for better attendance
4. And raises the salaries
5. Which motivates the Employee to have a better attendance record
6. Which eventually helps in reducing the Loss to the Company

11.2.9 Related Pattern and Measurability

11.2.9.1 Related Pattern: Traditional Model

AnyCorrectiveAction stable design pattern has the goal of Maintenance. That, however, is something that can be confused with, for example, betterment, prevention, correction, etc., as we saw earlier in the "Known As" section of the pattern documentation. Moreover, there are many ways that Maintenance through corrective action has been attempted and is being attempted. There exists, however, no such stable pattern that captures the essence of achieving Maintenance through employing corrective action in a way that is holistic and stable. Due to lack of such a stable pattern, this section illustrates a nonstable pattern approach to doing corrective action. Since, there exists no stable pattern on corrective action, presented here is a "traditional" approach (Figure 11.3) to the same, along with a comparison (as shown in Table 11.1) with the stable pattern-based approach,

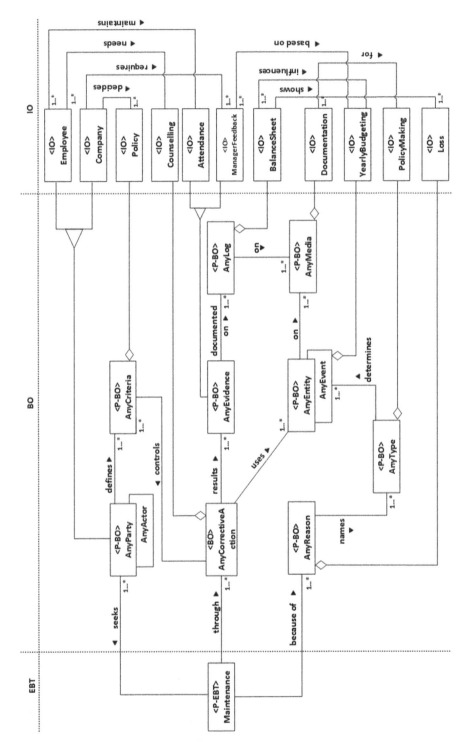

FIGURE 11.2   Application 1: company trying to reduce losses class diagram.

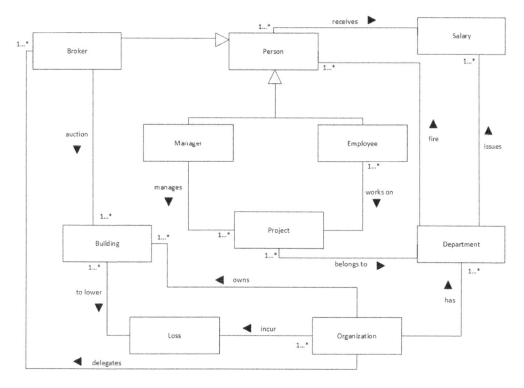

FIGURE 11.3 Traditional model of CorrectiveAction.

on the basis of how well does each of the methods satisfy one or more modeling adequacies to building an application that employs corrective action and does Maintenance.

11.2.9.2 Comparison
Comparison, as shown in Table 11.1.

11.2.9.3 Measurability
11.2.9.3.1   Quantitative
Quantitative Comparison of Traditional and Stability Models

Sl. No.	Criterion	Traditional Model	Stability Model	Indicator (Aspect w.r.t. Stability Model)
1	Number of aggregation relationships	1	4	Good
2	Number of inheritance relations	3	2	Good
3	Number of applications	1	Many	Good
4	Number of tangible classes	10	8	Good

Implications of the abovementioned indicators:

Number of Aggregation Relations and Inheritance Relations: As we can see here, the traditional model is not very flexible and adaptable to more than one application. The number of inheritance relationships is more in the traditional model than the number of inheritance relationships in the stability model, which makes it evident that the traditional model makes its class, especially, the subclasses that inherit from their superclasses very brittle. This makes the system difficult to maintain and has a lot of dependencies in it. If a superclass has to be modified for some reason, it affects all

TABLE 11.1

Comparison between Traditional Model and Stable Pattern Based on Modeling Adequacies

1	Adequacy:	Descriptive Adequacy

Descriptive adequacy refers to the ability to visualize objects in the models. Every defined object should be browsable, allowing the user to view the structure of an object and its state at a particular point in time. This requires understanding and extracting Meta data about objects that will be used to build a visual model of objects and their configurations. This visual model is domain dependent—that is, based on domain data and objects' Meta data.

Comments and Measures

Traditional Model

The traditional model class diagram describes the system in a way that shows which class is interacting with which other class and the relationship between them, but it fails to address the goals of the system and as to why this is happening, giving a very static view which does not show what is being done for what. **Only 50% adequate.**

Stable Pattern

The stable pattern clearly shows the various themes of the system and the BOs that realize them along with the IO classes pertaining to the system that is being implemented. It also has all the information that the traditional model had for describing the relations between the classes and the various conditions and constraints. It describes the system more effectively than what traditional model did. It provides a good "in a nutshell" view of the system; however, a textual description of the entire system is probably a better way to describe the system. So, as far as the class diagram of the stability model goes, we think that it is **90% adequate.**

Indicator (Stable Pattern):   **Good**

2	Adequacy:	Logical Adequacy

Logical adequacy refers to the representation tools that describe the pattern's participants' behavior, roles and responsibilities.

Comments and Measures

Traditional Model

Describes the classes and relationships, but only of one scenario that the pattern is catering to, so it is a very static view of the system and does not show which class is fulfilling which responsibility. **Only 10% adequate.**

Stable Pattern

Describes the classes and relations of the pattern as well as the application being catered to, by showing how each IO is connected to each other that is forming the application and how each of them is connected to a BO which realize the EBTs, giving a more logical understanding of how the system is functioning. However, it is still not a dynamic view. **90% adequate.**

Indicator (Stable Pattern):   **Good (Much Better)**

3	Adequacy:	Understanding Adequacy

Understanding adequacy relates to the easiness with which it is able to understand for anyone, how the system is built and what it is all about; how easy is it to understand.

Comments and Measures

Traditional Model

The class diagram is visually self-describing in a way that it shows the classes and the associations clearly. There are no crosslines and the flow is easy to understand. The system can be easily understood by this; however, only the static- and application-specific view of it. It does not still show the main aim and the concept of business that the system is made for. We would say that here the traditional Model class diagram is **about 30% adequate.**

Stable Pattern

The class diagram of the stability model is very effective, when we want to understand what the system is doing and how it is doing so. It has all the features that the traditional model provided, and also goes ahead to explain the aim(s) or the goals that the system is supposed to accomplish and the way the same is being done in the application that is in question. One can get to know how the concepts get implemented finally through the IOs or how the IOs are able to meet the goals that the system is supposed to achieve. It is **100% adequate.**

Indicator (Stable Pattern):   **Good**

(Continued)

TABLE 11.1 (*Continued*)

Comparison between Traditional Model and Stable Pattern Based on Modeling Adequacies

| 4 | Adequacy: | Analytical Thinking Adequacy |

Analytical thinking adequacy relates to the analytical thinking approaches and tools that are concentrating on analytical aspects of the system. It also includes the utilization of analysis patterns

Comments and Measures

Traditional Model	Stable Pattern
Traditional model class diagram is not at all analytically adequate. One cannot analyze the system in any way using this. The goal behind doing anything and where a class fits into the big picture of achieving one or more ultimate goals cannot be judged from it. **Only 10% adequate.**	Stable pattern class diagram is very adequate, when we consider analytical thinking adequacy, as it shows very clearly the exact place of any class in the system, what the system is built for, and how in the big picture a class fits in toward realizing the ultimate goal the EBT. It uses clear and appropriate use of stable analysis and stable design patterns to do this, making it **95% adequate.**
Indicator (Stable Pattern):	**Good (Huge Improvement)**

| 5 | Adequacy: | Systematic Adequacy |

Systematic adequacy relates to the systematic approaches that are utilized in modeling, such as bottom-up, top-down, and middle-out approaches. It also includes functional decomposition techniques, and the selection of the correct level of abstraction at each stage of analysis.

Comments and Measures

Traditional Model	Stable Pattern
The traditional model class diagram has no concept of taking a very systematic approach to modeling the system. It is a view of the system showing how things are and it does not go from one stage of analysis or abstraction to other. **10% adequate if at all.**	The stable pattern class diagram is very systematic. There are the ultimate goals of the system that are realized through the BOs and they are finally implemented through the IOs, going from top to bottom in a systematic approach to modeling the system. It clearly shows what the goals of the system (EBTs) are and how they are realized through respective BOs and then the IOs that hook up to the BOs. **100% adequate and very systematic.**
Indicator (Stable Pattern):	**Very Good (Way ahead of Traditional Model)**

| 6 | Adequacy: | Epistemological Adequacy |

Epistemological adequacy refers to tools for representing objects in the real world.

Comments and Measures

Traditional Model	Stable Pattern
The traditional model is an application-specific view of the system and is very static. There are no levels of abstraction anywhere except for the inheritance or such relationship, for example, roles of any person, etc. not much adequate **only about 10%–20% adequate.**	The stable pattern class diagram shows various levels of abstraction of the system, that is, the ultimate goals or the EBTs than coming down to the next level of abstraction, that is, the BOs which show how the EBTs are realized and then finally the IOs, so the application is abstracted out into different levels and it presents a much better view of the system, making it **90%–95% adequate.**
Indicator (Stable Pattern):	**Very Good (Much Better)**

(*Continued*)

TABLE 11.1 (*Continued*)

Comparison between Traditional Model and Stable Pattern Based on Modeling Adequacies

7	Adequacy:	Simplicity Adequacy

Simplicity adequacy relates to how simple your models will be.

	Comments and Measures
Traditional Model	**Stable Pattern**
The class diagram shows the classes that have been included in the use cases. The flow and associations are clearly mentioned in the form of arrows and lines This makes it simple to read and follow. It is **45%** **adequate.**	In terms of simplicity, the stable pattern class diagram shows the classes and the relationships between them. The direction arrows show the flow quite well. It is close to traditional model in terms of simplicity. The only thing that makes it better is the holistic view of the system, showing the EBTs, BOs, and IOs and also showing them in neatly segregated sections increasing the simplicity. **99%** **adequate.**
Indicator (Stable Pattern):	**Good (A Little Better)**

of its base classes, and higher the number of inheritance relationships, the more impact the system will undergo. The stability model reduces this and it has more number of aggregations in it, than the traditional model. This provides the system a flexibility of modifying just those parts which are necessary, without propagating the effect further into the system, thus making it more stable over time and easier to maintain. It also results in a significant reduction of Maintenance costs, which currently forms more than 80% of the cost of any software.

Number of Applications: Traditional model is a static model and it is the view of a particular scenario, and a specific application; it is impossible or very hard and too costly to adapt it to any other application and hence it just serves just one application. Stability model, on the other hand, has the ability to be modeled to any number of applications, because it has been modeled on the general concept of EBT and Maintenance in this case, and it can be applied to any other scenario as well; hence, this is a good feature that stability model offers.

Number of Tangible Classes: In a traditional model, the number of tangible objects used is 10, when compared with the 8 in the stability model. This speaks for itself in the lack of flexibility of the traditional model and the cost of Maintenance. Requirements always change over time and trying to adapt to the changing needs, the traditional model does not satisfy that need, since a large part of the system would have to be changed for that. Whereas in the stability model, the impact would be considerably lesser than that in the traditional model, since we just need to modify the IO classes and we are at ease to move forward.

11.2.9.3.2 Qualitative

Adaptability and Reuse: The SSM is highly adaptable to changes in business needs and within the system. The large number of inheritance relations in the traditional model makes it very difficult to maintain and change according to the changing needs. This problem is reduced in the stability model due to less number of inheritance relations. The changes and modifications to be made to a superclass are not propagated further into the system to the subsequent subclasses, since it does not use too many numbers of inheritance relations, hence making it easier to maintain and adapt and reuse. The traditional model also does not have the capability to be modeled to a different scenario, while the stability model owing to the approach of modeling the system on the lines of an EBT, that is, Maintenance in this case, and BOs, gives us the capability to be modeled to any other scenario, where Maintenance is required by AnyCorrectiveAction, thereby making it domain independent.

11.2.10 MODELING ISSUES, CRITERIA, AND CONSTRAINTS

11.2.10.1 Abstraction

To design this project, we have used the SSM, which consists of three parts. The EBT and the BOs form the essential parts of the model. They remain stable throughout any design and for any application. The IOs change for different applications. The EBT is the enduring business concept, it is the stable analysis model and using this ensures that the ultimate goal of the system or business as such is never compromised, and that the system is modeled using that goal in mind. The EBT is then realized using BOs, which are the workhorses for this EBT and these BOs provide hooks, various IOs to be attached to the pattern model according to the system that we are planning to design.

Here, the EBT is Maintenance, which serves as the goal for any such system and there are various BOs associated to it like AnyCorrectiveAction, AnyPolicy, AnyMechanism, AnyEvidence, AnyLog, etc. they realize the enduring concept of Maintenance and according to the application that we want, for example, reducing losses to an organization in this case, we attach the necessary IOs like employee, company, loss, policy, etc. to these BOs, thereby modeling the entire system that is wanted.

11.2.10.2 Modeling Heuristics

While modeling the system, we will need to follow certain modeling heuristics that enable a better system design. Few of them are as follows:

No Star: There should not be any macho classes in the system, since having a star in the system calls for a focus of control on one class and it gets overloaded, thereby making the system prone to failures because of it. In our system, classes such as AnyCorrectiveAction and AnyCriteria have three connections already in the pattern class diagram, so we need to make sure that while connecting the IOs, we do not overload these classes by having more than one new connection to it, as that would result in an unbalanced situation and a macho class or a star.

No Dangling: No dangling classes should be left in the system. Dangling refers to a situation, where there is just one connection to a class. This situation is not good, since it creates a possibility of loss of control, as one the execution reaches that class it cannot be brought back. In this system, we have AnyEvidence dangling in the pattern class diagram; thus, we will need to make sure that we have at least one IO that is connected to this BO, so that AnyEvidence is not left in a dangling state.

No Sequence: We should avoid sequences in the system, because they result in some situations, where the system may go into infinite loops without resulting into any tangible output. Figure 11.4 shows such a situation.

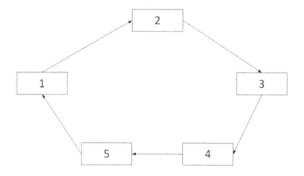

FIGURE 11.4   Sequence (must be avoided).

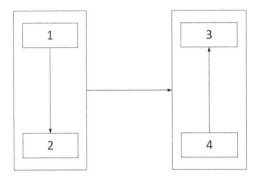

FIGURE 11.5 Alternate view of an existing sequence.

Moreover, if there is a sequence, then it may also mean that the selection of components and classes was not proper. Thus, when there is a sequence of say 8 classes, it may rather be clubbed into one whole class or 2 classes that are communicating with each other, as shown in Figure 11.5.

General Design: The system should not be designed, while keeping in mind just one domain or application. We are looking at designing a system, which can be reused readily and can be applied to any domain in a domain-independent fashion. Thinking about the possibilities in just a particular scenario will limit the capabilities of the pattern and it will defeat the purpose of stability modeling.

11.2.11   Design and Implementation Issues

11.2.11.1   Design Issues

11.2.11.1.1   Selecting EBT and BOs

To accurately model the stable design pattern of AnyCorrectiveAction, the first and the most essential job is to choose the correct EBT for it. There are instances and goals which can be confused with the ultimate goal of a corrective action, like instead of choosing Maintenance, one may get confused with betterment or rectification, but AnyCorrectiveAction is not used for those situations, which require betterment as such. Maintenance is the correct EBT for the system. Also, the related BOs are very important too. The BO AnyCorrectiveAction cannot realize the goal of Maintenance without the presence of the most appropriate and other relevant BOs in the system. AnyCorrectiveAction must be approved by AnyParty and must have AnyReason for it. The party is supposed to approve the corrective action. Then, AnyCorrectiveAction must produce AnyLog and AnyEvidence to support and sh   ow that the corrective action was indeed taken and accomplished, so that the issuing authority gets to know that it was done. Hence, the correct selection of the EBT and other BOs in the system is critical for the correct behavior of the pattern.

11.2.11.2   Implementation Issues

11.2.11.2.1   Using Command Pattern

While implementing, instead of using approaches like inheritance, etc., which make the system too brittle owing to the property of the methods of superclass being transferred to the subclass even if some of them are needed or not, we can use the command pattern to implement the system and this can be used while linking BOs to IOs.

In the command pattern, there is a command interface and a ConcreteCommand implements this interface. There is a receiver that contains the specific method's code and is called by the ConcreteCommand upon request from the invoker, who wants to invoke that specific method. This will allow us the flexibility to avoid the clients that are tied to specific implementation code, and they can just use the command to call a function without having to bother about its location of implementation details. On the service provider's end, it provides the system the flexibility to be maintained in a way that just the specific parts of the system are changed without affecting other classes in the system or the clients.

Here is the basic structure of the Command Pattern as shown in Figure 11.6.

In our pattern, the command pattern can be applied as follows:

```java
/*the Command interface*/
public interface Command{
    void execute();
}

/*Receiver class*/
public class AnyCorrectiveAction{

    public Counseling(){   }

    public void produceLog(){
       System.out.println("producing log");
    }
    public void resultEvidence (){
       System.out.println("making evidence");
    }
}
/*the Command for producing log*/
 public class ProduceLogCommand implements Command{

    private AnyCorrectiveAction myCorrAction;

   public ProduceLogCommand(AnyCorrectiveAction corrAction){
       this.myCorrAction =corrAction;
       }

   public void execute(){
     myCorrAction.produceLog();
   }
}
/*the Command for resulting evidence*/
public class ResultEvidenceCommand implements Command{

   private AnyCorrectiveAction myCorrAction;

   public ProduceLogCommand(AnyCorrectiveAction corrAction){
       this.myCorrAction =corrAction;
       }

   public void execute(){
     myCorrAction.resultEvidence();
   }
}
/*the Invoker class*/
public class Invoker {

    private Command evidenceCommand;
    private Command logCommand;
```

```
    public Invoker(Command evidenceCmd, Command logCmd){
        this.evidenceCommand=evidenceCmd;
        this.logCommand=logCmd;
    }

    public void resultEvidence(){
        flipUpCommand.execute();
    }

    public void produceLog(){
        flipDownCommand.execute();
    }
}
/*The Client*/
public class Client{

    public static void main(String[] args){
        AnyCorrectiveAction corrA = new AnyCorrectiveAction();
        Command ev=new ResultEvidenceCommand(corrA);
        Command lg=new ProduceLogCommand(corrA);

        Invoker correctiveAction=new Invoker(ev,lg);

        corrtiveAction.produceLog();
        correctiveAction.resultEvidence();
    }
}
```

11.2.12 BUSINESS ISSUES

11.2.12.1 Business Rules

Business rules are constraints that govern the way in which a business is carried out. Our system pertains to the ultimate goal of Maintenance, which is achieved through AnyCorrectiveAction and other related BOs. Here are some business rules that apply to this system:

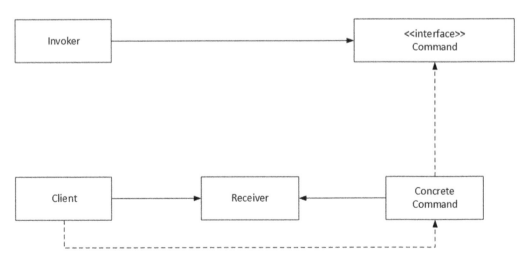

FIGURE 11.6   The command pattern. (From Command Pattern, n.d. (Online), available at Wikipedia, http://en.wikipedia.org/wiki/Command_pattern, accessed May 26, 2010.)

1. AnyParty can define one or more AnyCriteria. Only specific kind of parties can create AnyCriteria. They are government agencies for administering the country for its well-being. Scientists who discover and make theories or laws based on empirical vales and experiments, companies who make them for the employees to abide by for code of business and ethics.
2. AnyMechanism is controlled by AnyCriteria. No mechanism being employed for corrective can go beyond the boundaries set forth by one or more governing criteria.
3. There must be AnyEvidence and AnyLog which results from AnyCorrectiveAction, which is being taken that would serve as a proof to the authority that issued and approved it, of the fact that corrective action was taken.
4. AnyParty is the governing authority that observes the situation and approves that AnyCorrectiveAction be taken. AnyCorrectiveAction cannot be taken without the permission and approval of the governing authority.

11.2.13 KNOWN USAGE

- CorrectiveAction is a term which is more familiar in military. Many of the Military applications use this pattern to carry out many of its tasks; for example, there can be a situation which involves the military to reroute or reframe its strategy of attack. This has to be authorized by the commanding authority and the corrective action of realigning the military to organize a coordinated and effective plan of action will be taken.
- The pattern has also found its implementation in the business world. "A corrective action is a change implemented to address a weakness identified in a management system." Normally, corrective actions are implemented in response to a customer complaint, abnormal levels of internal nonconformity, nonconformities identified during an internal audit, or adverse or unstable trends in product and process monitoring such as would be identified by "Statistical Process Control (SPC)" [10].
- AnyCompany can use this pattern to correct any behavior that leads to some sort of nonconformity to its rules and policies.
- Another known usage of the pattern is in educational institutions like universities to take corrective actions in order to counter the problems like plagiarism or discrimination, etc.
- The pattern can really be applied to any situation or domain, where there is a need for Maintenance due to some reasons of nonconformance to the desired behavior. Though traditionally used in military applications, it can be exported over to other domains too.

11.3 TIPS AND HEURISTICS

- Betterment, as good as it may sound to be an EBT for AnyCorrectiveAction pattern, actually it is not. Maintenance is a better choice.
- To fit the EBT of Maintenance and to realize the pattern AnyCorrectiveAction, the other BOs in the system must be appropriate and in context with AnyCorrectiveAction, or else the true purpose of this may not be solved.
- There were certain BOs that were chosen earlier, but were discarded later on because they did not fit in the pattern or were redundant, plus, it was hard to find IOs for those BOs.
- While developing the IO, when the connections are not possible, it means that something is missing in the BO or the chosen EBT is not correct. This situation was faced, when a particular BO was never used in any application or an application could not be linked.

11.4 SUMMARY

Stability modeling approach toward modeling the system is a big leap ahead from the traditional way of modeling the system, because it provides us a pattern that can fit into one or many applications, in a domain-independent fashion. While in traditional modeling, we have developed a system with a specific purpose in mind and we have modeled just one application of a pattern, because it was not possible to model to any other scenario. But, the stability model keeps in mind the ultimate goal and it models a system, thus making it a flexible and a more maintainable system. Here, we saw that corrective action is a technique applied mostly in military. But, when we modeled it by using the stability modeling technique, we could use it and apply to other domains like business, as shown in the applicability part. It was a good learning experience and an eye-opener to realize the flaws that existed in the current approaches and to experience a better way of software modeling.

11.5 REVIEW QUESTIONS

1. What are the different types of maintenance?
2. How betterment is different than corrective action?
3. Why prevention cannot be used as a synonym for corrective action?
4. What are the differences between corrective action and correction?
5. Define AnyCorrectiveAction SDP.
6. What does CorrectiveAction mean?
7. Describe a few contexts in which corrective action is used in day-to-day life.
8. Write two to three scenarios, where AnyCorrectiveAction SDP is applicable.
9. Discuss the functional requirements for AnyCorrectiveAction SDP.
10. Discuss the nonfunctional requirements for AnyCorrectiveAction SDP.
11. Write six challenges of using AnyCorrectiveAction SDP.
12. Write a few constraints of AnyCorrectiveAction SDP.
13. Draw a class diagram for AnyCorrectiveAction SDP.
14. Explain and justify the participants in AnyCorrectiveAction SDP.
15. Write CRC cards for AnyCorrectiveAction SDP.
16. What are the consequences and challenges of AnyCorrectiveAction SDP?
17. Explain the application of AnyCorrectiveAction SDP in a scenario, where a company is trying to reduce losses. Draw a class diagram for it. Write any one use case for this application and draw a sequence diagram for the same. Write test cases for the same.
18. Draw traditional model for AnyCorrectiveAction.
19. Describe the adequacies used to evaluate any model.
20. Compare traditional CorrectiveAction model with AnyCorrectiveAction SDP over seven adequacies. Which one does you like more?
21. Compare both the models quantitatively in terms of number of aggregation and inheritance relationships, number of applications, and number of tangible classes.
22. Compare both the models qualitatively over adaptability and reuse.
23. Discuss the process of abstraction of AnyCorrectiveAction SDP.
24. Discuss the modeling heuristics with the help of diagrams for each of them. State the importance of each of them.
25. Discuss the design and implementation issues of AnyCorrectiveAction SDP.
26. What are the issues encountered while selecting EBT and BOs for AnyCorrectiveAction SDP?
27. Explain command pattern.
28. Why command pattern is used instead of inheritance?
29. Explain how command patterns are implemented with the help of a diagram.

30. Write some business rules for AnyCorrectiveAction SDP.
31. Discuss some of the known usages of AnyCorrectiveAction SDP.
32. What did you learn from this chapter?
33. Write your findings, tips, and heuristics.

11.6 EXERCISES

1. **Describe a scenario of an application of AnyCorrectiveAction taken by parents to improve their children's knowledge.**
   a. Draw a class diagram of the application.
   b. Document a detailed and significant use case.
   c. Create a sequence diagram of the created use case of b.
2. **Describe a scenario of an application of AnyCorrectiveAction taken by a software engineer to fix a bug.**
   a. Draw a class diagram of the application.
   b. Document a detailed and significant use case.
   c. Create a sequence diagram of the created use case of b.
3. **Describe a scenario of an application of AnyCorrectiveAction taken by a mechanic to fix a car.**
   a. Draw a class diagram of the application.
   b. Document a detailed and significant use case.
   c. Create a sequence diagram of the created use case of b.
4. **Describe a scenario of an application of AnyCorrectiveAction taken by a student to correct his assignment.**
   a. Draw a class diagram of the application.
   b. Document a detailed and significant use case.
   c. Create a sequence diagram of the created use case of b.

11.7 PROJECTS

1. Model an application of AnyCorrectiveAction taken by government to reduce crime rate in the country.
2. You are a part of a team that is maintaining an e-commerce web site. Model the application of AnyCorrectiveAction SDP taken by the team for the same.
3. Model an application of AnyCorrectiveAction taken by government to reduce pollution in the country.

REFERENCES

1. Sir David Paradine Frost (April 7, 1939–August 31, 2013), last modified on June 21, 2017, https://en.wikipedia.org/wiki/David_Frost
2. Corrective Action, n.d. (Online), available at Onelook.com http://onelook.com/?w=corrective+action&ls=a, accessed May 26, 2010.
3. M.E. Fayad and A. Altman. Thinking objectively: An introduction to software stability, *Communications of the ACM*, 44(9), 2001, 95.
4. H. Hamza, A. Mahdy, M.E. Fayad, and M. Cline. Extracting domain-specific and domain-neutral patterns using software stability concepts, *Lecture Notes in Computer Science*, vol. 2817, Springer, Berlin/Heidelberg, pp. 191–201, 2003.
5. Correction, n.d. (Online), available at MSN Encarta, http://encarta.msn.com/encnet/features/dictionary/DictionaryResults.aspx?refid=1861600494, accessed May 26, 2010.
6. Correction, n.d. (Online), available at YourDictionary.com, http://www.yourdictionary.com/correction, accessed May 26, 2010.

7. M.E. Fayad, H.A. Sánchez, and H.S. Hamza. A pattern language for CRC cards, in *Proceedings of Pattern Language of Programs' 2004 (PLOP'04)*, Monticello-Illinois, September 2004.

8. M.E. Fayad, H. Hamza, and H.A. Sánchez. A pattern for an effective class responsibility collaborator (CRC) cards, in *Proceedings of the 2003 IEEE International Conference on Information Reuse and Integration (IRI'03)*, Las Vegas, Nevada, pp. 584–587, October 2003.

9. E. Freeman, K. Sierra, B. Bates. *Head First Design Patterns*. Sebastopol, California, O'Reilly, 2004. https://en.wikipedia.org/wiki/Command_pattern, last modified on May 7, 2017.

10. Corrective Action, n.d. (Online), available at Wikipedia, https://en.wikipedia.org/wiki/Corrective_and_preventive_action, last modified on May 24, 2017.

12 AnyDebate Stable Design Pattern

> Debate is an important vehicle to aid the understanding of any concept, any situation, any policy, and any decision, in a very acceptable way.
>
> **M.E. Fayad**

Debate is a form of an interactive argumentation, carried out by the participating parties, in order to arrive at a better understanding of the factor(s) being discussed and debated. A debate generally involves two or more parties who present their points of view on the topic or factor of discussion, while a mediator, who can also be considered as the coordinator of the debate, mediates the whole ordeal. Debate has been one of the most effective forms of discussion, argument, and counterargument on any topic of interest. Its usage is applicable upon any concept in the world, be it science, nature, philosophy, day-to-day activities, law, civilization, just about anything, and hence a system that can allow facilitation of such a debate is a highly sought after software system, especially today. Now many debating and discussion platforms exist, both offline and online. Many of them focus on catering to various needs that any debating platform would be required to; however, they lack a holistic view of the requirements of such a system.

Presented here is an attempt to capture the core knowledge related to debate as a pattern, in a formal way, by considering numerous applications and domains, where a debate can be conducted, such that it is domain independent and stable over time, that is, the core knowledge related to the pattern of AnyDebate does not change with time, even if the technology, the domain of application, and the environment change. The AnyDebate stable design pattern, along with related patterns, captures the core requirement for any system that can be used to conduct debate.

12.1 INTRODUCTION

A debate is a discussion between sides with different views. Persons speak for or against something before making a decision. Debates are a means of encouraging critical thinking, personal expression, and tolerance of others' opinions. Today, debate still remains essential for democracy. Debates are conducted in governing assemblies, held in lecture halls and public arenas, presented in schools and universities, written in newspapers and magazine columns, heard on radio, or seen on the television. Like our predecessors in ancient Greece, people argue about what is best for their societies and shape the course of law, policy, and action.

A debate generally involves two or more parties presenting their points of view on the topic or factor of discussion, while a mediator who can also be considered as the coordinator of the debate, mediates the whole ordeal. Debate has been one of the most effective forms of discussion, argument and counter argument on any topic of interest. A debate's effectiveness can be witnessed in almost all spheres of life. One can see debates in the area of science, nature, philosophy, everyday activities, in courts of laws, and in any other field where discussions are usually held. Therefore, a system that allows integration of rounds of debate is considered very productive especially, in the area of software industry, which is the main topic of study in this book. For example, a debate can be held for the Presidential Elections of United States, when appointed candidates enter into a debate which is about the most current topics and this debate takes place in the presence of the citizens of United States. The result of this debate plays a vital role in the election of the President [2].

This work focuses on capturing the knowledge that is associated with any debate in a pattern, which is stable over time. After a careful analysis of the various domains and scenarios in which any debate can be conducted, we have tried to capture the core knowledge that is associated with any form of debate into a stable pattern, that is, something that remains unchanged over time, even when the application scenario and domain of application of the system changes. AnyDebate stable design pattern is a building block and the core of a software system that facilitates a debate. This chapter aims to isolate the core knowledge of debate as a concept is isolated from the application-specific logic to form something that is not specific to a particular application only, and that can be applied across application domains and in addition, present it as a stable design pattern to provide the reusable core for other applications sharing the same core concept. In order to achieve this goal, the AnyDebate pattern is designed based on the SSM [3–7]. The SSM provides a stable and reusable core for multiple applications sharing the same core knowledge [3]. SSM also helps finding the ultimate goals of debate and express them as the EBT. AnyDebate, as well as other participant patterns in turn for the capabilities or BOs that realize the ultimate goal, Understanding [8] in this case. The system can then be implemented in an application-specific context by customizing the BOs, by adding IOs to the periphery of the core, through hooking them up with the BOs. Thus, in this way, a whole system that can be used to conduct, capture, analyze a debate and aid understanding of the issue can be created.

This chapter describes a pattern structure for the pattern, showing how the Anydebate stable design pattern fits into the big picture, with other related patterns to satisfy the goal of "Understanding." CRC cards and application scenarios have been provided to aid the understanding of how the pattern can be applied. Also, discussed are various modeling, design, and implementation issues, associated business rules, known usage, etc. The pattern is also supplemented with a use case for a possible application with a detailed description of it including a complete class diagram and a sequence diagram.

12.2 ANYDEBATE DESIGN PATTERN DOCUMENT

12.2.1 AnyDebate Stable Design Pattern

Debate is a process of discussing a certain topic or factor(s) in an argument–counterargument fashion, where two or more parties present their opinions on the issues under question, while supporting or opposing one another's arguments and views on the matter. Parties, who participate in the debate, collectively come to a better understanding of the issue at hand, such that the understanding of the issue incorporates the various views that many parties may have about it. Debate is also defined as "a contestation between two or more persons, in which they take different sides of a question and maintain them, respectively, by facts and arguments; or, it is a discussion, in writing, of some contested point" [9]. Debate can be an important vehicle to aid the understanding of any concept, any situation, any policy, and any decision in a very acceptable way.

12.2.2 Known As

Debate is often confused with the word argument and used interchangeably in various situations. However, they always differ in usage.

Debate is generally used in a positive tone and suggests something constructive. However, an argument is usually used in a negative tone. Although both debate and argument require different point of view, and both require one person to explain his or her understanding on a topic, and the other person giving supporting argument on the topic. However, they still differ in many ways. The main difference is that there is no anger in a debate and argument generally has anger involved.

To debate is to discuss, consider, and deliberate. To argue, can be, to bicker, dispute, and squabble. Now, most debates could and often do include many of the characteristics of an argument, but debate in its purest form should be about providing a persuasive argument to explain and further your cause that can be appreciated by your opponent. In debating societies, the topic is decided and then the debaters draw lots to decide which side of the "argument" they will express. The most skillful debater will be adjudged to win. We guess what you are not doing is adding enough calm rational reasoning behind your "argument."

12.2.3 Context

The concept of debate can be used differently depending on the application domain in which it is used. Therefore, a careful study is needed in order to come up with the pattern that is common to every domain of field where debate exists. Although there can be certain exceptions where this pattern does not apply.

1. *Presidential Elections in United States*: Consider an example of Presidential Elections (AnyEvent), where the elected candidates take part in a political (AnyType) debate (AnyDebate) on the most current issues (AnyAspect) and debate their views (AnyView) and understanding (Understanding) on them, in the presence (AnyRule) of the citizens (AnyParty) of United States and the candidate, who wins here has a high probability of winning (AnyConsequence) the elections to become President of United States. This debate is broadcasted to the citizens of United States by using the Internet, TV (AnyMedia), and other media.
2. *Patent Application*: The patent examiners (AnyParty) debate (AnyDebate) on patent issuance (AnyAspect) about patent application (AnyEvent), which are published in the latest publications (AnyMedia). The outcome (AnyConsequence) of the debate is noted (AnyView) and understood (Understanding) by the government (AnyParty) and the patent application is issued or rejected (AnyRule) accordingly.

In this project, we have taken an example of Presidential Elections, where the elected candidates, debate on the most current issues in the presence of the citizens of United States, and the candidate who wins here, has a high probability of winning the elections to become the President of United States. Another example in our discussion is that the patent examiners debate on patent issuance about patent application, which is published in the latest publication. The outcome of this debate is noted by the government and the patent application is issued or rejected accordingly. In both these examples, we have used the same pattern that we have developed.

12.2.4 Problem

As we have already discussed that the debate concept is applicable in the various applications irrespective of the domain. Thus, the main problem is to create a universal debate pattern that will be applicable in multiple application contexts. The other problem is that the way in which this concept is applicable in various applications. Thus, a general pattern with common characteristics will help in understanding the problem more clearly, irrespective of the application domain. We also have to deal with the scalability and flexibility factors, which are also important for some applications. For any debate, the principal requirements are that the system must have two or more parties discussing a certain topic of interest and presenting their views. The system should allow capturing the information and knowledge thus generated through this process. Debate also needs a mediator or conductor to ensure smooth functioning of the debate. It should be a solution that addresses AnyDebate that is possible in any context on any topic of interest.

12.2.4.1 Functional Requirements

1. *Understanding*: The system must be able to generate enough understanding on the subject matter that is being debated, the ultimate requirement of the system is to do that and capture the knowledge that aids in making the understanding of the known concept better. Understanding has a unique ID, a type, and a level. It enables decisions, increases knowledge, and makes people aware.

2. *AnyParty*: The debate must have participant parties that come together to discuss and interact with each, get views, and express views about the topic of discussion to collectively arrive at a better understanding of the concept that the debate is about. AnyParty has a unique ID, name, type, and role. AnyParty can be a member of a group and perform a particular activity. AnyParty plays one or more roles, can switch roles, can join or leave a group, and can participate in multiple activities. AnyParty has a capability to learn.

3. *AnyEntity*: There has to be a topic of discussion, and it has to be debated among the participants. If there no such topic of interest, then there is no question of debating anything. AnyEntity has a unique ID, a name, and a description. It triggers discussion, leads to debate, and demands focus.

4. *AnyRule*: There would be certain rules that would govern the way in which the debate is conducted. These rules may decide the boundaries of the debate that need to be followed throughout the debate, so that the event goes on smoothly, while the required objectives are met. AnyRule has a unique ID, a name, a type, and a description. It also brings discipline, imposes constrains, and specifies boundaries.

5. *AnyEvent*: There has to be an event that hosts the debate that is being conducted. Without a proper event as such, the debate would lose its significance. There will not be a proper schedule and no concrete results would be captured. AnyEvent has a unique ID, a period, time, and venue of occurrence. It starts, occurs, and ends.

6. *AnyView*: The participating parties should have some views about the topics being discussed, as without having any opinion, they cannot contribute to the debate that is being held. AnyView has a unique ID, a kind, a description, and level. It reflects thinking, touches issue, and provides vision.

7. *AnyConsequence*: The debate should produce some results that extend the body of knowledge or they should refine it. If there are not any consequences as a part of the discussion/debate, there is no need of the debate. AnyConsequence has a unique ID, a reason, effect, and a level. It affects, happens, and follows action.

8. *AnyAspect*: AnyAspect is the aspect of the topic on which debate has to be done. It has a unique ID, a description, and a title. It gives clarification, decides the direction of the debate, and defines the scope and boundaries of the debate.

9. *AnyType*: AnyType is used to classify AnyParty. AnyType has a unique ID, a name, and a classification property based on which the type is decided. AnyType operates on AnyParty, provides labels, and names them. It can also be used to change the type of AnyParty.

10. *AnyMedia*: AnyMedia is used to record and to broadcast debate, so that maximum people can watch it. AnyMedia has a unique ID, a state, and a type.

11. *AnyDebate*: AnyDebate is used to have a conversation and discussion on a particular topic and then draw conclusions and understanding out of that discussion. AnyDebate has a unique ID, a topic, description, and a title. It starts, ends, and provides understanding to the listeners and participants.

12.2.4.2 Nonfunctional Requirements

1. *Informative*: The debate should be informative in nature. The exchange of views and arguments made should bring out the topic's essence in a good way. The discussion should enhance the knowledge related to the area of discussion and enable a viewer to augment his/her knowledge about the area of interest, while getting to know the multifaceted view of the issue being debated.

2. *Usefulness*: The debate must be useful in the sense, if there is no business related to the topic of discussion and no one really cares about it, or if that concept is very trivial in nature, there is no need for any discussions and debates on it.

3. *Effectiveness*: The debate should also be effective. This means that the method of debating, the views being expressed, and the mechanisms of capturing them should be proper. It should be understandable, comprehendible, capturable and disseminatable.
4. *In Context*: AnyDebate should be within the context of the topic being discussed. It should not cross the boundaries and it should not go out of the way. For example, if the topic being discussed is education, then the discussion should focus on education only and should not discuss politics.
5. *Interactive*: AnyDebate should be interactive in the sense that both the parties debating should express their views. If only one party is discussing its views, then it will be like one-way communication and people will not get to know the flip side of the coin. Thus, AnyDebate is interesting only if both the parties defend their stand very well by interaction.
6. *Relevant*: AnyDebate should be relevant, as there is no point discussing on a topic that is of no use. For example, if during Presidential Elections, if both the candidates discuss the current issues of the country, then it would be relevant and people would like it; but, if they will discuss the best movie of the year, then no one would be interested in it.

12.2.5 CHALLENGES AND CONSTRAINTS

12.2.5.1 Challenges

Challenge ID	0001
Challenge Title	Multiple Parties Involved in Debate
Scenario	Debate in college
Description	If there are two candidates expressing their views on a particular topic, then it would be easier to understand what they want to say, but if there are 1000 candidates involved in debate, then it would be really difficult to figure out who wants to say what.
Solution	There should only be a limited number of candidates that can participate in a debate. If there are many participants, then they should be grouped together to form a team.
Challenge ID	0002
Challenge Title	Multiple Rules for Debate
Scenario	Debate Competition in School
Description	In a school, when there is a debate competition, some rules are set such as candidates have to stick to topic, or a certain time limit for discussion, etc. If there are a few rules, then candidates can remember them, but if there are 1000s of rules, then candidates may find it challenging to concentrate on the topic and discussion and may get distracted by thinking about not breaking the rules.
Solution	Only required rules should be enforced and some freedom should be given to candidates.
Challenge ID	0003
Challenge Title	Multiple Aspects for Debate
Scenario	Debate on any topic
Description	When candidates are debating on a particular topic that is very vast like a country, there are innumerable aspects that they can focus on. They can focus on any one of the problems that the country might be facing at that time, on the progress of the country or on any other issue. It there are five aspects that need to be discussed, then that would be alright, but if all the aspects of the topic need to be discussed, then it becomes challenging to express the views on such a wide topic.
Solution	Two to three aspects of a particular topic should be assigned to candidates so that they can focus only on those.

12.2.5.2 Constraints

1. One or more AnyParty achieves understanding.
2. Understanding is achieved using one or more AnyDebate.

3. AnyParty follows one or more AnyRule.
4. One or more AnyRule influences AnyDebate.
5. AnyDebate is done on one or more AnyAspect.
6. AnyAspect results in one or more AnyConsequence.
7. AnyDebate has one or more AnyType.
8. AnyType names one or more AnyEntity.
9. AnyType names one or more AnyEvent.
10. AnyEntity/AnyEvent is on one or more AnyMedia.
11. AnyEntity/AnyEvent results in one or more AnyView.
12. AnyDebate shows one or more AnyMedia.

12.2.6 Solution

12.2.6.1 Pattern Structure and Participants

The solution here explains the concept of debate. Figure 12.1 shows the class diagram for the debate pattern.

12.2.6.2 Class Diagram Description

1. Understanding is the EBT of this pattern and it is done by one or more AnyParty (BO).
2. Understanding (EBT) is done using one or more AnyRule (BO) through AnyDebate (BO).
3. AnyParty (BO) obeys AnyRule (BO) and considers one or more AnyFactor (BO).
4. One or more AnyFactor (BO) affects AnyDebate (BO), which leads to AnyConsequence (BO).

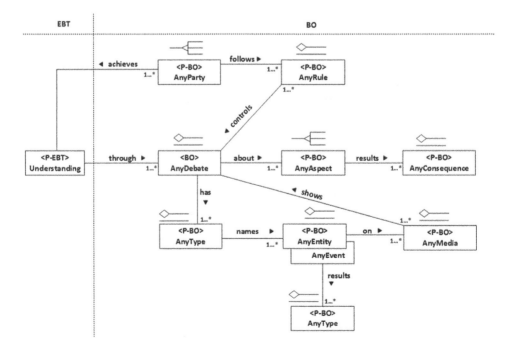

FIGURE 12.1 Class diagram for AnyDebate stable design pattern.

5. AnyDebate (BO) has AnyType (BO) about AnyEntity (BO), which results into one or more AnyView (BO).
6. One or more AnyEntity (BO) is within AnyEvent (BO) on AnyMedia (BO).

12.2.6.2.1 Classes

Understanding: The ultimate goal of the AnyDebate stable design pattern and the related BOs that form the pattern structure is to aid the understanding related to the topic of debate. This class represents that goal and captures the knowledge/understanding of the participants or the debating event as such, related to the concept under discussion.

12.2.6.2.2 Patterns

AnyDebate: This class represents a topic on which two people present their own views. This represents the debate and it is a central part of the software.

AnyParty: This class represents the legal entity that participates in the debate. The participants, as well as the mediator/conductor of the debate, are a type of AnyParty. The functions of these formal entities are satisfied by this class and are a generalization of any such users of the system.

AnyType: The debate can be of any type. This class is the common ground for the types of debates that can be and it serves the functions of determining the type of the debate and satisfying the related functions.

AnyEntity: This class represents all the topic(s) of interest, about which the debate is being conducted.

AnyEvent: This represents the event that hosts the debate on the topic of interest. AnyEvent is thus the facilitator of such debating undertaking.

AnyMedia: It also represents the medium on which the proceedings of the debate as well as the knowledge are generated out of it, and that betters the understanding of the entity under discussion.

AnyView: Captures the view presented by the participating party on the entity under discussion.

AnyRule: This class represents the rules that govern the process of AnyDebate that is being conducted.

AnyFactor: This class represents the factors that affect a debate.

AnyConsequence: This class represents the result of the debate after discussion between two or more people on a topic.

12.2.6.3 CRC Cards

Responsibility	Collaboration	
	Client	Server
Understanding(Understanding)(EBT)		
To give knowledge	AnyParty	enableDecision()
	AnyDebate	increaseKnowledge()
	AnyRule	makeAware()
Attributes: level, concept		
AnyDebate(AnyDebate)(BO)		
To engage in argument	Understanding	exploreTopic()
	AnyFactor	testAnalyticalSkill()
	AnyConsequence	precedesConclusion()
	AnyType	
Attributes: debateID, debateTime, debatePlace, topic, numberOfPeople, type		

Continued

	Collaboration	
Responsibility	**Client**	**Server**
AnyFactor(AnyFactor)(BO)		
To help in making decision	AnyDebate	influence()
	AnyParty	helpReasoning()
		causeAction()
Attributes: factorName, factorDescription, factorCategory		
AnyConsequence(AnyConsequence)(BO)		
To affect	AnyDebate	affect()
		happen()
		followAction()
Attributes: consequenceDescripton, reason, effect, consequenceLevel		
AnyView(AnyView)(BO)		
To reflect thinking	AnyEntity	reflectThinking()
		touchIssue()
		provideVision()
Attributes: viewKind, viewDescription, issueLevel		
AnyEvent(AnyRule)(BO)		
To happen	AnyEntity	occur()
	AnyMedia	start()
		end()
Attributes: eventId(), eventDescription, eventName, period, venue		

12.2.7 CONSEQUENCES

The consequences are related to the challenges and constraints for the system and more so with the challenges at large. The only aim to design a debate pattern is to derive a pattern that should be applicable in all the situations, whatever the domain of the context it is. The above-mentioned debate pattern fits all the domains, where any type of debate is performed on any topic. Since this pattern is applicable in all the domains, it remains stable in all the situations. The other objective is to make the pattern general by taking out only common things from the different domains. This helps in building a generic pattern. Since, the pattern can be also used in different applications, it is adaptable. The best part of debate design pattern is that the pattern is derived with stability in mind. It captures the enduring knowledge of business and its capabilities.

12.2.7.1 Flexibility

A good thing about the debate pattern is that it is very flexible and it can alter according to the topic of the debate. As the topic changes, the consequence of the debate can be altered easily.

12.2.7.2 Reusability

The debate pattern is a stable pattern and can be used in wide variety of applications. It can be reused in many different scenarios spread across many different fields.

12.2.8 APPLICABILITY WITH ILLUSTRATED EXAMPLES

Election debate illustrated in Figures 12.2 and 12.3.

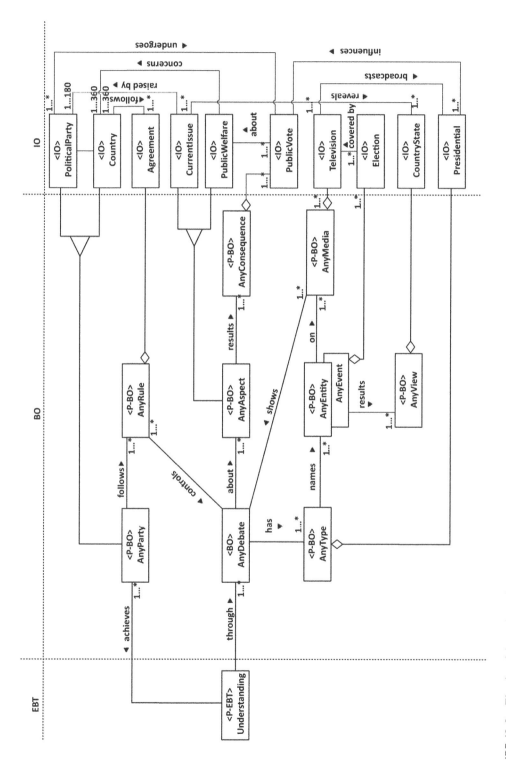

FIGURE 12.2 Election debate class diagram.

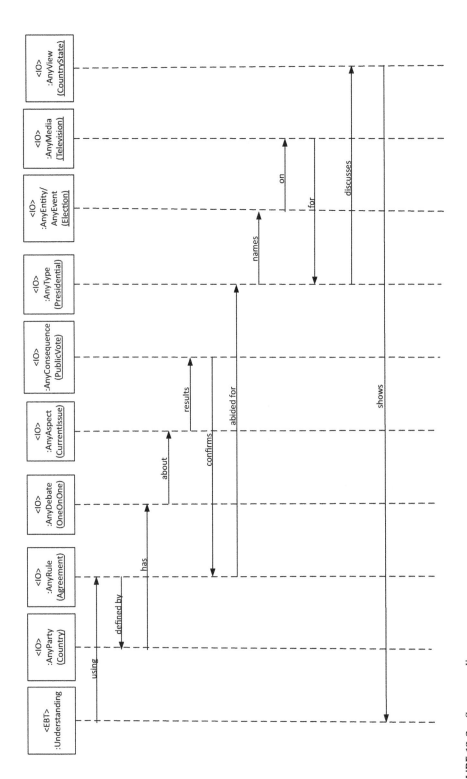

FIGURE 12.3   Sequence diagram.

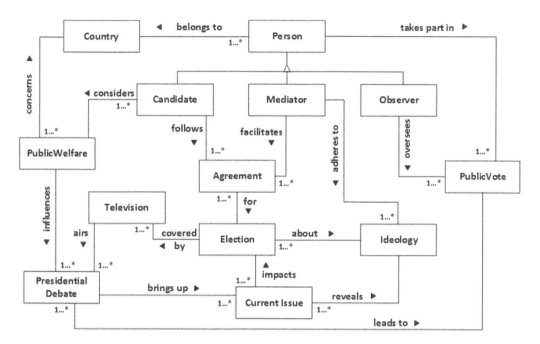

FIGURE 12.4   Traditional model class diagram for presidential debate.

12.2.9   Related Pattern and Measurability

12.2.9.1   Related Pattern

In order to explain the benefits of stable pattern, a comparison is made between the Meta model, which is a traditional model, and stability model in this section. Find printed below a Meta model class diagram of an application, presidential debate, which is designed using traditional model (Figure 12.4).

12.2.9.2   Measurability

12.2.9.2.1   Qualitative Measure

Traditional model is not stable as it is tightly coupled with IOs and does not allow the model to scale. IO is specific to domains and varies from application to application. If there is any requirement change, the model has to be redesigned in order to meet the latest changes. This makes the traditional model hard to be reused during the requirement changes. Thus, the traditional model involves high development and maintenance cost.

The stability model helps in achieving the stable patterns. These patterns are not designed to meet a specific application and are loosely coupled with the IOs. Thus, when there is a requirement change, only the IOs are changed, leaving the EBTs and BOs unchanged. This involves the reduced development and maintenance cost because most of the elements can be reused from the existing stable pattern and it leads to less amount of maintenance. Since the IOs are not tied up with the stable pattern, this allows the pattern to be easily scalable. The features of stability model such as stability, scalability, reusability, maintainability, and simplicity make it far better as compared to traditional model.

12.2.9.2.2   Quantitative Measure

The number of classes in the traditional model is always greater than the number of classes in the stability model. The number of operations is directly proportional to number of classes. So, the stability model will contain fewer numbers of operations than traditional model.

12.2.10 MODELING ISSUES

12.2.10.1 Abstraction

A model can be used for describing something and is considered as a layer of abstraction. It should show the associations between the entities, their behavior, roles, and responsibilities. The AnyDebate stable design pattern is abstracted in all the possible ways. However, certain issues that were encountered while we tried to remodel this pattern are listed below.

The problem at hand was to design a pattern that suites all the applications that involve any debate of any type. Then, we created following EBT candidates: convincing, discussion, confusion, consensus, proving, clarity, and agreement. Debate is neither convincing the candidates nor proving something to the opponents, and it is much more than just a discussion. The parties are not confused on the topic they are debating on, so, clarity and convincing need not be achieved through a debate. Clarity is an attribute of debate; it describes the quality of the debate. There is no agreement required between the parties involved in the debate. Thus, the abovementioned EBTs do not match with the AnyDebate stable design pattern. After careful examination of all the possible goals and factoring in subgoals, that lead to a significant ultimate goal, we realized that Understanding a concept is the ultimate goal that any debate may plan to satisfy. Goals such as discussion, agreement, and consensus became a part of Understanding and it encompasses these terms to lead to a much bigger objective. Therefore, Understanding turns out to be the correct EBT for this pattern.

12.2.10.2 No Dangling

Any model has certain common capabilities, which are to be carefully associated with each other. Their relationships are to be appropriately identified. Lack of attention in this issue will cause many problems like dangling. Dangling classes are those classes that lock the control of the program. This scenario is explained with an example as shown in Figure 12.5.

In the above example, state 4 is the class which cannot get the control back to the other states. So, state4 is called a "Dangling Class" and it may cause a lock situation to impede flow of execution at any point or may result in an indefinite hold. Such classes are disabled classes and cause a deadlock. This situation is sometimes termed as "Life Lock." These classes have no interfaces or classes to interact with and thus cannot do anything and halt the entire system from functioning.

To avoid this situation, any model should avoid dangling classes. The above example is corrected and is illustrated as shown in Figure 12.6.

The above diagram is shown to illustrate how a proper balance among the capabilities can reduce the dangling classes. This model is well balanced and thus ensures efficient functionality

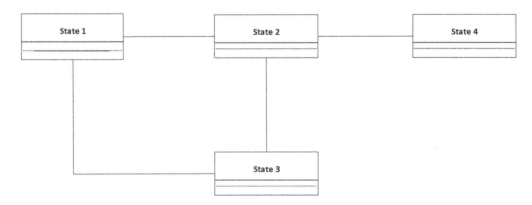

FIGURE 12.5 Example of dangling class.

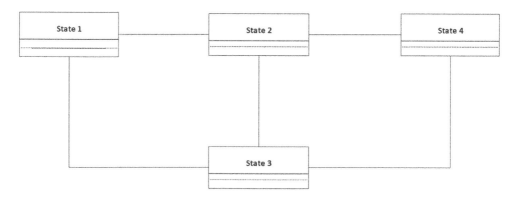

FIGURE 12.6 Example of corrected dangling class.

of the system. AnyDebate stable design pattern is modeled in such a way that it avoids the dangling classes.

12.2.11 DESIGN AND IMPLEMENTATION ISSUES

For the Understanding EBT, the necessities of any party or actor are carefully considered. When there is a topic of debate, it needs understanding. Thus, we came to the conclusion that AnyDebate is indeed not an EBT, but a BO. We also believe that the ultimate goal of the AnyDebate class is in reference with the Understanding as the ultimate goal. Hence, we can also say that any party or any actor debates on a topic for the Understanding.

In the case of implementation issues for our BOs, we also found that the cases of generalization was for the AnyParty, AnyEntity, AnyRule, AnyFactor, and AnyView classes. AnyDebate can be about AnyEntity and AnyEntity results in AnyView. AnyFactor affects AnyDebate.

12.2.11.1 Delegation

Inheritance is present in all the software patterns. The EBTs are supported by certain BOs like AnyParty. AnyParty is inherited by an organization, country, or human involved in the pattern. Inheritance cannot make the pattern reusable sometimes, as inheritance lets us stick to the roles specified in the application. To avoid this issue, the design pattern should ideally support delegation and thus give flexibility to the pattern. This is clearly illustrated as shown in Figure 12.7.

The relationship between the classes mentioned above is an "OR" relationship. That means the number of classes associated with "class1" is just one. As the design does not allow any flexibility to add more number of classes in future, and as it does not allow for "on-the-fly" generation of classes, it is considered as a static model and thus is not preferred here. Figure 12.8 explains how delegation differs from inheritance.

Delegation is a better way of design, than inheritance, as the relationship between the classes is "And" relationship. Also, delegation allows "on-the-fly" generation of classes and is considered as dynamic way. However, the important aspect to be identified here is, all the subclasses (class2, class3, class4) are tightly coupled with the Delegatee (class1). This restricts a certain level of flexibility of the user, as the user has to be tightly coupled with the design rules. These issues are addressed with aggregation which is discussed below.

Aggregation is considered as the best of the other when compared to the designs discussed above, as it gives complete flexibility for the user to add components to the containers (Figure 12.9). The design is open for any number of classes to be attached to it. There is no tight coupling between the owner's design and user's application. This makes the design extremely user-friendly and reusable.

Based on all the issues mentioned above, appropriate capabilities (BOs) are chosen to have aggregations with the application-specific objects (IO).

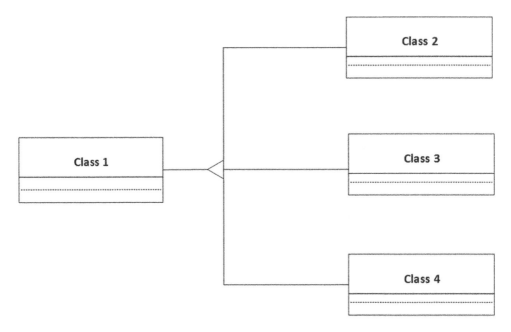

FIGURE 12.7 Example showing design of a static inheritance.

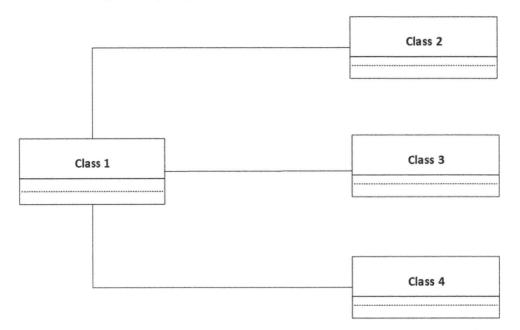

FIGURE 12.8 Example showing the design of delegation.

12.2.11.2 Common Interface

The AnyRegistry stable design pattern has several BOs that have important operations relevant to this pattern. Ideally, a design pattern supports an interface that contains the declarations of all the significant methods of the application. This way, it creates more levels of abstraction and provides more flexibility and reusability of the pattern. The applications that follow this pattern can implement the interface and avoid massive refactoring, when something changes.

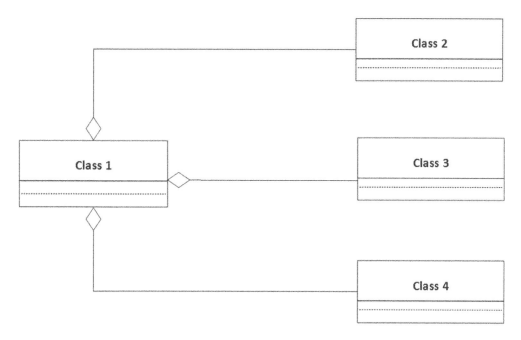

FIGURE 12.9   Example showing the design of aggregation.

12.2.12   TESTABILITY

If AnyDebate stable design pattern can be used, as it is without changing the core design and by only plugging IOs for infinite number of applications, then AnyDebate pattern can be said to be highly testable. This chapter describes an application scenario in the *Applicability* section, where IOs are hooked up to the core infrastructure, or the core pattern, through the BOs, thus realizing the application without having to rework the whole system. Similarly, more such applications can be built by simply changing the IOs.

Testing is focused at the core of the system, that is, the enduring part of the system is tested for proper functionality. The EBT and the BOs are tested rigorously to ensure that the system core is rock solid and remains stable over time. The test cases are designed at the time of documenting the use cases; this ensures that the system behaves as it is supposed to, fulfilling the required requirements. This can be seen in the use case under the *Applicability* section of this chapter. Focusing on the enduring core of the software application saves on a lot of redundant work, which not only ensures that the system is stable over time but also ensures that the required goals are indeed met, thereby making the system highly testable [10].

12.2.13   BUSINESS ISSUES

12.2.13.1   Business Rules

1. AnyParty or/and AnyActor discuss a topic.
2. AnyDebate needs Understanding on a topic.
3. AnyDebate should have certain constraints that restrict the debate topic.
4. AnyDebate needs to be identified for each domain.
5. AnyEntity specifies the topic on which the debate is done.
6. Debate can be done by any organization, country, human, separately, or together, depending on the application.

The pattern is stable, scalable, and adaptable as it can be applied in many domains.

12.2.13.2   Business Integration

AnyDebate stable design pattern is designed in such a way that it can easily be used in many business models. In order to connect this pattern to a particular domain, domain-specific IOs need to be connected to the BOs in the pattern. Thus, any business model can be easily integrated with the stable pattern.

12.3   TIPS AND HEURISTICS

1. AnyDebate generates Understanding.
2. The debate can be about AnyEntity.
3. The debate can be done by AnyParty, which involves human, organization, country, and political parties, or by AnyActor which involves human.
4. The class AnyRule is included in order to restrict the debate.
5. This pattern can be used for almost any type of debate.
6. When a debate is done, it can be for one or more entity.
7. EBT holds the thread of control that is the sequence should start from and end in the same EBT.
8. The flow in the sequence diagram should follow the snake pattern; this facilitates the easy understanding of the concept.
9. EBTs interact only with BOs.
10. BOs interact with EBTs and IOs.
11. IOs can interact only with BOs.
12. Number of interactions between the objects in the sequence diagram should be within the range of interaction in the class diagram.
13. Aggregation, inheritance, and delegation should be employed properly in the class diagram.
14. Aggregation is better than delegation and delegation is better than inheritance.
15. Test cases should be included for all BOs and EBTs in the use cases and should not be included in the IOs.
16. Application scenario should reflect the real-world problem and the case studies selected for supporting the pattern should be intelligent.
17. There should not be any direct connections between the actors or roles in the system.
18. BOs should be properly selected, so that it will be suitable to all the similar applications.
19. EBTs must represent the goals of the pattern.
20. Identifying the correct EBTs is very important.

12.4   SUMMARY

Described in this chapter is a stable design pattern, applicable in numerous contexts and application scenarios, whatever the domain of the situation may be. The above AnyDebate stable design pattern fits in all the domains, where any type of debate is done for any entity. Since this pattern is applicable in all the domains, it remains stable in all the situations and over time, even though the application scenario, topic of the debate, participants, venue, event, etc., and the whole surroundings of the system change with time. The other objective was to make the pattern general by abstracting out the common infrastructure involved in applications from different domains. This helped us in building a generic pattern that can be used in different applications, and that is adaptable to changing application context without much effort needed for the same. It is highly reusable and can form the basis of software that facilitates any debate on any topic in many application contexts.

12.5   REVIEW QUESTIONS

1. Explain what is debate.
2. What is the difference between a debate and an argument?

3. Describe a few contexts in which debate is used in day-to-day life.
4. Write two to three scenarios where AnyDebate SDP is applicable.
5. Discuss the functional requirements for AnyDebate SDP.
6. Discuss the nonfunctional requirements for AnyDebate SDP.
7. Write four challenges of using AnyDebate SDP.
8. Write a few constraints of AnyDebate SDP.
9. Draw class diagram for AnyDebate SDP.
10. Explain and justify the participants in AnyDebate SDP.
11. Write CRC cards for AnyDebate SDP.
12. What are the consequences AnyDebate SDP in terms of following:
    a. Flexibility
    b. Reliability
13. Explain the application of AnyDebate SDP in a scenario of Presidential Debate during a Presidential Election. Draw a class diagram for it. Write any one use case for this application and draw a sequence diagram for the same. Write test cases for the same.
14. Explain the application of AnyDebate SDP in the issue of patent applications. Draw a class diagram for it. Write any one use case for this application and draw a sequence diagram for the same. Write test cases for the same.
15. Draw traditional model for AnyDebate.
16. Compare traditional AnyDebate model with AnyDebate SDP. Which one of the two do you like more?
17. Compare both the models quantitatively in terms of number of classes, cost estimation, coupling among classes and constraints.
18. Compare both the models qualitatively.
19. Discuss the process of abstraction of AnyDebate SDP.
20. Describe the concept of no dangling.
21. Discuss the design and implementation issues of AnyDebate SDP.
22. Discuss the concept of inheritance and delegation with the help of a diagram. Specify when to use delegation and when to use inheritance.
23. Discuss the testability of AnyDebate SDP.
24. Write some business rules for AnyDebate SDP.
25. Discuss some of the known usages of AnyDebate SDP.
26. What did you learn from this chapter?
27. Write your findings, tips, and heuristics.
28. Explain the term debate?
29. What is the usage of the AnyDebate stable design pattern?
30. Can the term debate be used in any other context than what you thought of?
31. Can AnyDebate design pattern be used interchangeably with any other design patterns from this book? Explain.
32. What problem does the AnyDebate design pattern solve?
33. What are the challenges faced in implementing the AnyDebate design pattern (AnyRule)?
34. What are the challenges faced in implementing the AnyDebate design pattern (AnyType)?
35. What are the constraints faced in implementing the AnyDebate design pattern (AnyRule)?
36. What are the constraints faced in implementing the AnyDebate design pattern (AnyParty)?
37. What are the constraints faced in implementing the AnyDebate design pattern (AnyType)?
38. Discuss briefly the functional requirements of AnyDebate design pattern (AnyEntity)
39. Discuss briefly the nonfunctional requirements of AnyDebate design pattern (AnyEntity)
40. Discuss briefly the functional requirements of AnyDebate design pattern (AnyRule)

41. Discuss briefly the nonfunctional requirements of AnyDebate design pattern (AnyRule)
42. Discuss briefly the functional requirements of AnyDebate design pattern (AnyAspect)
43. Discuss briefly the nonfunctional requirements of AnyDebate design pattern (AnyAspect)
44. Explain AnyDebate pattern model (AnyParty) with the help of class diagram and CRC cards.
45. Explain AnyDebate pattern model (AnyRule) with the help of class diagram and CRC cards.
46. Explain AnyDebate pattern model (AnyType) with the help of class diagram and CRC cards.
47. What are the design and implementation issues for the given AnyDebate design pattern (AnyEntity)?
48. What are the design and implementation issues for the given AnyDebate design pattern (AnyMedia)?
49. What are the design and implementation issues for the given AnyDebate design pattern (AnyConsequence)?
50. Provide some patterns related to the AnyDebate design pattern (AnyType).
51. Provide some patterns related to the AnyDebate design pattern (AnyParty).
52. Provide some patterns related to the AnyDebate design pattern (AnyRule).
53. Explain usage of AnyDebate design pattern (AnyParty) with two examples other than the ones provided in this chapter.
54. Explain usage of AnyDebate design pattern (AnyRule) with two examples other than the ones provided in this chapter.
55. Explain usage of AnyDebate design pattern (AnyEntity) with two examples other than the ones provided in this chapter.
56. How does traditional model differ from the stability model? Explain using the AnyDebate design pattern model (AnyParty).
57. How does traditional model differ from the stability model? Explain using the AnyDebate design pattern model (AnyType).
58. How does traditional model differ from the stability model? Explain using the AnyDebate design pattern model (AnyMedia).
59. List some of the business issues encountered for the AnyDebate design pattern (AnyMedia).
60. Explain procedure for testing the AnyDebate design pattern.
61. Discuss some of the real-time usages of AnyDebate design pattern (AnyMedia).
62. Discuss some of the real-time usages of AnyDebate design pattern (AnyConsequence).
63. Discuss some of the real-time usages of AnyDebate design pattern (AnyRule).
64. What are the lessons learned by you from AnyDebate pattern.
65. List some of the domains in which AnyDebate design pattern (AnyParty) can be applied.
66. List some of the domains in which AnyDebate design pattern (AnyAspect) can be applied.
67. List some of the domains in which AnyDebate design pattern (AnyMedia) can be applied.
68. What is the tradeoff of using the AnyDebate pattern?
69. List some advantages of using AnyDebate design pattern (AnyMedia) in real applications.
70. List some advantages of using AnyDebate design pattern (AnyConsequence) in real applications.
71. List some advantages of using AnyDebate design pattern (AnyEntity) in real applications.
72. Can you think of any scenarios, where AnyDebate design pattern will fail? Explain each scenario briefly.
73. Describe how the developed AnyDebate design pattern would be stable over time.
74. List some of the testing patterns that can be used to test AnyDebate design pattern.
75. Can you think of any other goal which is not covered by AnyDebate design pattern?
76. Briefly explain how AnyDebate design pattern (AnyType) supports its objective.
77. Briefly explain how AnyDebate design pattern (AnyMedia) supports its objective.
78. Briefly explain how AnyDebate design pattern (AnyConsequence) supports its objective.

79. Examine the functional requirements of all the patterns involved – Are there any missing requirements? Discuss them.
80. Examine the nonfunctional requirements of AnyDebate design pattern – Are there any missing requirements? Discuss them.
81. Try to list few more business rules for the AnyDebate pattern.

12.6   EXERCISES

1. Describe a scenario of an application of AnyDebate in a school debate competition.
    a. Draw a class diagram of the application.
    b. Document a detailed and significant use case.
    c. Create a sequence diagram of the created use case of b.
2. Describe a scenario of an application of AnyDebate in the process of defending PhD Theses.
    a. Draw a class diagram of the application.
    b. Document a detailed and significant use case.
    c. Create a sequence diagram of the created use case of b.

12.7   PROJECTS

1. You are a part of a team that is organizing a reality show and many contestants are participating in it. Based on their auditions, you have to select a few of the contestants. There are three other judges. All four of you debate over each contestant before making a final decision of selecting or not. Draw a class diagram for this application by using AnyDebate SDP. Write three use cases and draw sequence diagrams for each of them.
2. Use the AnyDebate pattern and other patterns in this book to model a system that can be used in courtroom settings. Two parties debate in order to come to an understanding on the subject at hand, (e.g., a guilty innocent verdict). During court proceedings evidence and discussion are presented to the courtroom. The system needs to log these proceedings, so they can be looked up later. Draw the class diagram, develop CRC cards, and draw the sequence diagram of the system.
3. Online debating formats are meant to allow debaters to engage in short debates using instant messaging or video conferencing software. These debates will have one debater representing the "affirmative" and another debater presenting the "negative." While online debates are not meant to replace face-to-face communication, they are a way to bridge geographic distances and to allow for discussion between people, who might not otherwise have a chance to meet. Draw the class diagram, develop CRC cards, and draw the sequence diagram of the system.
4. Cross-Examination (Policy) Debate: Like other forms of debate, Cross-Examination Debate focuses on the core elements of a controversial issue. Cross-Examination Debate develops important skills, such as critical thinking, listening, argument construction, research, note-taking, and advocacy skills. Cross-Examination Debate is distinct from other formats (with the exception of two-team Parliamentary Debate) in is use of a two-person team, along with an emphasis on cross-examination between constructive speeches. While specific practices vary, Cross-Examination Debate typically rewards intensive use of evidence and is more focused on content than delivery. Draw the class diagram, develop CRC cards, and draw the sequence diagram of the system.
5. Public Forum Debate offers students a unique opportunity to develop on-their-feet critical thinking skills by situating them in contexts not unlike the U.S. political talk shows. Public Forum debaters must anticipate numerous contingencies in planning their cases and must

learn to adapt to rapidly changing circumstances as discussions progress. Public Forum's open-ended cross-examination format encourages the development of unique rhetorical strategies. Public Forum debates should be transparent to lay audiences, while providing students with real-world public speaking skills, through the discussion of contentious ideas. Draw the class diagram, develop CRC cards, and draw the sequence diagram of the system.

REFERENCES

1. Debate (Online), available at Wikipedia, http://en.wikipedia.org/wiki/debate, accessed May 28, 2010.
2. United States Presidential Election Debates (Online), available at Wikipedia, http://en.wikipedia.org/wiki/U.S._presidential_election_debates, accessed May 28, 2010.
3. M. Fayad and A. Altman. Introduction to software stability, *Communications of the ACM*, 44(9), 2001, 95–98.
4. M. Fayad. Accomplishing software stability, *Communications of the ACM*, 45(1), 2002, 111–115.
5. M.E. Fayad. How to deal with software stability, *Communications of the ACM*, 45(4), 2002, 109–112.
6. H. Hamza. A foundation for building stable analysis patterns, Master thesis, University of Nebraska-Lincoln, Lincoln, NE, 2002.
7. M.E. Fayad, H.A. Sanchez, and S.K. Singh. Knowledge maps—Fundamentally modular approach to software architecture, design, development and deployment, in *19th Annual Conference on Software Engineering and Data Engineering (SEDE)*, San Francisco, June 2010.
8. Understanding (Online), available at Merriam-Webster Dictionary, http://www.merriam-webster.com/netdict/understanding, accessed May 28, 2010.
9. Debate (Online), available at Bouvier's Law Dictionary, http://www.constitution.org/bouv/bouvier_d.htm, accessed May 28, 2010.
10. M.E. Fayad, H.A. Sanchez, S.G.K. Hegde, A. Basia, and A. Vakil. *Software Patterns, Knowledge Maps, and Domain Analysis*, Auerbach Publications, Boca Raton, FL, Taylor & Francis Catalog #: K16540, December 2014, ISBN-13: 978-1466571433.

Section III

SDPs' Mid-Size Documentation Template

This part consists of five chapters.

Each chapter presents the detailed documentation pattern that consists of the following sections: (1) the name and type of the pattern, (2) context, (3) the functional and nonfunctional requirements, (4) solution, and (5) applicability which has five scenarios and one detailed application.

1. AnyHypothesis Stable Design Pattern (Chapter 13)
2. AnyConstraint Stable Design Pattern (Chapter 14)
3. AnyInteraction Stable Design Pattern (Chapter 15)
4. AnyTranslation Stable Design Pattern (Chapter 16)
5. AnyArchitecture Stable Design Pattern (Chapter 17)

13 AnyHypothesis Stable Design Pattern

Truth in science can be defined as the working hypothesis best suited to open the way to the next better one.

Konrad Lorenz [1]

Hypothesis is a well-known concept in the domain of science, logic, legal studies, and arts that deals with the issue of proposition or a set of propositions. They also provide explanations for the existence of some specific group of phenomena: asserted as temporary conjecture to lead investigations into any subject or be accepted as most probable in the presence of established facts. The main idea of this chapter is to develop a stable pattern on hypothesis that can be easily applied to all possible domains. One of the most significant advantages of developing AnyHypothesis pattern is its robustness and stability that eventually leads to a solid solution where a developer need not work repeatedly every time he or she needs hypothesis in an applicable domain. This particular pattern may be applied to diverse areas of study like law, science, technology, arts, logic, and other areas where one needs to argue forcefully by providing hypothesis and possible solutions to them. The AnyHypothesis pattern could be considered as an important solution that deals with inner core to find a valid proof for Supposition; incidentally, the main aim of creating hypothesis is to provide a series of Supposition.

13.1 INTRODUCTION

Hypothesis is an important English word that has its roots in Ancient Greece. It is an explanation for a set of observations. It could also be a mere assumption or guess. It is also a tentative, testable answer to a scientific question and leads to one or more predictions that can be tested by experimenting. Hypothesis always needs considerable evaluation and assessment before it is considered as truthful. In other words, hypothesis is largely a general term whose meaning could be used in almost all domains of daily life.

Hence, it is almost difficult to imagine an area of interest without Supposition, and it can never be proved without Supposition and it is an offshoot of Supposition. Therefore, the common goal of AnyHypothesis is Supposition. The Hypothesis stable design pattern consists of a set of one EBT and innumerable IOs. Many hypothesis scenarios are included in this chapter.

13.2 PATTERN DOCUMENTATION

13.2.1 NAME: ANYHYPOTHESIS STABLE DESIGN PATTERN

The stable design pattern of AnyHypothesis describes hypothesis as an explanation of some phenomena based on propositions and in what manner AnyActor or AnyParty defines AnyCriteria to create AnyHypothesis. With all stable design patterns, the BO needs to support an everlasting and ultimate purpose, which is known as an EBT. The ultimate goal of AnyHypothesis is Supposition. Any is used in front of hypothesis to prove that this document supports the design for any hypothesis which exists. Regardless of the context (AnyContext), it determines the hypothesis (AnyHypothesis) which is about an entity (AnyEntity) or event (AnyEvent). Additionally, the word "any" is attached to the beginning of all BOs because a BO is externally stable while internally unstable. This means that regardless of which hypothesis is describing any event or entity, this model will remain stable throughout all eternity in terms of describing the purpose of Supposition.

13.2.2 CONTEXT

Several scenarios describe AnyHypothesis stable design pattern. In this section, we will describe three scenarios that support the stable design pattern proposed.

13.2.2.1 Understanding Current Age of Obesity

It is a known fact that both obesity and calorie consumption have drastically increased in the last couple of decades. People (AnyParty) and even pets (AnyActor) can suffer from a number of health-related issues like mobility difficulty, diabetes, and other symptoms related to obesity (AnyCriteria). In the past, three scientists suggested that obesity is related to calorie scarcity and stress due to mild cold may no longer exist because people can eat whenever they like and stay at a temperature as comfortable as desired during cold months (AnyHypothesis) [2]. By reviewing past history and the behavior of food habits (AnyIndicator), the scientists were able to conclude that the changes in our food storage system (AnyEvent) and introduction of food manufacturing companies deciding what foods we consume (AnyEntity) was the leading cause to modern obesity. The scientists did not conduct a study to test the theory (AnyLevel), but one could be forwarded, by comparing a control group existing now in current conditions (AnyContext) with a study group that survived in similar calorie availability and temperature environment millions of years ago.

13.2.2.2 Finding Cure for Cancer

Scientists (AnyParty) usually study mice for their projects (AnyActor), specifically brain cell behavior and functionality research, and such initiatives are always supported by different cancer research foundations (AnyEntity). In the course of one of the research experiments, it was discovered that tumor brain cells behave similarly to stem cells (AnyEvent) [3]. Based on the brain cells behavior (AnyContext), it was also observed that mouse brain tumors are partially composed of cells that initially act as tumor cells, but eventually lose this ability (AnyIndicator). Scientists believe that if they continue to study these cells, new treatments can be discovered to cure tumor brain cells (AnyHypothesis). This study is still in the initial stages of research on mice (AnyLevel). As long as scientists continue to see the brain cell behavior amongst brain cell tumors (AnyCriteria), this insight can be used to prepare a new cancer treatment.

13.2.2.3 Creating a New Branch in Physics

Albert Einstein (AnyParty) published the theory of General Relativity in 1907 [4]. Based on a series of mathematical formulas and calculations (AnyCriteria), and the study of gravity (AnyActor), Einstein proposed that space-time is curved (AnyHypothesis). This theory is supported by incorporating it into other future theories (AnyLevel). By comparing the proposed mathematical values with those of results of tests (AnyIndicator), CERN (AnyEntity) was able to determine the validity of the theory. By building on the currently available knowledge in physics (AnyContext), further discovery of mathematical results can be comprehended to understand the physical world (AnyEvent).

13.2.3 PROBLEM

The main problem of the Hypothesis stable design pattern is to design a valid solution to describe and support the system. In order to do so, a SSM for Supposition must be developed. This design introduces the BOs which are externally stable and internally unstable requirements. As we created a stability model, this design will be stable eternally, since the requirements for Supposition will never change. The classes that vary from situation to situation are known as the IOs, and they will be added to the design later, when applying the Supposition SSM to concrete applications.

13.2.3.1 Functional Requirements

1. *AnyActor*: Any creature or system involved with a given hypothesis to help substantiate a theory. AnyActor participates in various activities that includes hypothesis. AnyActor has a *name, id and category*. It has operations such as *interact(), look(), plan(), and conclude()*.

2. *AnyParty*: Any organization involved in developing and studying plausible ideas to conclude a Supposition. It represents the hypothesis maker. It models all the parties that deal with *AnyHypothesis*. Party can be a person, researcher, scientist, philosopher, or a business entrepreneur. AnyParty starts using AnySource like hypothesis, deduction, logic, and induction to find the *AnyHypothesis* for any entity. Any Party has a name, a type and an authority level. In addition, any party plays a specific role. AnyParty also initiates a hypothesis process, explores possibilities, deduces, inducts, debates, evaluates, comprehends, and collects evidence and achieves Supposition and eventual truth.

3. *AnyCriteria*: The basis used to support a Supposition. A Supposition will need a solid ground to be supported by a reason. AnyCriteria represents a class that is used to suppose something. It has attributes such as *standard, principle, proof, and judgment*. It has operations such as *theorize() test(), confirm(), verify() and assess(), and suppose()*.

4. *AnyHypothesis*: The precise statement used to uphold a given Supposition. To state something as a hypothesis, one may need to design a method to state problems, questions, and assumptions. When a hypothesis is formed, it has to be subjected to a series of test to test its validity.

5. *AnyLevel*: A basis to indicate at which point of the process a given theory is supported by applicable tests and results. A detailed social research study may involve subjecting a theory to a series of research methodologies, and to deduct /induct to find a series of results. A researcher may find out results and accompanying findings sometime during a certain phase of the experiment.

6. *AnyIndicator*: Events or ideas, which support a given theory and its relevance to a specific situation. AnyIndicator supports AnyHypothesis under AnyContext to provide a suitable environment to find out the truth.

7. *AnyContext*: The overall environment in which the Supposition is held in and contributes to the existence of the idea. It is the BO of the pattern and it has operations like updateRelevantEvent(), determineImportance(), and valueField().

8. *AnyEvent*: A specific situation, which encourage a discovery or propagation of a specific Supposition. This class represents any viewable occurrence. An event is something that happens in the near future. It has attributes such as *name, occasion, type, and outcome*. It has operations such as *appear(), happen(), and arrange()*. It has also many attributes like field timePeriod location, time and era.

9. *AnyEntity*: An organization or a person used to help produce, develop, and support or disprove the validity of a proposed Supposition. It has many entities like entityID, field investment, and Source organization leader. It also has many operations like addEmployee(), updateTestProcess(), and appendHypothesis(). AnyEntity can be an organization or an individual who make hypothesis and debate with strong Supposition. Any Entity has a unique Id, a title, and a type. It contains Supposition, argues for it, and always involves party with it.

13.2.3.2 Nonfunctional Requirements

1. *Knowledge*: A new set of knowledge will be created by developing new Suppositions. Although Suppositions look like true, further analysis and discovery should made

regarding a specific phenomenon or event that leads a general belief in the hypothesis. In addition, by generating new ideas, fresh and critical knowledge will be obtained as theories are either proved or disproved. Regardless of the outcome, after conducting studies in newer areas and domains, different ways of viewing different scenarios are created and probed. In fact, hypotheses will promote knowledge in every possible aspect of life.

2. *Measurable*: In order to ensure that a given Supposition is reproduced multiple times, it must be measurable and quantifiable. If not, then a Supposition can never be supported by repetitive events supporting the idea. Therefore, Suppositions should be measurable and they can be divided into different areas of specific calculation, which ensures that events supporting theories can be reproducible anytime and it can be determined whether similar results were encountered as expected.

3. *Progress*: Suppositions involve hypotheses that propose different ideas, which many not always be obvious or perceptible. Therefore, by studying related events that help produce outcomes that support a given Supposition, always provide visible progress in the field under study. By creating, studying, and developing hypotheses, humankind is pushing forward toward greater understanding of any theoretical topic. By achieving a better view of a specific field, other ideas and past theories can be further supported or disproved, which helps the momentum in the specific field of study. Since, all hypotheses contain this characteristic of fresh ideas, studying them in detail helps in developing the field further.

13.2.4 SOLUTION

Class Diagram Description as shown in Figure 13.1.

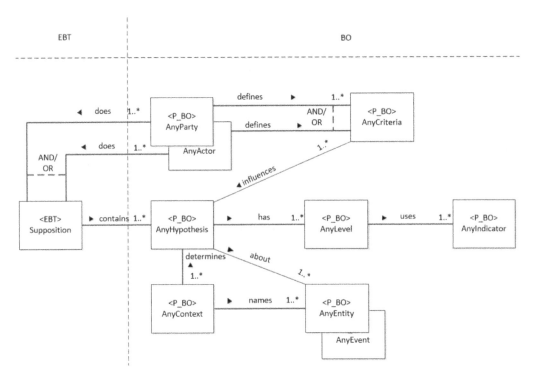

FIGURE 13.1   Stable design pattern of AnyHypothesis.

This diagram shows the EBT of Supposition. The diagram first begins in the EBT stage of the stability model and moves toward the BOs, which make up the design. AnyParty, AnyActor, or both define AnyCriteria, which influences AnyHypothesis used to support the Supposition. AnyHypothesis has AnyLevel, which uses AnyIndicator. Additionally, AnyContext names the AnyEntity or AnyEvent that the AnyHypothesis is referring.

13.2.5 DIFFERENT APPLICATIONS OR APPLICABILITY (AS SHOWN IN TABLE 13.1)

Case Study 1: Understanding Current Age of Obesity as shown in Table 13.2.
  Scenario: Understanding Current Age of Obesity

TABLE 13.1
Five Applications of AnyHypothesis Stable Design Pattern

EBT	BOs	Understanding Current Age of Obesity	Finding Cure for Cancer	Creating a New Branch in Physics	Allocating Enough Memory for Process	Hypothesis of Earthquakes
Supposition	AnyActor	Pets that have trouble to walk	Mice	Gravity	Operating system process	The land
	AnyParty	People suffering with health-related diseases	Scientists	Expert physicists	User running software application	Humans living on the land
	AnyCriteria	Health-related issues caused by obesity	Brain tumors have stem cell foundation	Mathematical computation	Operating system state	Richter scale
	AnyHypothesis	Obesity linked to those who keep overnourished and warm	Stem cells can help cure brain tumors	Space-time is curved	Process will allocate only enough memory to allow other processes to run	Earthquake occurring when seismic instrument intercepts ground waves
	AnyLevel	Number of tests conducted	Beginning stages of devising treatment option	Accepted into later theories	Number of total applications running	Location distance from epicenter
	AnyIndicator	Past history and study of food	Results in preliminary tests	Comparison of proposed values and results of tests	Process able to get memory as expected	Sudden oscillations in readings
	AnyEntity	Food manufacturing company	Company supporting testing	CERN	Developer program	Seismology department
	AnyEvent	Changing the storage of food	Discovering tumor cell behavior	Deeper mathematical investigation	Running new process	Damaged homes
	AnyContext	Abundance of food and lack of sleep	Behavior of brain cells	Current state of physics	All processes running on system	History of damage

TABLE 13.2

Determine Test Candidate

ID		Determine Test Candidate	
Title		1	
Actors		**Roles**	
AnyParty		Company, Person	
Class	**Type**	**Attribute**	**Operation**
Supposition	EBT	idea	supportHypothesis()
		field	purposeOfIdea()
		hypothesisID	defendTheory()
		suppositionID	
		support	
AnyParty	BO	hypothesis	increaseInvestment()
		organizationID	addEmployee()
		numEmployee	developTest()
		field	
		support	
AnyActor	BO	roleID	followCommand()
		supportValue	listenToPurpose()
		location	supportIdea()
		importance	
		interactionCount	
AnyHypothesis	BO	listAuthor	listReason()
		year	giveSupport()
		area	offerExplanation()
		listTest listResult	
AnyCriteria	BO	value	applyFieldMeasurment()
		fieldName	determineValue()
		measurmentToolvariance	occurrenceRatio()
		relevance	
AnyLevel	BO	listIndicator	increaseCredibility()
		levelID	addIndicator()
		support	appendLevelWeight()
		listField	
		accurateMeasure	
AnyIndicator	BO	value	addIndicator()
		weight	updateWeight()
		listSupport	increaseSupportFactor()
		occurrencePattern	
		occurrenceDifficulty	
AnyContext	BO	field	updateRelevantEvent()
		timePeriod	determineImportance()
		location	valueField()
		time	
		era	
AnyEntity	BO	entityID	addEmployee()
		field	updateTestProcess()
		investmentSource	appendHypothesis()
		organization	
		leader	

(Continued)

TABLE 13.2 (*Continued*)
Determine Test Candidate

Class	Type	Attribute	Operation
AnyEvent	BO	length date listContributor difficulty relevance	measure() increaseOccurrence() changeIndicator()
HealthIssue	IO	name cause side-effects symptom history	addSymptom() findCause() determineLevel()
ObesePatient	IO	weight sex height BMI lifeExpectancy	changeDiet() decreaseTemp() intakeCalorie()
HeatAndIntakeMonitor	IO	time pace location temperature method	addReader() installThermometer() increaseSpeed()
TemperatureHabit	IO	currentTemp averageTemp tooLow delta numDays	decreaseTemp() increaseInterval() adjustAverage()
Test	IO	ID length numSections numSubjects numControl	addSubject() adjustQuestion() appendInstruction()
History	IO	averageTemp numYears sinceDifference expectedResults source	addSource() updateData() modifyComparision()

Context: As the name implies, AnyHypothesis stable design pattern can be applied to different situations and contexts. Here, we will consider the topic of obesity trends in America and a possible cause for the recent health epidemic. Only recently have obesity and its related negative health issues been seen, causing such havoc in societies leading to increased death toll. Due to this quick and drastic change within humans, scientists produce theories to explain the sudden shift. One important hypothesis is the absence of cold winter and availability of an abundant quantum of calories. By comparing past living status of people during hunter-gatherer days, and the status of modern society, the scientists have been able to develop a theory to explain the sudden increase of obesity in a large population of people. This is a mere hypothesis, as further testing would be needed to prove the validity and correctness (Figure 13.1).

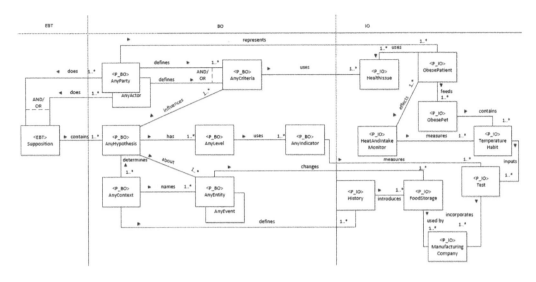

FIGURE 13.2   Software stability pattern for understanding obesity.

13.2.5.1   Class Diagram (as Shown in Figure 13.2)
Use Case Description:

1. AnyParty being the ObesePatient needs to determine whether AnyCriteria of HealthIssues is met or not.
2. AnyHypothesis is tested by getting results from HeatAndIntakeMonitor, which records the current temperature and the number of calories consumed by the patient.
3. AnyContext is determined with TemperatureHabit, which measures the average temperature over a period.
4. AnyLevel is measured by the scope and validity of the test used.
5. The test helps to indicate, if the ObesePatient is qualified to fall under the patient category. If yes, additional tests are used to obtain daily statistics to compare to history.
6. AnyEvent is defined by changes in history and the way humans now live differently.
7. Results to compare AnyCriteria with test results and history are obtained.
8. If TemperatureHabit determines average temperature is always warm during the winter on a patient that qualifies for the examination, then AnyHypothesis determining the reason for the increase is obesity can be applied.
9. Must repeat process to find as many qualifying patients as possible.

Sequence Diagram as shown in Figure 13.3.

13.3   SUMMARY

This document discusses the software stability model of Hypothesis. Since, this can be applied to all hypotheses that exist in the world the SSM was used. By introducing both EBTs and BOs along with IOs, a model can be developed, which is applicable to all situations. This ensures that it may be reused a number of times, regardless of the concrete situation in which it is applied. When developing the model, we must keep in mind both things that are specific to the problem at hand, along with the underlying concepts, which are applied to all situations regardless of the specific content. By doing this extra step, the software stability model is introduced, and the produced model can be applied to all applicable scenarios.

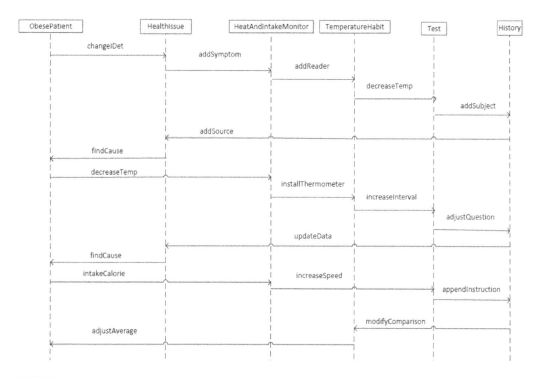

FIGURE 13.3 Sequence diagram for understanding obesity.

13.4 REVIEW QUESTIONS

1. Define the AnyHypothesis stability design pattern.
2. Hypothesis is an important term in English. List five instances when we use it on a periodical basis.
3. List a few application domains for the AnyHypothesis design pattern.
4. Hypothesis always leads to Supposition. What might be the intermediate steps before supposing something?
5. AnyHypothesis as a stable design pattern is domain and application specific—say True or False.
6. List any two problems, while formulating the AnyHypothesis design pattern.
7. List all the participants in the AnyHypothesis design pattern.
8. Illustrate the class diagram for the AnyHypothesis design pattern.
9. List three areas where you can apply AnyHypothesis patterns.
10. List some functional requirements other than those given in this chapter.
11. List some nonfunctional requirements that are covered in this chapter.
12. List two main advantages of using the AnyHypothesis stable design pattern.
13. List two applications of the AnyHypothesis design pattern.
14. Briefly describe why Supposition is the most appropriate EBT for use in the AnyHypothesis design pattern.
15. Why do you think that the AnyHypothesis design pattern would be stable and robust over time?
16. Can you list some of the research issues related to the AnyHypothesis design pattern.
17. List one scenario that will not be covered by AnyHypothesis design pattern.
18. List three use cases for this pattern. List those that are not covered in this chapter.
19. List some of the related design patterns used in creating the AnyHypothesis stable design pattern.

13.5 EXERCISES

1. Hypothesis versus assumption stable patterns
   a. What are the differences between stable hypothesis pattern and assumption pattern?
   b. Create the assumption stable pattern.
   c. Name and describe three contexts for each of them.
   d. Show how to apply the two patterns including with one context of your choice and including a use case, the application of each pattern with your select context.
2. Hypothesis versus proposition stable patterns
   a. What are the differences between stable hypothesis pattern and proposition pattern?
   b. Create the proposition stable pattern.
   c. Name and describe three contexts for each of them.
   d. Show how to apply the two patterns including with one context of your choice and including a use case, the application of each pattern with your select context.
3. Create five applications of AnyHypothesis patterns in a table as shown in Table 23.1 (Chapter 23).
4. Hypothesis versus approximation stable patterns
   a. What are the differences between stable hypothesis pattern and approximation pattern?
   b. Create the approximation stable pattern.
   c. Name and describe three contexts for each of them.
   d. Show how to apply the two patterns including with one context of your choice and including a use case, the application of each pattern with your select context.
5. Hypothesis versus justification stable patterns
   a. What are the differences between stable hypothesis pattern and justification pattern?
   b. Create the justification stable pattern.
   c. Name and describe three contexts for each of them.
   d. Show how to apply the two patterns including with one context of your choice and including a use case, the application of each pattern with your select context.
6. Create two applications of two contexts of AnyHypothesis stable design pattern and provide the following for each one of the applications:
   a. Context
   b. Application diagram using the Hypothesis stable design patterns.
   c. Detailed use case.
   d. A sequence diagram for the application

13.6 PROJECTS

Use Hypothesis in one of the following and create:

a. Hypothesis Testing
b. Alternative Hypothesis
Three contexts for each one of them
Stable architecture pattern for each of them
An application of each one of them

REFERENCES

1. Konrad Zacharias Lorenz (November 7, 1903–February 27, 1989), last modified on June 24, 2017, https://en.wikipedia.org/wiki/Konrad_Lorenz
2. R.J. Cronise, D.A. Sinclair, and A.A. Bremer. Th-e "Metabolic winter" hypothesis: A cause of the current epidemics of obesity and cardiometabolic disease, *Metabolic Syndrome and Related Disorders*, 12(7), 2014, 355–361.
3. P. Dirks. Brain tumor stem cells: The cancer stem cell hypothesis writ large. *Molecular Oncology*, 4(5), 2010, 420–430.
4. C.M. Will. August 1, 2010. Relativity, Grolier Multimedia Encyclopedia, accessed May 02, 2015.

14 AnyConstraint Stable Design Pattern

The more constraints one imposes, the more one frees one's self. And the arbitrariness of the constraint serves only to obtain precision of execution.

Igor Stravinsky [1]

A constraint is a well-recognized English word that plays the most important component of a physical, financial, or social restriction. Derived from an intransitive verb, constrained, it is something that uses threat or force to prevent, restrict, and dictate the action or thoughts. The main goal of this chapter is to design and create a stable pattern on the term *constraint* that is applicable to use in all possible domains that relate to constraint. Three positive issues that govern patterns on constraints are its robustness, stability, and extensibility. By developing this pattern, developers need not work repeatedly every time they need constraint in applicable domain. This approach not only saves time but also helps them realize the intrinsic goal of making the pattern robust and testable over time. AnyConstraint stable patterns could be widely used in diverse areas of applications such as law, science, arts, business, personal development, academics, and any other areas of interest that relate to constraint. This will help developers argue with conviction to provide lasting solutions to different domains. Hence, the developed pattern on constraint also helps developers find out the core of the problem as a proof for restriction that is the essential themes of the pattern [2–7].

14.1 INTRODUCTION

Restriction is an important word in English language that has its roots in the verbal form, *constrained*. In a common usage, constrain is the state of being restricted or confined within a particular perimeter. Edith Wharton described the word as embarrassed, reserved, or reticent. Restriction is also meant to be a compulsion, force, or restraint. Inhibition also mimics the word constraint, while another usage indicates repression or control of natural feelings or impulses. Innumerable social constraints may force a person to keep continued silence, while serious constraints may force to drop its plan to go big with its marketing and advertisement plans.

On a personal note, a person may not feel any constraint in speaking about an issue of controversy. In a corporate domain, a certain type of constraint may inhibit the potential creativity that many employees feel in their line of jobs. In academics, many students may decide to attempt only certain numbers of questions in an exam because of the severe time constraint. In business, the best way for a CEO to control ballooning expenses constraint is to cut and manage unnecessary expenses. A young learner may face severe constraints while learning basic math, while a college student may find that severe constraints are hampering while learning advanced computation.

In the domain of software programming, a complex algorithm may need additional improvements because of severe space constraints. All these examples demonstrate that the word constraint has a broader dimension and unlimited usage. It is almost impossible to find an aspect of life where we do not use the term *constraint*, while AnyRestrint stable pattern can never be designed and formulated without learning more about limitation that is closely related to the English word *constraint*. Every stable pattern has a common goal and the ultimate goal of restraint is limitation. An important offshoot of AnyConstraint is limitation.

14.2   PATTERN DOCUMENTATION

14.2.1   NAME: CONSTRAINT STABLE DESIGN PATTERN

AnyParty, who wants to impose limitation through AnyConstraint on AnyEntity, can describe stable design pattern of a constraint. So, the EBT for constraint is "Limitation." If any party wants to impose any limitation, then they will define any criteria. Limitation has many types, which name the entity. AnyCriteria or AnyRule influences AnyConstraint.

14.2.2   CONTEXT

Constraint stable design pattern can be applied to any domain, where limitation is applied or forced. Constraint limits an entity to produce a certain output that can be positive or negative depending on the situation. It restricts certain activity by imposing noticeable limitations. Constraints are usually negative forces that can stop someone to take some actions. Constraints also impose severe restrictions by the way of unforeseen circumstances and scenarios. Some of the scenarios, where constrains are applied are highlighted below:

1. *Political Constraint*: National and international politics may force severe constraints in the way a country is governed or the manner in which international policies are created. A political party has a definite plan and a working plan of action to govern a country efficiently. However, they will need to follow some rigid rules and detect constraints that may affect the way in which the country is governed. Any governing or policy decision taken by any party must follow certain rules and they should consider all types of constraints.
2. *Business Constraint*: A business firm will face innumerable constraints and hindrances while creating a workable plan for increasing turnover. In other words, business constraint always plays a significant role in the growth of a business. Some of the business constraints may include physical limitation, fiscal limitation, time limitation, resource limitation, policy limitation, and most of these constraints play an important role and they invariably affect the business goals.
3. *Learning Constraint*: Not all university students are created equal especially in their ability to score better grades in the examinations. Some are very good and they may never face any type of constraints. On the contrary, some of them may face innumerable constraints that might impede their ability to score better marks. Some of these constraints are learning disabilities, absence of memory power, writing disability, dyslexia, and deep fear of taking examinations. Some students may even face financial constraints that may affect their ability to pay fees on time.

14.2.3   PROBLEM

To design a pattern for any given scenario is a truly complicated and time-consuming task. There are many such scenarios, which force some constraints in the process of their implementation. The basic solution is to design a stable design pattern that can be applied to similar kind of scenarios that define constrains. Constraint stable design pattern is reusable and easily implementable for any kind of problems that are similar.

14.2.3.1   Functional Requirements

1. *Limitation*: A set of rules or restriction is applied on the system. Limitation may defer the action taken for the execution of the problem. Limitation also causes a delay in the execution. The law enforcement limits the activity of any party or actor performing any task or

application. Limitation is one of the most important aspects in politics, business, and in other fields. It actually shapes the process according to the limitations.

2. *AnyConstraint*: Constraint is some factor, which affects the execution of an application or an entity that exists within the system. Sometimes, a constraint is a positive factor for the task. For example, constraint in politics forces the politician to consider it and take any decisions only for the welfare of citizens. Constraints in business force the management to improve and fine-tune the product to sell more and enhance turnover.

3. *AnyParty*: It is a party or an organization that imposes a limitation through constraints. There should be at least one party or an actor to impose the limitation. In politics, there are one or more political party, which usually impose limitation on any of the entity. In business, the board of directors or the company is the party.

4. *AnyCriteria*: Any party defines a set of rules, which is to be followed for taking any decision. Criteria define the boundary of an activity. It can be rules that are defined by an organization or an actor. A criterion is must for the activity to be completed in a desired way. If the criteria are not defined, then the execution will result in many problems and the outcome will not be as desired as we wish it to be.

5. *AnyType*: Limitation has different types. It can be a time limit, resource limit, or limitation in legal matters. There are many entities, which apply the time limit. For example, the budget should be created and requested to the Congress within the time limit set for the activity. In business, some products must be sold and disposed off within the time limit. If the resource is limited, then the design of the product should also be limited according to the availability of resources.

6. *AnyEntity*: It is a part of a system or an element of the system that performs given tasks for the execution of the problem given. For example, creating a budget in politics is an entity of politics.

14.2.3.2 Nonfunctional Requirements

1. *Understandable*: Constraints that is defined by any party should be easy to understand and comprehend. There should not be any complexity in studying the problem. Following up on the constraint while planning is very important, so understanding of the constraint becomes easy. Constraints could have many ramifications in a given area of study. Some constraints need an in-depth study and debate to understand its complexities.

2. *Applicable*: Constraint defined for a given problem domain should be applicable to the problem. If the constraint is not applicable, then it is meaningless. Constraint should be positive to improve the problem and not to make the problem worst, so that the execution is delayed. Constraint should be applicable to all areas of study so that it becomes easy to use for any numbers of applications.

3. *Flexible*: Sometimes, the requirement of the problem changes, so the constraint should also be changeable and it should not be very rigid. If the constraint is very rigid, then it might create a problem in the management process of the problem. Flexible constraint allows the application to be executed without any delay. An inflexible constraint is the one that makes understanding of limitation difficult and challenging. For example, if a given problem in the study of algorithm is inflexible to comprehend, then a developer may find it difficult to create and make a pattern around the problem.

4. *Necessity*: A constraint could be a necessity while creating a stable pattern. It is almost an important necessity, as all constraints are complex to understand and study. It is a big necessity to find the core of the problem to find the most appropriate EBT for the pattern. A constraint becomes a deep necessity when a developer is trying to use the right application of the meaning of the problem related to a specific constraint.

14.2.4 Solution

Class Diagram Description as shown in Figure 14.1:

- AnyParty impose Limitation on any entity
- Limitation is imposed through AnyConstraint
- Limitation has one or more type
- AnyType names one or more Entity
- AnyEntity is within AnyConstraint
- AnyParty define one or more AnyCriteria

14.2.4.1 More Applications (as Shown in Table 14.1)

1. *Political constraint*: Every political activity has many constraints. A political party (*AnyParty*) faces many problems while ruling a country. The president analyzes, evacuates, and prepares a budget (*AnyEntity*) and submits it to the Congress for approval. He may need to submit the budget within the given time limit (*AnyConstraint*). The party defines the rule for the governance (*AnyRule*).

2. *Business Constraint*: A company (*AnyParty*) may need to launch new products (*AnyEntity*) to increase the turnover and revenue. The manager (*AnyActor*) proposes a new product and the company will check the corporate rules (*AnyRule*) and approves the product if everything is in order. Then, the marketing and promotion (*AnyConstraint*) of the product should be carried out in an effective manner to increase the sales of the product.

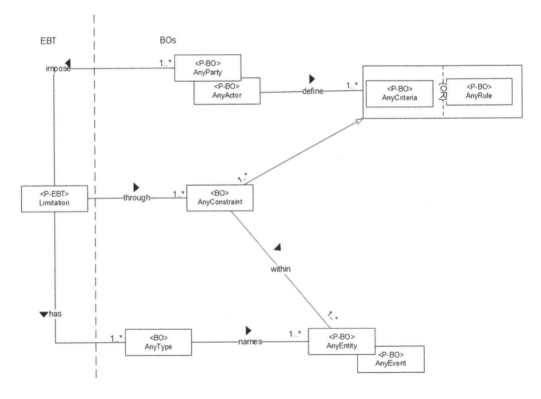

FIGURE 14.1 Constraint stable design pattern class diagram.

TABLE 14.1

AnyConstraint Stable Design Pattern Applications

EBT	BO	Political IO	Business IO	Financial IO	Education IO	Environmental IO
Limitation	AnyParty/ AnyActor	Political party, president	Company management, manager	Country, citizen	Student	Human, animals
	AnyCriteria/ AnyRule	Political criteria, rules	Business criteria	Financial rules	Complete education	Environmental cycle
	AnyConstraint	Time limit	Marketing	Recession	Less finance	Natural calamity
	AnyEntity/ AnyEvent	Budget	Advertisement	Development process	Books, materials	Weather, air, water
	AnyType	Resource, time	Finance, man power	Growth, earning	Personal, professional	Long term, short term

3. *Financial Constraint*: Every country (*AnyParty*) faces big financial constraints on the road to development. The overall economy of the country is affected by an occasional bout of recession (*AnyConstraint*) and this will impede and restrain the development process (*AnyEntity*).

4. *Education Constraint*: There are many constraints in the education and academic sphere. A student (*AnyActor*) might have family problems, which stop him from availing quality education. Meager finance (*AnyConstraint*) may also affect the ability of a student to get an admission in a college. Books and materials (*AnyEntity*) cannot be purchased without the required financial assistance. Any personal problem (*AnyType*) may also cause educational constraints.

5. *Environmental Constraint*: Environment affects humans (*AnyActor*) and all living animals and plants. The ensuing effect can be long term or short term (*AnyType*). The environment follows an environmental cycle (*AnyCriteria*), but natural disasters such as earthquake, typhoons, cyclones, flood, and drought (*AnyConstraint*) may occur periodically and it will eventually affects and disturb the life cycle.

14.2.4.2 Applicability

14.2.4.2.1 Case Study #1: Political Constraint

14.2.4.2.1.1 Scenario: Political Constraint in Budget Process AnyConstraint stable design pattern can be applied to many domains, where limitation is applied. There are many limitations in politics. Any political party (AnyParty) imposes limitations through constraints (AnyConstraint) by following some criteria (AnyCriteria). Constraint exists within any of the political matters (AnyEntity). There are many types of limitations (AnyType) such as financial limitation, developmental limitation, management limitation, and others. Let us say, the U.S. President (AnyActor) creates a budget (AnyEntity) and submits it to the Congress (AnyParty). The budgetary process must follow the set and predefined rules (AnyCriteria) and it should be submitted within the time limit (AnyConstraint).

14.2.4.2.1.2 Class Diagram As shown in Figure 14.2.

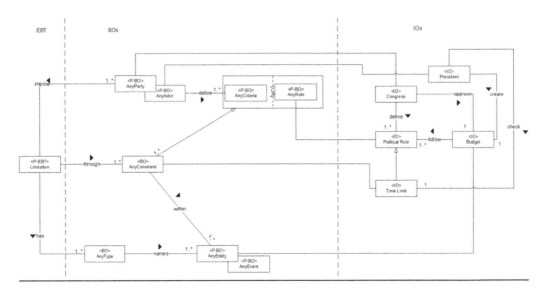

FIGURE 14.2   Stable class diagram for political constraint scenario.

14.2.4.2.2   Use Case

Use Case ID: UC#1

Use Case Name: Pass federal budget

Actors			Roles	
1. Political Party			1. Political Party (Congress)	
2. Person			2. Person (President)	

Class	Type	Attributes	Operations
Limitation	EBT	limitationID, limitationTitle, limitationDescription	applyRestriction() defineType()
AnyParty	BO	partyName, partyID, partyLocation, partyMemberID, partyMemberName	imposeLimitation() defineCriteria()
AnyActor	BO	actorID, actorName, actorDesignation,	imposeLimitation() followCriteria()
AnyCriteria	BO	criteriaID, criteriaName	listCriteria()
AnyConstraint	BO	constraintID, constraintName	listConstraint() appyLimitationOnEntity()
AnyEntity	BO	entityID, entityName	listEntity()
Person (President)	IO	presidentName	createBudget() checkTimeLimit() requestApproval()
PoliticalParty(Congress)	IO	partyName, partyLocation, partyMembers	approveBudget() defineRule()
Budget	IO	budgetType, budgetDetail	followRule()
Rule	IO	ruleID, ruleDetail	listRule()
Time Limit	IO	time, date	showTime()

Continued

Use Case Description

1. Limitation applies restriction on an entity (AnyEntity)
   TC: Who restricts an Entity?
2. Limitation define different types (AnyType)
   TC: What are all the types of limitations?
3. AnyParty imposes limitation on entity
   TC: Who imposes limitation?
4. AnyParty define criteria which can be applied to political criteria
   TC: Who defines criteria?
5. AnyActor checks the criteria defined by AnyParty
   TC: Who checks the criteria?
6. AnyConstraint lists the constraint applicable to politics
   TC: Who lists the constraints?
7. The President creates the budget and he also checks the time limit
   TC: Who creates the budget?
8. Political party defines rule
   TC: What does political party define?
9. The President request a budget approval from the Congress
   TC: What does President request?
10. The Congress approves the budget
    TC: Who approves the budget request?
11. Budget follows the rules
    TC: What rules are followed by budget?

14.2.4.2.3 Sequence Diagram

Sequence Diagram as shown in Figure 14.3.

14.3 SUMMARY

Software stability model is more stable and reusable, when compared with the traditional model. When the system is modeled in a traditional way, the system becomes very specific. However, when

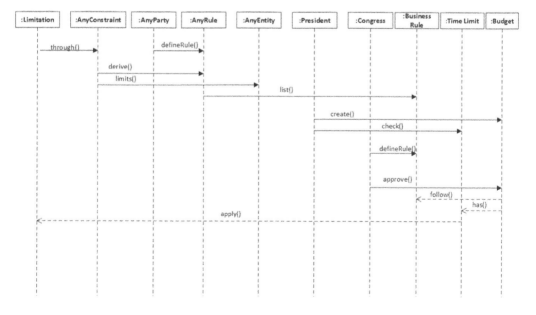

FIGURE 14.3 Sequence diagram of political constraint scenario.

the stability model is used, it can be applied to many similar domains. Using the software stability model is cost efficient and it makes the system stable, versatile, repeatable, and reusable. In this chapter, we have demonstrated the use of constraint as a topic for creating a stable model. With its wide applications under different context, it is almost challenging to find the core of the given problem. In other words, a big challenge may crop up while a developer is trying to find the EBT for the term constraint.

14.4   REVIEW QUESTIONS

1. Define the AnyConstraint stability design pattern.
2. Why do you consider AnyConstraint as a challenging pattern?
3. List a few application domains for this design pattern.
4. List four challenges that one might face while formulating the AnyConstraint design pattern.
5. AnyConstraint as a stable design pattern is domain and application specific—say True or False.
6. List any two disadvantages while formulating the AnyConstraint design pattern.
7. List all participants in the AnyConstraint design pattern.
8. Illustrate the class diagram for the AnyConstraint design pattern.
9. List one tradeoff or giveaway while using this pattern.
10. Document the CRC cards for the AnyConstraint BO.
11. The AnyConstraint pattern involves the use of other patterns too. Briefly explain.
12. List two main merits of using the AnyConstraint stable design pattern.
13. List four applications of the AnyConstraint design pattern.
14. Briefly describe why limitation is the most appropriate EBT for use in the AnyConstraint design pattern.
15. Why do you think that the AnyConstraint design pattern would be stable and robust over time?
16. List some of the research issues related to the AnyConstraint design pattern.
17. Why the AnyConstraint design pattern provides extensibility and robustness over a period?
18. List two modeling challenges that can crop up, when modeling the stable design pattern for AnyConstraint.
19. List one scenario that is not covered by this pattern.
20. List four test cases to test the participants of the AnyConstraint design pattern.
21. List some related design patterns used in formulating the AnyConstraint stable design pattern.

14.5   EXERCISES

1. Constraint versus pressure stable patterns.
   a. What the differences between stable constraint pattern and pressure pattern?
   b. Create the pressure stable pattern.
   c. Name and describe three contexts for each of them.
   d. Show how to apply the two patterns including one context of your choice and including a use case, the application of each pattern with your select context.
2. Constraint versus motivation stable patterns.
   a. What the differences between stable constraint pattern and motivation pattern?
   b. Create the motivation stable pattern.
   c. Name and describe three contexts for each of them.
   d. Show how to apply the two patterns including one context of your choice and including a use case, the application of each pattern with your select context.

3. Create five applications of AnyConstraint patterns in a table as shown in Table 23.1 (Chapter 23).
4. Constraint versus necessity stable patterns.
   a. What the differences between stable constraint pattern and necessity Pattern?
   b. Create the necessity stable pattern.
   c. Name and describe three contexts for each of them.
   d. Show how to apply the two patterns including one context of your choice and including a use case, the application of each pattern with your select context.
5. Constraint versus repression stable patterns.
   a. What the differences between stable constraint pattern and repression pattern?
   b. Create the repression stable pattern.
   c. Name and describe three contexts for each of them.
   d. Show how to apply the two patterns including one context of your choice and including a use case, the application of each pattern with your select context.
6. Create two applications of two contexts of AnyConstraint stable design pattern and provide the following for each one of the applications:
   a. Context.
   b. Application diagram using the constraint stable design patterns.
   c. Detailed use case.
   d. A sequence diagram for the application.

14.6 PROJECTS

Use constraint in one of the following and create:

a. Constraint Satisfaction
b. Constrained optimization
c. Biological constraints
Three contexts for each one of them
Stable architecture pattern for each of them
An application of each one of them

REFERENCES

1. Igor Fyodorovich Stravinsky (June 17 [O.S. June 5], 1882–April 6, 1971), last modified on June 24, 2017, https://en.wikipedia.org/wiki/Igor_Stravinsky
2. W. Sun and Y.X. Yua. *Optimization Theory and Methods: Nonlinear Programming*, Springer, p. 541, 2010, ISBN 978-1441937650.
3. J.J. Leader. *Numerical Analysis and Scientific Computation*, Addison-Wesley, Boston, MA, 2004, ISBN 0-201-73499-0.
4. R. Dechter. *Constraint Processing*, Morgan Kaufmann, Burlington, MA, 2003, ISBN: 1-55860-890-7.
5. K. Apt. *Principles of Constraint Programming*, Cambridge University Press, Cambridge, UK, 2003, ISBN 0-521-82583-0.
6. E. Freuder and A. Mackworth, eds. *Constraint-Based Reasoning*, MIT Press, Burlington, MA, 1994.
7. F. Thom and S. Abdennadher. *Essentials of Constraint Programming*, Springer, Berlin/Heidelberg, Germany, 2003, ISBN 3-540-67623-6.

15 AnyInteraction Stable Design Pattern

For good ideas and true innovation, you need human interaction, conflict, argument, debate.

Margaret Heffernan [1]

Interaction is a relationship word. An action occurs, as two or more entities exert an effect on each other. The concept of interaction is easy to understand when we learn more about the basic idea of two-way effect. Interconnectivity is a close word to interaction and it involves interactions on many micro levels. Interaction has many meanings under different contexts. Interaction involves a reciprocal action or effect or influence or it may also be understood as reciprocation. Interaction is a well-known word in many spheres of life although it has a wider usage in science domain especially in the field of physics. It is also well recognized in personal or professional communication.

The main objective of this chapter is to create a stable pattern on the English word, *interaction*. It is assumed that this pattern will be applicable for extended use in all possible domains related to interaction. This pattern is unique because of its three-way advantage to a user: robustness, stability, and extensibility. Developers need work repeatedly every time when they need interaction as a pattern in different domains. With this pattern, developers may save time, energy, and effort needed to create it, while realizing the core issues related to the intrinsic goal of making the patter testable over time also becomes very easy. The AnyInteraction pattern could be used in a number of scenarios and diverse areas of application like science, personal relationship, sociology, legal matters, business, and academics. Hence, developers may provide lasting solutions to different applications of interaction. Finding out the core of the problem also becomes easier; incidentally, it will also provide an ingenious way to understand the word interaction that is the "Essential Business Object" (EBT) for the pattern.

15.1 INTRODUCTION

In modern English, interaction is an extremely important word. In its common usage, interaction is a close action that occurs when two or more entities apply an effect on each other. An interaction is a result of two or more individual effects that result in some ensured result. An interaction could be extremely personal as in the case of a married couple, who interact continuously to maintain a harmony in the relationship. An interaction could also be the action between a teacher and student that results in a mutually productive association; incidentally, the student will get the maximum benefit here. In sports, an interaction will occur when both the coach and the player discuss ways to improve sporting performance.

In science domain concerning physics, interaction is interpreted in many different ways. One such area is the effect that one type of particle has on another. An example is the study of atoms. In a mathematical expression, it specifies the nature and strength of this close association. In the domain of business, all team members of a team should interact with each other to ensure the success of a project. In theater arts like drama, an interaction between different characters will ensure better performance on the stage. Likewise, the term *interaction* can occur in many domains and in diverse contexts. Although, this word is applicable for any domains, its usage may differ slightly depending on the domain in which it occurs.

It is almost difficult to search for an area in our lives where we fail to use this term. AnyInteraction stable pattern is a unique platform that manages any domain with utmost flexibility and ease. However, a developer must ensure that the term *interaction* is understood in a proper manner without confusing the term's vastness in term of usage and application. Every pattern is closely associated to a common goal and eventual goal of the term *interaction* is in fact, interaction.

15.2   PATTERN DOCUMENTATION

15.2.1   PATTERN NAME: INTERACTION (STABLE ANALYSIS PATTERN)

This chapter mainly focuses on the stable analysis pattern for interaction. This chapter also wraps all the possible scenarios for the EBT "Interaction." AnyParty or Actor can involve in an interaction process. This interaction can be carried out in a convenient manner by using an apt mechanism, so that every party can participate in the interaction. AnyParty, who is intended to participate in interaction, sets any protocol. These protocols have the capability to influence the mechanism by which interaction is carried out. The event in which the interaction is done makes use of the mechanism. Interaction should be done within context making sure it is not carried away form the intended context. The context of the interaction determines the type of interaction, such as vocal based, video based or even gesture based. The context of the interaction also aims at the AnyData that is used for an interaction process. AnyMedia may be used for recording these data.

15.2.2   CONTEXT

Interaction stable analysis pattern can be applied across various domains that are holding different contexts. These interactions are held within time duration and they specify a protocol for the interaction itself. These protocols influence the mechanism of interaction.

1. *Classroom Interaction*: A classroom is the place, where a professor (AnyParty) plans to deliver his/her lecture (AnyEvent). This lecture has duration, which can last for 3 hours (AnyDuration) based on the subject (AnyContext) and professor's teaching (AnyMechanism) style. She/he makes use of the PPT and Projector (AnyMedia) to explain the key points (AnyData) of the concept. A student is expected to attend the class and note down the notes (AnyProtocol) that is delivered by the lecturer. The lecture can completely remain a vocal interaction (AnyType) or the professor may even ask students to watch the prerecorded subject videos (AnyMedia) using which they can revise the concepts.
2. *Project Meeting*: A project meeting (AnyEvent) is a common feature of all projects. The manager (AnyParty) of the project schedules a meeting to discuss the project status (AnyContext) by sending a mail (AnyMedia) and by specifying the location and the time (AnyDuration) of the meeting. The team members (AnyParty) of the project should attend the meeting (AnyProtocol) with all the required project material (AnyData). The manager addresses (AnyType) all the team members and tracks the sprint (AnyMechanism) status by looking at the burn down charts.
3. *Gaming*: Gaming is the one activity where a player (AnyParty) involves in playing games (AnyEvent). The player makes use of the joystick (AnyMedia) for playing. This can be a gesture-based (AnyType) game in the case of Xbox video games. The gamer will need to follow certain set of rules (AnyProtocol) to play (AnyContext) the game. The game will remain active for certain time (AnyDuration), within which the user is expected to show his or her skill (AnyMechanism) and collect all the points (AnyData) to complete the game as a winner.

15.2.3 PROBLEM

The main problem that a developer comes across while creating a stable interaction pattern that can be applied to various contexts, is finding the actual EBT. Once this theme is found out, we can build any number of applications, where there is a scope of interaction. The ultimate goal is to make the pattern reusable and extensible for any type of scenarios, where Interaction is applicable as AnyParty can participate during interaction. Developing such a pattern requires basic requirements that can be characterized into functional and nonfunctional requirements. These requirements for the "Interaction Stable Analysis Pattern" are discussed below:

15.2.3.1 Functional Requirements

1. *AnyParty*: AnyParty is the legal user that participates during the interaction process. For an instance, a party can be a lecturer in case of classroom interaction. A party holds or looks at certain protocols for the interaction. A party uses a mechanism for interacting. For example, a professor uses teaching as a mechanism in case of classroom interaction. A round of interaction always requires two AnyParty entities.

2. *AnyActor*: AnyActor represents the person who involves in an interaction. An actor even has certain set of protocols that is needed to be followed in the interaction process. An actor plays a main role in interaction. AnyActor sets AnyProtocol to involve in a round of interaction.

3. *AnyProtocol*: Protocols are a set of rules that are designed by the party, who participates in interaction. These protocols are taken care by the other parties who are involved in interaction. AnyProtocol is the official procedure or a system of rules that play in any occasion. For an instance, a student is expected to take down the notes in a round of classroom interaction AnyProtocol is influenced by AnyMechanism.

4. *AnyMechanism*: AnyMechanism is the standard procedure that is followed during the interaction process. AnyParty involves in interaction by following the mechanism prescribed for interaction. A mechanism can be influenced by protocols defined by the party and can change the entire course of action. AnyMechanism uses AnyEntity to influence AnyProtocol.

5. *AnyDuration*: AnyDuration represents the time interval in which the interaction is conducted. This duration can be varying depending on the mechanism that is chosen for interaction. For example, in the case of a project meeting, the duration of the meeting can last very long. In contrast, the classroom interaction can last long for a period of 3 hours. Therefore, for every interaction, the mechanism will have a different duration.

6. *AnyContext*: The context determines the situation where the interaction will occur. It restricts the interaction to a specified set of boundaries, so that it may not be deviated. For example, in case of classroom interaction, the context of interaction will be the subject the professor teaches.

7. *AnyType*: AnyType specifies the kind or the manner in which the interaction is held. There can be many types of interaction such as vocal, gesture, touch based, and others. For example, in the case of classroom interaction, a professor may heavily depend on the vocal type of interaction with the students.

8. *AnyEvent*: AnyEvent refers to the entities for which the interaction is taking place. It specifies the importance of interaction. For example, the project meeting itself is an event in the context of a business. An event may vary from scenario to scenario and within many contexts.

9. *AnyMedia*: AnyMedia is used to record any kind of data that is used in the interaction process. It can also help the party with interaction. Media can be of any type such as Internet, newspaper, projector, joystick, and others.

10. *AnyData*: AnyData refers to the content on which the interaction is carried out. These data can be textual in case of reports or it can be in a digital format in case of videos. A party makes use of the data to interact. The Data are recorded by the media.

15.2.3.2 Nonfunctional Requirements

1. *Informative*: Information is a quality factor, which determines the usefulness of the interaction. It determines how fruitful the interaction is. AnyInteraction process can be informative as long as it sticks to its context. For example, a classroom interaction is very informative than an open interaction as it sticks strictly to the subject. Interaction becomes finer and more meaningful when the quality of the content under interaction provides enough statistics and data.

2. *Understanding*: Understanding is one of the prime quality factors for interaction. Every interaction can last for its intended duration only, if it is understandable by two parties that are participating in the interaction. Interaction process is a two-way traffic; when one actor fails to understand the other party, the entire process becomes useless. Both or more parties should understand what other parties are debating.

3. *Relevant*: The interaction process must be always relevant to the entity that it operates. The relevant nonfunctional requirement specifies the bonding between the interaction and the context on which it is carried out. To avoid any deviation from the context, the interaction should be more relevant.

4. *Useful*: It is one of the quality factors for interaction. To get the best out of interaction, it should be useful for all parties who interact with each other. Usefulness determines the extent to which the interaction can be utilized. The degree of usefulness depends on the topic of interaction, its relevance, and applicability.

5. *Expressive*: For an interaction to be successful, the entire process should be expressive. It is a person's responsibility to keep the interaction expressive and attractive. When an actor who involves in a round interaction is expressive and positive in the process, interaction becomes more powerful and hard-hitting.

6. *Helpful*: For a round of interaction to be successful, it should be helpful to all parties who interact in the process. Helpfulness is gauged by factors like subject matter's relevance, usefulness, its practicality, benefits, and its application to today's life. Helpful quality factor always focuses on the future aspects of the interaction. It refers to the gaining factors from the process of interaction.

7. *Synergy*: There should be a perfect synergy between different participants of interaction process. Synergy always sets the future tone for harmony. Synergy is typified by participants' helping attitude and concern for others. Harsh and bitter interaction always leads to bitterness, anger, frustration, quarrels, and future animosity. Interactions could be synergetic when a participant respects and honors others who participate in the process of interaction.

8. *Cooperation*: When there is enough cooperation, a round of interaction becomes extremely successful. Cooperation in an interaction process could be in the form of guidance, help, or even assistance. Cooperation is possible only when all participants in the interaction know their role and responsibility.

15.2.4 SOLUTION

A round of interaction should produce some definite result that must lead to practical solutions. A teacher, who interacts successfully with his students, will provide many solutions to the latter especially in the subject matter under study. Without the availability of solutions, we can never call interaction as successful and productive. Productive interactions always provide successful and effective solutions.

Class Diagram Description as shown in Figure 15.1: One or more parties participate in interaction

1. Multiple protocols can be set by AnyParty
2. Interaction is done through AnyMechanism

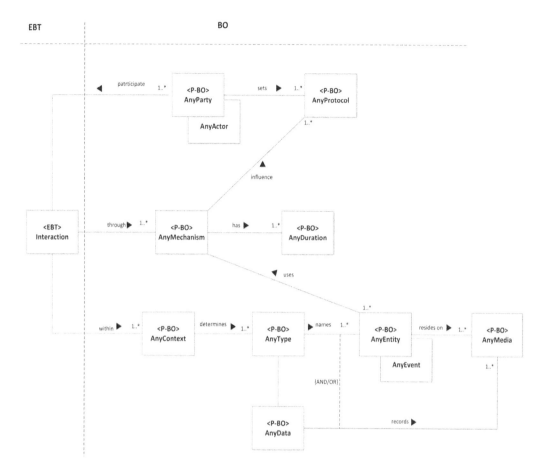

FIGURE 15.1 Interaction stable analysis pattern class diagram. <P-EBT> = Pattern EBT, <P-BO> = Related Pattern BO, BO = Business Object, EBT = Enduring Business Theme.

3. AnyMechanism has a specified duration
4. AnyProtocol can influence AnyMechanism
5. Interaction should be within AnyContext
6. AnyContext determines AnyType of interaction
7. AnyType of interaction names AnyEntity
8. AnyEntity resides on AnyMedia
9. AnyEvent resides on one or more Media
10. AnyMedia records AnyData
11. One or more Entity uses AnyMechanism

15.2.5 APPLICABILITY

Case Study #1: Interaction through Social Network as shown in Table 15.1.

Interaction stable analysis pattern can be utilized in many applications. Let us consider an example application of social network, where interaction stable analysis pattern can be applied.

An interaction is very common on Facebook (AnyParty). A user (AnyParty) needs to login (AnyProtocol) to enter Facebook and to enable chatting and messaging (AnyMechanism). These (AnyContext) can be done for any time (AnyDuration) by using chat option (AnyType). The Messaging system produces text (AnyData), which is recorded in the ChatBox (AnyMedia).

TABLE 15.1

Chat on Facebook

Use Case Id #: UC-1

Use Case Name: Chat on Facebook

Actor			Roles	
1. AnyParty			User, Facebook	

Class	Type	Attributes	Operations
Interaction	EBT	type	enableCommunication()
		context	
AnyParty	BO	partyName	doesInteraction()
		type	defineProtocol()
		size	
AnyProtocol	BO	type	makesCondition()
		name	influenceMechanism()
		validity	
AnyMechanism	BO	name	determinesDuration()
		type	generateInteraction()
		efectiveness	
AnyDuration	BO	interval	notify()
		frequency	
AnyContext	BO	scope	validateDomain()
		type	
AnyType	BO	name	revealType()
		usage	
Any Entity	BO	name	useMechanism()
		description	
AnyData	BO	type	showType()
		length	
AnyMedia	BO	name	recordData()
		id	
		type	
User	IO	name	login()
		purpose	performInteraction()
		address	
Facebook	IO	size	provideAuthentication()
		url	
Login	IO	userName	restrictsLogin()
		password	
Chatting	IO	type	specifyMechanism()
		duration	
Messaging	IO	type	sendMessages()
		content	
Time	IO	interval	obtainInterval()
		frequency	
Text	IO	length	displayData()
		type	
ChatBox	IO	size	captureText()
		capacity	

(Continued)

TABLE 15.1 *(Continued)*

Chat on Facebook

Use Case Description and Test Cases:

1. Interaction is done by AnyParty
   TC: What causes interaction?
   Who interacts with whom?
2. Interaction should be within the context
   TC: Who specifics the context?
   What is the scope of the context?
3. AnyParty sets the Protocol for Interaction
   TC: What kind of protocols can be applied?
   Who should follow the protocol?
4. Interaction is done through a mechanism
   TC: What kind of mechanism is involved?
   How many steps are involved in mechanism?
   How long will the mechanism work?
5. AnyMechanism specifies AnyDuration
   TC: Is there any constraint on duration?
6. AnyContext determines the Type of Interaction
   TC: Is there any criteria for Type?
7. AnyData is recorded by AnyMedia
   TC: What type of media can be used?
   What kind of data is recorded?
   Is there any default data?
8. User make use of Facebook for interaction
   TC: Does everyone use Facebook?
   Is Facebook the only way for interaction?
9. Facebook has authentication
   TC: What kind of authentication does Facebook provide?
   Is authentication required every time?
10. User should provide his/her credentials to login
    TC: What do credentials include?
    Is there any restriction on credential format?
11. Login enables chatting
    TC: What kind of chatting does it provide?
12. Chatting can be used for messaging
    TC: What messages can be sent?
    Who receives messages?
13. Messaging produces text
    TC: What is the format of the text?
    What length of text does it produce?
14. Text is recorded by ChatBox
    TC: What is the capacity of ChatBox?
    Is there any regulation on content?

15.2.5.1 Class Diagram

Class diagram as shown in Figure 15.2.

15.2.5.2 Sequence Diagram

Sequence diagram as shown in Figure 15.3.

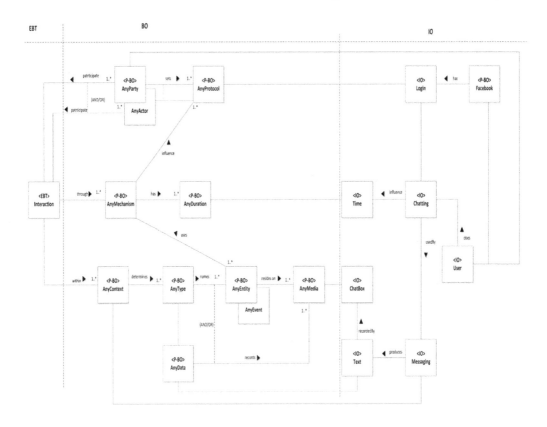

FIGURE 15.2 Class diagram for Case Study #1. <P-EBT> = Pattern EBT, <P-BO> = Related Pattern BO, BO = Business Object, EBT = Enduring Business Theme.

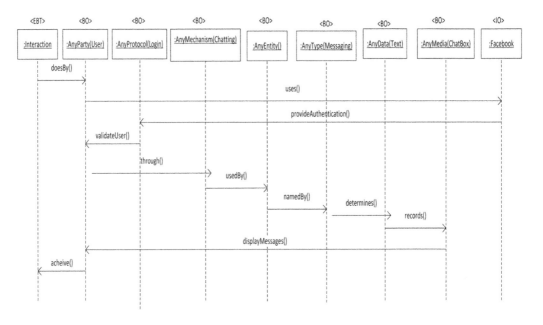

FIGURE 15.3 Sequence diagram for Case Study #1.

15.3 SUMMARY

The term interaction is widely used as a process for communication between two entities that eventually result in an effect. Hence, the term interaction is too general and wide enough which makes difficult for a developer to find the most appropriate meaning for this chapter's context. However, a careful examination will reveal the most appropriate theme for the pattern developed in this chapter. Interaction seems to the most relevant EBT for AnyInteraction pattern. This chapter also establishes that the pattern developed is robust, flexible for any usage, extensible, and practical for any purpose. The main goal of this chapter is to create pattern that can be used widely under any context and in any domains of everyday usage.

15.4 REVIEW QUESTIONS

1. Define the AnyInteraction stability design pattern.
2. Why do you consider interaction as the most appropriate EBT for the pattern?
3. List a few application domains for the AnyInteraction design pattern.
4. List four challenges while formulating the design pattern.
5. AnyInteraction as a stable design pattern is domain and application specific—say True or False.
6. List any two constraints and pitfalls while formulating the AnyInteraction design pattern.
7. List all participants in the AnyInteraction design pattern including BOs.
8. Illustrate the class diagram for the AnyInteraction design pattern.
9. List three advantages of this pattern.
10. List English terms similar to interaction and explain why they have been left out of this chapter.
11. The AnyInteraction pattern involves the use of other patterns too. Briefly explain.
12. Explain the basics of the term interaction with reference to its relevance to this chapter.
13. List two applications of the AnyInteraction design pattern.
14. Provide some examples of nonfunctional requirements for this design pattern.
15. Why do you think that the AnyInteraction design pattern would be stable over time?
16. Can you list some of the existing research issues related to the AnyInteraction design pattern?
17. Why the AnyInteraction design pattern provides extensibility and applicability over a period.
18. List two or three modeling issues that you will come across when modeling the stable design pattern for AnyInteraction.
19. List one specific scenario that will not be covered by this design pattern.
20. List three test cases to test all participants of the AnyInteraction design pattern.
21. List some of the implementation issues while developing AnyInteraction pattern.

15.5 EXERCISES

1. Interaction versus feedback stable patterns
   a. What is the difference between stable interaction pattern and feedback pattern?
   b. Create the feedback stable pattern.
   c. Name and describe three contexts for each of them.
   d. Show how to apply the two patterns including one context of your choice and including a use case, the application of each pattern with your selected context.
2. Interaction versus MInterface stable patterns
   a. What is the difference between stable interaction pattern and interface pattern?
   b. Create the interface stable pattern.
   c. Name and describe three contexts for each of them.
   d. Show how to apply the two patterns including one context of your choice and including a use case, the application of each pattern with your selected context.

3. Create five applications of AnyInteraction patterns in a table as shown in Table 23.1 (Chapter 23).
4. Constraint versus stable patterns
   a. What is the difference between stable interaction pattern and reflection pattern?
   b. Create the reflection stable pattern.
   c. Name and describe three contexts for each of them.
   d. Show how to apply the two patterns including one context of your choice and including a use case, the application of each pattern with your selected context.
5. Interaction versus emergent stable patterns
   a. What is the difference between stable interaction pattern and emergent pattern?
   b. Create the emergent stable pattern.
   c. Name and describe three contexts for each of them.
   d. Show how to apply the two patterns including one context of your choice and including a use case, the application of each pattern with your selected context.
6. Create two applications of two contexts of AnyInteraction stable design pattern and provide the following for each one of the applications:
   a. Context.
   b. Application diagram using the interaction stable design patterns.
   c. Detailed use case.
   d. A sequence diagram for the application.

15.6   PROJECTS

Use interaction in one of the following and create:

a. Interaction frequency
b. Interaction design
c. Social interaction
Three contexts for each one of them
Stable architecture pattern for each of them
An application of each one of them

REFERENCE

1. Margaret Heffernan (born 1955), last modified on June 13, 2017, https://en.wikipedia.org/wiki/Margaret_Heffernan

16 AnyTranslation Stable Design Pattern

> Without translation, I would be limited to the borders of my own country. The translator is my most important ally. He introduces me to the world.
>
> **Italo Calvino [1]**

In the history of English language, many words and phrases are considered as classical and legendary with their own interpretations and meanings. Out of such compelling gems, the word translation seems to be the most widely used and practiced. The most widely used definition of translation as per Merriam-Webster Online Dictionary is "words that have been changed from one language into a different language" [2]. Translation is a critical concept that deals with the primary issue of converting or changing something from one form to another. In the context of this chapter, it also means understanding something that has been converted from one form to another.

The main goal behind this chapter is to create a stable software model that could be applied to all applicable and available domains. The resulting pattern will give us a working solution to AnyTranslation, which is the main theme of this chapter. This will help developers to stop trying finding solution to a recurring problem every time they need the requirement of translation in their domains. In other words, this chapter envisages creating a stable software pattern that is not only robust but also reusable across a variety of domains.

16.1 INTRODUCTION

Translation is the act or process of translating something into a different language to allow readers of the translated language to understand the meaning of original text. In other words, when something is translated into another language, the passage of popular culture of the translated language to the other ensures deeper understanding and comprehension of language, dialects, stories, anecdotes, history, traditions, customs, and other hitherto unknown facts. Hence, the main theme of the word translation is "understanding." However, translation is a generic concept that presents the real essence of the problem that is presented in this book; in a way, translation is an enduring concept.

As a diverse range of applications may impose different needs for completing translation, it may not make any sense to develop a unified framework that consists of all translation issues within such applications. Nevertheless, different applications may still share some portions of translation needs although they differ in some aspects. The main goal of this chapter is to create a conceptual, stable software pattern that can capture and elaborate the core knowledge that exists within the domain where translation concept is visualized; additionally, this model is thought to work best in any application and contexts.

A typical stable pattern has just one ultimate goal. The ultimate goal or EBT of translation is Understanding. In the meanwhile, AnyTranslation as a stable pattern could be the best depiction of understanding something. AnyTranslation solves any type of problem that tries to understand something which could be in the form of business communication, academic research, translating the content of text or even in personal communication.

16.2   PATTERN DOCUMENTATION

16.2.1   Name: AnyTranslation Stable Design Pattern

The Stable design pattern for AnyTranslation describes the term translation that is used by AnyParty or AnyActor, who wants translate AnyMedia. The EBT that describes the ultimate goal of "AnyTranslation" is "Understanding." If any party wants to recreate anything in a language that is understood by anyone, they use translation approach. The Understanding has AnyContext. The ultimate goal of AnyTranslation is to recreate everything written in the documents so the actor or party has to follow certain rules (AnyRule). AnyTranslation produces AnyOutcome. Once the translation is performed, they are published in AnyMedia, such as books and digital records in order to convey the meaning of the text to others. At the end of translation, a checklist or log (AnyLog) is generated to verify if the translation is achieved successfully.

16.2.2   Context

As explained before, translation operates in a wide array of situations and scenarios. AnyTranslation stable design pattern can be applied to any domain, where a number of written records in one language, has to be converted to another language by following established rules. Some areas where translation helps in achieving communication are illustrated below:

1. *Business Communication*: Multinational businesses operating in multiple countries, many of which do not speak English will need translation (AnyTranslation) to operate business operations smoothly. The translation process will produce a series of corporate results (AnyOutcome). This may include documents (AnyMedia) like employee manuals, legal contracts descriptions, market research notes, presentation slides, financial statements, electronic communications, and social media messages. The documents are of multiple types (AnyType). A translation must ensure that they must satisfy defined rules (AnyRule), such that the translation is meaningful and accurate. A translation is generally circulated to employees (AnyParty) or organization (AnyParty) by e-mail (AnyMedia). At the end of translation, a checklist/log (AnyLog) is created to verify if the translation is executed successfully.
2. *Academic Research*: Any researcher (AnyActor) or a group of researchers (AnyParty) may have a written document (AnyMedia) in an unknown language, where they want to translate into some other language (AnyTranslation). This process will have some outcome (AnyOutcome) and translation is carried out according to certain rules (AnyRule). The actor or party may have any reason (AnyReason) for translation. Once translation is completed, one can make a checklist (AnyLog) of completed translations to make sure if all the translations are done in a proper manner.
3. *Government Agencies*: Every country (AnyParty) will have a government agency for translation (AnyActor), which handles correspondence and communication by translating (AnyTranslation). AnyActor will translate all the communication in print or in voice to produce translated materials (AnyOutcome) by following set rules (AnyRule), and by placing some ethical and time constraints (AnyConstraint). The translated documents are published for public or private record-keeping purposes (AnyMedia) in their web site or in a digital archive. As soon as the translation is completed, the agency takes a log (AnyLog) containing the document details and a summary of it.

16.2.3   Problem

Translation stable design pattern will be used to solve different problems based on changes resulted because of Understanding process. The model is designed in such a way that the same design pattern can be used for different applications to make the system stable and robust. This is achieved by including EBT, BOs, and IOs. The ultimate goal of AnyTranslation is "Understanding." As the BO "AnyTranslation" is related to some kind of understanding in the system, "Understanding" is considered as EBT of the pattern translation. IO varies based on application so it can be considered based on particular applications.

16.2.3.1   Functional Requirement

1. *Understanding*: Understanding can be defined as a reason for AnyTranslation to convert text from one language to another. AnyParty and AnyActor are responsible for understanding by providing AnyRule for AnyTranslation.
2. *AnyActor*: AnyActor are the users of the system or application. AnyActor is any person who is responsible for the Understanding by providing AnyRule for AnyTranslation. AnyActor is responsible for following AnyRule for AnyTranslation in a given application.
3. *AnyParty*: AnyParty is a user such as organization who is responsible for following AnyRule. AnyParty will have understanding for AnyTranslation.
4. *AnyRule*: AnyRule can be defined as a set of laws that AnyParty needs to be followed before defining AnyOutcome for AnyTranslation.
5. *AnyContext*: AnyContext can be defined as a scenario for Understanding. AnyContext is the part of AnyTranslation. Without AnyContext, AnyTranslation has no value and cannot exist.
6. *AnyOutcome*: AnyOutcome is defined as consequence or result of the something. It is derived from AnyTranslation.
7. *AnyTranslation*: AnyTranslation results in the AnyOutcome. AnyOutcome of change in the system is based on AnyTranslation.
8. *AnyEvent*: AnyEvent is something that happens in an organization. AnyTranslation is based on AnyEvent of the system. AnyEvent is related to AnyMedia, which records the change in the system.
9. *AnyType*: AnyType can be considered as AnyType of AnyEntity. AnyContext names AnyType of the system.
10. *AnyMedia*: AnyMedia is media such as documents, books, web sites, and Internet, which is used by AnyEvent or AnyEntity, to publicize about AnyTranslation, which broadcasts AnyOutcome.
11. *AnyEntity*: AnyEntity is something that makes something happen in an organization. AnyTranslation is based on AnyEntity of the system. AnyEntity is related to AnyMedia that records the change in the system.
12. *AnyLog*: AnyLog is used to store details from AnyMedia.

16.2.3.2   Nonfunctional Requirement

1. *Effective*: Effectiveness is the measure of how AnyTranslation results in a change for AnyActor. Effective application or system produces a desired or intended result. While initiating AnyTranslation, it is very much important to consider the effect of AnyTranslation, so that AnyOutcome is positive. For example, if a reader fails to understand the contents of translated text, then the entire translation process is thought to be ineffective.
2. *Realistic*: A translation process should be realistic and meaningful. The outcome of the translation process should be positive and benefitting to the person who wants to understand something. Unrealistic translation may lead to confusion, lack of understanding, and disagreement.
3. *Convincible*: Convincible is the term for which AnyParty or AnyActor should be convinced about AnyContext that is proposed by AnyTranslation. Application for AnyTranslation should be convincing in AnyOutcome, so that AnyMedia should be able to use AnyLog. Translation should be able to convince someone with conviction. If the translation process

is erroneous or faulty, then the reader/listener who reads or listens to translated materials may not believe in the authenticity of the converted material.

4. *Positive*: Positive translation ensures deeper understanding and comprehension. Positivity in the context of translation means a process that can result in expectation, encouragement, motivation, and a deeper urge to learn more.

5. *Analysis*: Translation should lead to deeper analysis of data, fact, and statistics. The person who reads or listens to a translated material should be able to analyze the content in an appropriate manner. In other words, if a person or entity analyses any translated materials in a convincing manner, then the entire translation process is said to be complete and foolproof.

6. *Explanation*: Translation should provide explanation for translating something. It should explain why something is translated and what would happen if a person understands something from a translated material. Explanation could be in the form of a note or a brief or even a memo.

7. *Apprehension*: A person who tries to understand a translated material should become apprehensive by the latter. He or she should not be anxious or panic while trying to understand translated materials. In other words, translation should be simple and straightforward without excessive focus on the deeper and elaborate practical details of translation process.

8. *Clarification*: It is an action of making a statement or situation less confused and more comprehensible. In other words, translation should be able to make the entire process easy, flexible to comprehend, and less complicated.

9. *Simplification*: Translation should make the process of understanding simple and easy. Simplified translation process will make understanding quicker and faster. Complicated translated materials are difficult to understand, time consuming, and they need additional effort too.

16.2.4 SOLUTION

Solution as shown in Figure 16.1.

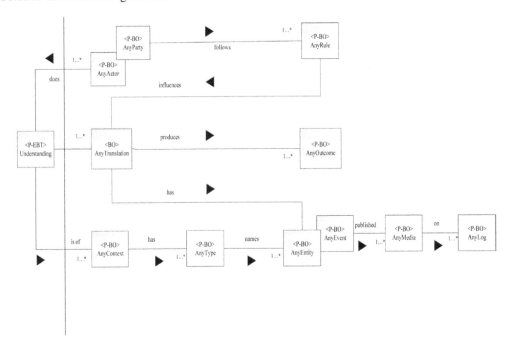

FIGURE 16.1   Class diagram of translation stable design pattern.

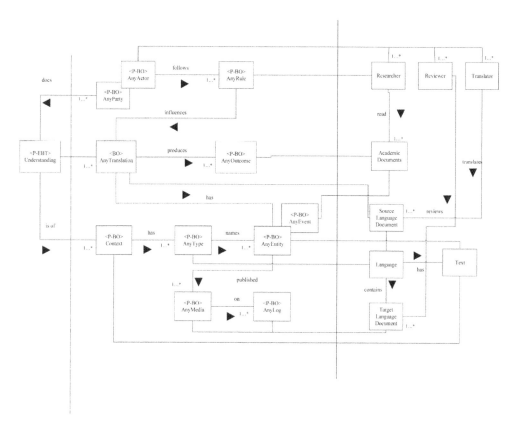

FIGURE 16.2   Class diagram for translation for academic research.

16.2.4.1   Applications

16.2.4.1.1   Translation for Academic Research

The second case study relates to the translation of written documents for an academic research. A researcher finds a native speaker/translator to translate some text. The translator and researcher agree on the delivery time for translated text, while the translator agrees with the period and starts the translation process. The researcher also finds a second native speaker/reviewer to carry out a review of the translated text. The translator then sends the completed translation to the researcher. The researcher dispatches the source text and the actual translation on to the reviewer who carries out the review. The reviewer sends the review back to the researcher. The researcher resends the review on to the translator, who in turn creates the final version. The translator sends the final version to the researcher. The researcher will use the translation for academic purposes (as shown in Figure 16.2).

16.2.5   Sequence Diagram

Sequence diagram as shown in Figure 16.3.

16.3   SUMMARY

In this chapter, we proposed a possible and stable solution for the translation stable analysis pattern. The pattern is a conceptual model that can be used to analyze and understand the basic translation concept. The pattern is extremely generic, since it captures just the core knowledge of the translation concept, which makes the pattern applicable to different domains and applications and reusable under any type of situation. In addition, the pattern also generalizes the basic properties of translation, thus making the model adaptable for any kind of application.

Use Case #	Use Case 2		
Use Case Title	Camp for Money Donation		
Actor	**Roles**		
AnyActor	Researcher, Translator, Reviewer, Document		
Classes	**Type**	**Attributes**	**Operations**
Understanding	EBT	1. name	1. understandForTranslation()
		2. type	2. bringsPerception()
		3. outcome	3. leadsToOutcome()
		4. description	
AnyActor	BO	1. id	1. readText()
		2. name	2. translateText()
		3. role	3. reviewText()
		4. purpose	4. studyText()
			5. followsRules()
AnyParty	BO	1. id	1. readText()
		2. name	2. translateText()
		3. role	3. reviewText()
		4. purpose	4. studyText()
			5. followsRules()
AnyRule	BO	1. name	1. governsTranslation()
		2. theme	2. helpsTranslators()
		3. description	3. establishConvention()
		4. context	
		5. id	
AnyTranslation	BO	1. id	1. convertDocument()
		2. name	2. compareText()
		3. result	3. finalOutput()
		4. description	
		5. outcome	
AnyOutcome	BO	1. id	1. analysis()
		2. name	2. comparison()
		3. result	3. finalOutput()
		4. description	
		5. context	
AnyContext	BO	1. id	1. initiateChange()
		2. name	2. translationType()
		3. type	3. understand()
		4. status	
AnyType	BO	1. id	1. change()
		2. typeName	2. operateOn()
		3. properties	3. nameContext()
AnyEvent	BO	1. id	1. initiateChange()
		2. eventName	2. causeUnderstanding()
		3. eventType	3. recordOnMedia
		4. status	
		5. position	
		6. states	
		7. type	
AnyEntity	BO	1. id	1. initiateChange()
		2. entityName	2. causeUnderstanding()
		3. entityType	3. recordOnMedia()
		4. status	
		5. position	
		6. states	
		7. type	

Continued

Classes	Type	Attributes	Operations
AnyMedia	BO	1. id	1. capture()
		2. mediaName	2. store()
		3. mediaType	3. display()
		4. status	
		5. sector	
AnyLog	BO	1. id	1. nameLog()
		2. logName	2. replaceLog()
		3. logType	3. removeLog()
		4. logReferences	4. searchLog()
		5. logCriteria	5. openLog()
		6. logSize	6. closeLog()
		7. logLocation	7. editLog()
		8. logPath	
Researcher	IO	1. name	1. studiesText()
		2. id	2. understandsText()
		3. degree	3. acceptsTranslation()
		4. address	
		5. zipcode	
		6. number	
Translator	IO	1. name	1. translatesText()
		2. id	2. helpsResearcher()
		3. degree	3. followsConventions()
		4. address	
		5. zipcode	
Reviewer	IO	1. name	1. reviewsTranslation()
		2. details	2. givesComments()
		3. description	3. editsText()
		4. number	
Document	IO	1. name	1. containsText()
		2. id	2. givesKnowledge()
		3. language	3. helpsRecord()

Use Case Description:

1. Understanding is used for converting a source language into target language. (TC: How source language is converted to target language?)
2. AnyActor participates in the translation process and plays a role of translator to translate text. (TC: What role does AnyActor play?)
3. In the process of translating text, some rules and criteria need to be followed. (TC: What needs to be followed during translation?)
4. AnyActor carries out the idea of proposing and organizing language translation. (TC: What is the role of the Financial Manager?)
5. The researcher gives documents in source language to the translator to convert it to the desired target language. (TC: What is the role of the Financial Manager?)
6. AnyOutcome of this translation is the result of understanding. Translation from source language to target language helps the researcher to understand the text. (TC: What does translation do?)
7. AnyMedia is used to communicate between the translators and researchers the result of the translation. (TC: What does AnyMedia do?)
8. All the data collected by AnyMedia are stored in AnyLog. AnyLog can be opened, removed, searched in order to check for the history of the translation process. (TC: What does AnyLog do?)
9. The researcher also sends the text to a qualified reviewer to check, if the translation is correct. (TC: What does publisher do?)
10. The reviewer makes comments and points out errors in the translated text, which is then sent to the first translator, as well as the researcher. (TC: What does reviewer do?)
11. The entire information of the translation is gathered and stored in AnyLog. (TC: What is stored on AnyLog?)

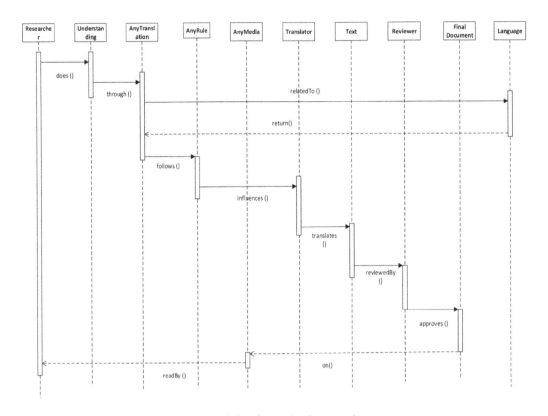

FIGURE 16.3   Sequence diagram for translation for academic research.

16.4   REVIEW QUESTIONS

1. What do you mean by the term "translation"? Can you use it in any other context?
2. Find out all terms that are same as "trust." Can you use them interchangeably?
3. What are the capabilities to carry out translation? Describe each one of them.
4. Draw and describe the class diagram for stable translation pattern? Explain the participants and justify their existence in the pattern.
5. Develop two case scenarios where you can apply the concept of translation. Try to fit these scenarios with the translation pattern.
6. Try to create a use case and interaction diagram for each of the scenarios you thought of in the above question.
7. Draw a class diagram for the application of translation SAP in providing translation for a classic book in English to Chinese. Describe and elaborate it. Write a use case for it and its description. Draw a sequence diagram for the use case and explain it.
8. Draw a class diagram for the application of translation SAP in business communication in a large corporation. Describe it. Write a use case for it and its description. Draw a sequence diagram for the use case and explain it.
9. List some major differences between the translation pattern described here and the traditional pattern.
10. List some design and implementation issues faced when implementing the translation pattern. Highlight each issue.
11. Give some important commercial applications where the translation pattern is used.
12. Mention some lessons that you have learned from this chapter.

13. Define and describe the translation stable analysis pattern.
14. The translation stable analysis pattern can be applied and extended to any domain (True or False).
15. List three important challenges in formulating the translation analysis pattern.
16. What are the classes and patterns involved in defining the stable pattern for "translation"?
17. Document the CRC card for the translation EBT.
18. What is the tradeoff of using this stable pattern?
19. What are the anticipated design issues for the translation EBT, when linked to the design phase?
20. Mention all the BOs that were synthesized for this pattern. Elaborate on each one of them.

16.5 EXERCISES

1. Translation versus interpretation stable patterns
   a. What is the difference between stable translation pattern and interpretation pattern?
   b. Create the interpretation stable pattern
   c. Name and describe three contexts for each of them.
   d. Show how to apply the two patterns including one context of your choice and including a use case, the application of each pattern with your selected context.
2. Translation versus explanation stable patterns
   a. What is the difference between stable interaction pattern and explanation pattern?
   b. Create the explanation stable pattern.
   c. Name and describe three contexts for each of them.
   d. Show how to apply the two patterns including one context of your choice and including a use case, the application of each pattern with your selected context.
3. Create five applications of AnyTranslation patterns in a table as shown in Table 23.1 (Chapter 23).
4. Translation versus stable patterns
   a. What is the difference between translation stable pattern and association pattern?
   b. Create the association stable pattern.
   c. Name and describe three contexts for each of them.
   d. Show how to apply the two patterns including one context of your choice and including a use case, the application of each pattern with your selected context.

16.6 PROJECTS

Use translation in one of the following and create:

a. Dynamic Translation
b. Brainstorming
c. Translation Language
Three contexts for each one of them
Stable architecture pattern for each of them
An application of each one of them

REFERENCES

1. Italo Calvino (October 15, 1923–September 19, 1985), last modified on May 31, 2017, https://en.wikipedia.org/wiki/Italo_Calvino
2. Merriam-Webster Dictionary. Encyclopædia Britannica Online. 2015. https://www.merriam-webster.com/, retrieved June 24, 2015.

17 AnyArchitecture Stable Design Pattern

We shape our buildings; thereafter they shape us.

Winston Churchill [1]

Architecture is a basic concept that plays an important role in many areas of life, technology, engineering, and development. In other words, it has a diverse number of applications in both business and technical architecture, with its highly sophisticated domain of understanding, needs careful evaluation and an in-depth study. In the domain of architecture, professionals that work different under differing situations have provided several definitions. For example, architecture in the realm of civil engineering is different from the one that is used in computer hardware engineering. The role of architecture, in the sphere of software pattern designing, suggests that it may not be practicable to develop a unified framework of patterns that encompass different architectural aspects within various applications; this is due to a demand for different requirements for ensuring architecture that are enforced by different applications.

With a goal-driven approach, this chapter intends to create a stable software model that is easy to apply to all domains where the term *Architecture* is likely to play its role. The main theme of this chapter, creating a robust AnyArchitecture pattern, is believed to provide a solid working solution to all recurring problems that may arise, while using the term *Architecture*; in fact, it will allow developers avoid working repeatedly to find similar solutions to all similar problems. In this chapter, we propose to introduce the concept of Architecture stable pattern that aims to provide a conceptual model that encompasses the main theme of Architecture concept and also includes the true requirements and ultimate solution. As the level of abstraction of this pattern is quite high, its application across a wide array of applications is ensured and guaranteed.

17.1 INTRODUCTION

Speaking specifically, architecture is the art and science of building. In an engineering context, architecture is the essential art of designing and building structures and buildings. Architecture also means a coherent form or structure and this definition could be used in many different forms. For example, when one means that a novel lacks architecture, it is thought to be devoid of a proper story or a meaningful plot. Another example is someone commenting on a garden that its design lacks proper architectural fundamentals. In computer hardware technology, architecture is the manner in which the components of a computer or computer systems are integrated, formed, and organized. Some of the most common synonyms of the word architecture are frame, configuration, framework, infrastructure, shell, and skeleton.

However, the main theme of the term architecture remains the samev—"creativity." Creativity is an enduring concept and this is taken as the essential enduring business theme (EBT) for this chapter on AnyArchitecture stable design pattern. However, architecture is still a generic term that presents many aspects and issue of different meanings that surround the word. A wide array of applications may demand different needs for ensuring creativity. The main objective of this chapter is to form a conceptual and stable software pattern that can seize and capture the core knowledge of architectural concept that usually exists with a domain where a developer can visualize and comprehend it [2–4].

17.2 PATTERN DOCUMENTATION

17.2.1 NAME: ANYARCHITECTURE STABLE DESIGN PATTERN

The subject chosen for the pattern is architecture. The EBT of this project is Creativity. It is based on the term architecture, which can be defined as the art of creating something that is of strategic or aesthetic importance. Architecture can be applied to create different types of objects both physical and virtual. AnyActor or AnyParty use their creativity in order to generate AnyArchitecture based on AnyCriteria and AnyContext. AnyParty or AnyActor, who wants support from others, may also define their criteria and reason. The ultimate goal of creativity can be expressed through AnyArchitecture.

17.2.2 CONTEXT

Architecture has many things to do with planning, designing, and constructing form, space, and ambience to reflect functional, technical, social, environmental, and aesthetic considerations. It requires creative manipulation and coordination of materials and technology, and of light and shadow. Often, conflicting requirements must be resolved. Various applications, in which architecture is an enduring concept, are

1. *Architecture of a Vehicle*: An automobile (AnyEntity) platform is a shared set of common design (AnyDesign), engineering, and production efforts, as well as major components, over a number of outwardly distinct models and even types of automobiles, often from different, but related marques. It is a standard norm in the automotive industry to reduce the costs associated with the development of products by basing placing products (AnyCriteria) on a smaller number of platforms. This approach further allows car making companies (AnyParty) to create distinct models from a design perspective on similar underpinnings. Each type of architecture can be created with different purposes in mind (AnyCriteria). There are different kinds of architecture for different types (AnyType) of vehicles that have to follow different semantics (AnyRule).

2. *Architecture of a Building*: It requires the creative manipulation and coordination of materials and technology, and of light and shadow. Often, conflicting requirements must be addressed and resolved. The practice of architecture (AnyArchitecture) also encompasses the pragmatic aspects of realizing buildings and structures, including scheduling, cost estimation and construction administration. Documentation stored on a computer (AnyMedia) produced by architects (AnyActor), typically drawings, plans and technical specifications, defines (AnyRule) the structure and/or behavior of a building (AnyEntity) or other types (AnyType) of system that is to be or has been constructed.

3. *Architecture of an Organization*: An Organizational architecture has two very different (AnyType) connotations. In one sense, it literally refers to the organization and its built environment, and in another sense, it refers to architecture in a metaphorical mode, as a structure that creates the business operations of the organization. Organizational architecture or organizational space is the influence of the spatial environment on humans (AnyActor) in and around organizations (AnyEntity). Organizational architecture (AnyOutcome) or organization design is the creation of roles, processes, and formal reporting relationships in an organization (AnyParty). Organization design can be defined narrowly (AnyContext), as the process of reshaping (AnyEvent) organization structure and roles, or it can be more effectively defined as the alignment of structure (AnyArchitecture), process, rewards, metrics, and talent with the strategy of the business.

4. *Architecture of Software*: Software architecture is the high-level structure of a software system, the discipline of creating such structures, and the documentation of these structures. It is the set of structures that are needed to describe the software system (AnyArchitecture), and it comprises of different software elements (AnyEntity), the existing relations between them,

and the properties of both elements and relations. The architecture of a software system is a metaphor, analogous (AnyContext) to the architecture of a building. Software architecture choices include specific structural options from possibilities in the design of software. For example, the systems that controlled the space shuttle launch vehicle have the requirement of being very fast and very reliable (AnyRule) in principle. Therefore, an appropriate real time computing language (AnyCriteria) would be chosen. Similarly, multiple redundant independently produced copies of a program running on independent hardware and crosschecking results would be the software-system architecture choice (AnyType) to satisfy the need for trust and reliability. Software architecture is about making fundamental structural choices, which are costly to change once implemented, i.e., which are used to (AnyMedia) "house" the more changeable elements of the program, for example, an operating system.

5. *Architecture of a Mobile Device*: In the past, computers needed to be disconnected from their internal network, if they needed to be taken or moved anywhere. However, mobile architecture allows maintaining this connection whilst during transit or movement of PC from one place to the other. Each day, the number of mobile devices (AnyEntity) keeps increasing, while mobile architecture is the part of technology that is needed to create a rich, connected user experience. Currently, there is a visible lack of uniform interoperability plans and implementation. In addition, there is a lack of common industry view on architectural framework. This increases cost and slows down third party mobile technology development. An open approach is required among industries to achieve same results and services (AnyArchitecture). A consortium of companies (AnyParty) are designing products and services based on open, global standards, protocols, and interfaces and that are not locked to proprietary technologies. Using stable architecture pattern, we can establish uniformity and a standard set of rules (AnyRule) in order to reduce the costs and maintenance.

17.2.3 PROBLEM

Architecture is a term with a wide array of usage under different contexts and scenarios. Hence, designing an architecture stable pattern that is easily adaptable and extendible is a challenging task. Another notable problem is the perceived lack of abstraction of architecture concepts under different circumstances. Hence, any possible solution should focus on providing an answer to the generality issue of architecture, which upon completion, would provide an opportunity to extend the resulting pattern to any type of applications. Every pattern needs two specific requirements while designing it. These two requirements are functional requirements and nonfunctional requirements.

1. *Functional Requirements*: Functional requirement is interpreted as "What the system is supposed to do?" They consist of an amalgamation of functions of the system and its components. The BOs define the functional requirements of the system in the software stability model. Functional requirements can be technical details, behavior of model, and associated calculations.

2. *Nonfunctional Requirements*: Nonfunctional requirements are interpreted as "How a system is supposed to be?" It is termed as a standard that is used to judge the action of the system. It neglects behavior and focuses on operation of the model. EBTs define the nonfunctional requirements of the system in software stability model.

17.2.3.1 Functional Requirements

The essential business theme for architecture is creativity. Architecture is way of representing creativity by any actor or party for any event or entity. The context for creativity may vary from party to party and from actor to actor. Creativity can be represented by using any type of media.

BOs: Stable Design Patterns

1. *AnyParty*: AnyParty can perform the act of architecture. AnyParty can request someone to do the architecture. AnyParty should plan the architecture and may ask some other party to do the architecture for them by negotiating the terms and conditions of architecture.
2. *AnyActor*: AnyActor can also perform architecture. AnyActor can also request someone else to do the architecture. AnyActor specifies AnyCriteria whether he/she will do the architecture or not. If AnyActor on behalf of a company plans AnyArchitecture, then he/she specifies AnyCriteria based on the company policy.
3. *AnyCriteria*: AnyParty or Actor will provide their creativity for AnyEvent of AnyEntity based on some criteria. In addition, each party and actor has criteria for creating the architecture. AnyCriteria represents the set of constraints AnyParty or AnyActor will follow.
4. *AnyType*: There are different types of architecture like mobile, building, software, and cars. Type depends on the purpose of the architecture and the type of audiences it wants to address. AnyType is determined by AnyArchitecture and will name the event or entity to be created.
5. *AnyArchitecture*: There is always a way to design architecture. Everything needs to be planned and all the arrangements need to be done on time and in a proper manner. Architecture can be of various types depending upon one or more criteria.
6. *AnyMedia*: For most of the architecture, there is a requirement of any type of media like radio, newspaper, television, Internet, etc. It depends on the type of architecture that is to be built for AnyParty or AnyArchitecture is targeting. AnyMedia is determined by AnyEvent or AnyEntity and creates AnyOutcome each time it is used.
7. *AnyEvent*: Architecture can be performed for various purposes and can be done for any event or service. Architecture is commonly done for many events like concerts, theater, and product launches, etc. AnyEvent is named by AnyType and uses AnyMedia for the purpose of architecture.
8. *AnyEntity*: An entity is the object for which endorsement is done. Entity can be any new product or any existing product. AnyEntity is named by AnyType and uses AnyMedia for the purpose of endorsement.
9. *AnyContext*: The architecture depending upon AnyType has AnyRules on which the architecture depends. There are different rules that apply for AnyArchitecture. Different architectures lead to AnyOutcome.

17.2.3.2 Nonfunctional Requirements

1. *Practice*: It serves as a run-through and helps acquiring or maintaining proficiency for an activity. Practice is the act of rehearsing a behavior repeatedly, or engaging in an activity continuously, for the purpose of improving or mastering it, as in the phrase "practice makes perfect." Architectural procedures need immense practice and without it, one can never anticipate the expected outcome.
2. *Discovering*: Discovery is the act of detecting something new or something "old" that had been unknown. With reference to science and academic disciplines, discovery is the observation of new phenomena, new actions, or new events and providing new reasoning to explain the knowledge gathered through such observations with previously acquired knowledge from abstract thought and everyday experiences. Every architectural endeavor should lead to a new discovery of a technique, method, or procedure. A proper architectural procedure will help developers in discovering new ways of creating a stable pattern.
3. *Understandable*: Architecture should be easily understandable and comprehend because if it is not easy to understand, then the purpose of the architecture will not be realized and it will be a waste of time. If people are not able to understand the process of architecture, they cannot apply it in practice.

4. *Legal*: There are certain rules and regulations established for the endorsements that each organization is bound to follow. So, if the rules are violated, the organizations may be stopped or prevented. In other words, the entire process of using the principle of architecture should be proper and legal so that all norms and regulations of creating a stable pattern are met and endorsed.

5. *Descriptive*: Architecture should be descriptive and wholesome, so that the message it wants to convey is easily and effectively conveyed to the audience. If it is not descriptive, then the audience might misunderstand or misinterpret the entire concept, which will make it ineffective and fruitless. The goal of the architecture will not be achieved under such a circumstance.

6. *Attractive*: Any architecture's success depends on how well it is able to influence and attract the audiences. Therefore, it should be always very eye catching and attractive. The issue of attractiveness is largely influenced by the way in which the principles of aesthetics and neatness are followed. For example, a stable pattern is said to be complete and perfect, when different issues of pattern making are placed together in a perfect harmony.

17.2.4 Solution

* <<P-BO>> - related Pattern BO * BO = Business Object * EBT = Enduring Business Theme Class Diagram Description as shown in Figure 17.1.

- **AnyParty** requests for creativity to create the architecture.
- **AnyContext** has one or many types and creativity has many contexts.
- **AnyEntity** is stored on any media which contains one or many logs.
- **AnyArchitecture** has one or many forms which have one or many aspects.
- **AnyRule or AnyCriteria** is followed by AnyParty.

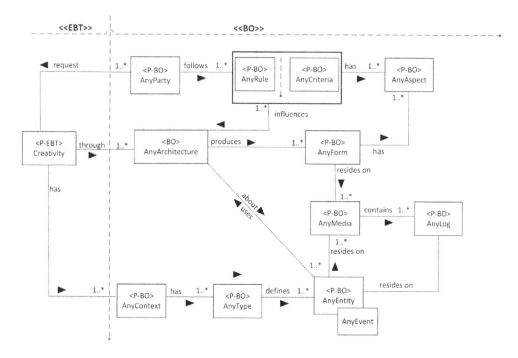

FIGURE 17.1   AnyArchitecture stable design pattern.

17.2.5 Scenarios

Scenario 1: AnyArchitecture Stable Design Pattern Applied to Software
Context: Software architecture is the high-level structure (AnyEntity) of a software system, the discipline (AnyRule) of creating such structures, and the documentation of these structures. It is a set of structures needed to discuss the software system (AnyMedia), and it comprises of different software elements, current relations between them (AnyContext), and the properties of both elements and relations. The architecture (AnyArchitecture) of a software system is a metaphor, which is also analogous to the architecture of a building. Software architecture choices include specific (AnyAspect) structural options culled from different possibilities in the design of software. For example, the systems (AnyParty) that controlled the space shuttle launch vehicle have the requirement of being very fast and very reliable in principle. Therefore, an appropriate real time computing language would be chosen naturally. Similarly, multiple, redundant, and independently produced copies of a program running on independent hardware and crosschecking ensuing results, would be software system architecture's natural choice to satisfy the need for reliability and trust. Software architecture is about making fundamental structural choices, which are costly to change once implemented, that is, which are used to "house" the more changeable elements of the program, for example, an operating system.

Scenario 2: AnyArchitecture Stable Design Pattern Applied to Construction
Context: Architecture has something to do with planning, designing, and constructing form, space, and ambience to reflect functional, technical, social, environmental, and aesthetic considerations (AnyCriteria). It requires the creative manipulation and coordination of materials and technology and of light and shadow. Often, conflicting requirements must be (AnyType) resolved. The practice of architecture also encompasses the pragmatic aspects (AnyAspect) of realizing buildings and structures (AnyEntity), including scheduling, cost estimation, and construction administration. Documentation produced (AnyEvent) by architects (AnyParty), typically drawings, plans, and technical specifications, defines the structure and/or behavior of a building or other kind of system that is to be or has been constructed.

Scenario 3: AnyArchitecture Stable Design Pattern Applied to E-Commerce
Context: Electronic commerce, commonly known as e-commerce, is trading in products or services by using computer networks, such as the Internet and Wi-Fi. Electronic commerce derives its inspiration from technologies (AnyArchitecture), such as mobile commerce, electronic funds transfer, supply chain management, Internet marketing, (AnyEvent) online transaction processing, electronic (AnyEntity)data interchange (EDI), inventory management systems (AnyType), and automated data collection systems. Modern electronic commerce typically uses the (AnyCriteria) World Wide Web for at least one part of the transaction's life cycle, (AnyAspect), although it may also use other technologies such as e-mail too.

Scenario 4: AnyArchitecture Stable Design Pattern Applied to Database
Context: The design of a DBMS depends on its architecture (AnyArchitecture). It can be centralized, decentralized, or hierarchical (AnyType). The architecture of a DBMS can be seen as either single tier or multitier. An n-tier architecture divides the whole system into related, but independent n modules (AnyEntity), which can be independently modified, altered (AnyEvent), changed, or replaced. In 1-tier architecture, the DBMS is the only entity, where the user directly sits on the DBMS and uses it. Any changes done here will directly be done on the DBMS itself. It does not provide handy tools for (AnyCriteria) end users. Database designers and programmers normally prefer using single-tier architecture. If the architecture of DBMS is 2-tier, then it must have an application through which the DBMS can be accessed. Programmers use a 2-tier architecture, where they access the DBMS by means of an application. Here, the application tier is entirely independent of the database in terms of operation, design, and programming.

Scenario 5: AnyArchitecture Stable Design Pattern Applied to Organization
Context: According to most authors and developers, organizational architecture is a metaphor. Like traditional architecture (AnyArchitecture), it shapes the organizational (some authors would say the informational) space where life will take place. It also represents a concept that implies a connection (AnyType) between the organizational structures with other systems inside the organization, in order to create a unique synergistic system (AnyEntity), which will be more than just the sum of its parts. Conventionally, organizational architecture consists of a formal organization (organizational structure), an informal organization (organizational culture), many business processes and strategies, and the other important human resources because an organization is a system of (AnyParty) people. The goal of organizational (AnyContext) architecture is to create an organization that will be able to create values for present and future customers, thereby optimizing (AnyEvent) and organizing itself into a systematic entity. Some under organizational architecture understands building blocks that are mandatory for the growth of the organization.

17.2.6 Applications

17.2.6.1 Case Study
Company tries to create architecture for a Mobile.

In this case study, a communications company plans to create architecture of a mobile. The company knows that if they create a stable mobile architecture, they will be able to produce a quality product by reducing the cost of maintenance and improve upon the version to create an even better one.

17.2.6.2 Case Study Class Diagram
Class Diagram Description as shown in Figure 17.2.

1. AnyParty is the one, who requests creativity to be performed on AnyArchitecture in order for it to be built.
2. AnyArchitecture has any context and follows one or many criteria, which can take on one or more forms.
3. AnyMedia is an infrastructure on which the architecture resides. It also contains one or many logs.
4. Architect is the one who uses his creativity, in order, to create the architecture.
5. Mobile is the form of the architecture to be created.

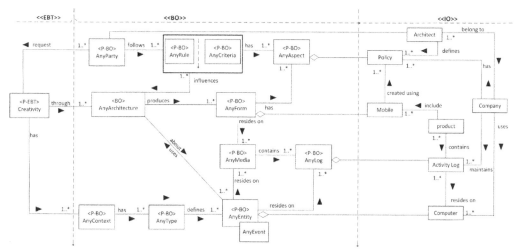

FIGURE 17.2 Case study class diagram.

6. A company can have one or many different products, like mobiles and other products.
7. Computer is one of the media, on which the architecture is believed to be resided.
8. AnyRule or AnyCriteria have to be followed while creating AnyArchitecture.

Case Study: Use case as shown in Table 17.1.
Use Case Description:

1. AnyParty provides creativity and has criteria based on AnyCriteria.
   TC: Who provides creativity? Criteria are based on what?
2. Creativity is achieved through AnyArchitecture. A company seeks creativity in order to create a new product.
   TC: Why is creativity needed? How is creativity achieved?
3. AnyParty defines AnyCriteria. AnyParty and company both define their policies that are determined by AnyContext of creativity.
   TC: Who defines AnyCriteria? AnyParty does what?

TABLE 17.1

Create an Architecture for a Mobile

Title	Plan architecture		
Id	1.1		
Actor	Roles		
AnyParty	Company, Architect		
Class	**Type**	**Attribute**	**Operation/Interface**
Creativity	EBT	degree, restriction	providedBy()
			achievedBy()
AnyContext	BO	validity, significance, stimulus	identifyCorrelation()
			createContext()
AnyCriteria	BO	name, type, clauseNumber, domain	checks()
			assistContext()
AnyEntity	BO	entityName, entityType, status	belongsToCompany()
			createArchitecture()
AnyLog	BO	name, size, media, logNumber, timestamp	record()
			nameLog()
AnyArchitecture	BO	name, type, context, purpose	useMedia()
			nameType()
AnyMedia	BO	mediaName, mediaType, capability	createLog()
			storeArchitecture()
AnyType	BO	Id, name	determineParty()
			nameEntiy()
Architect	IO	Name, occupation, age	createProduct()
			evaluateCriteria()
Policy	IO	text, clause, name	identifyCriteria()
			validateContext()
Company	IO	Name, location, objective	defineCriteria()
			planArchitecture()
			approachArchitect()
Product	IO	Name, Type, manufactureDate()	existOnRecord()
			determinePurposeOfArchitecture()
Company Records	IO	recordName(), type, purpose	maintainData()
			createRecord()
			saveRecord()

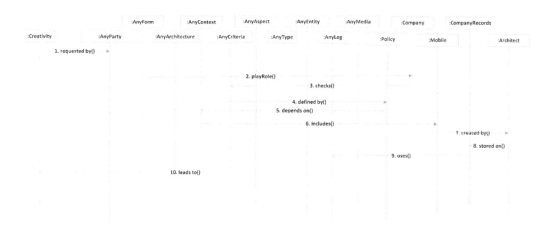

FIGURE 17.3 Case study sequence diagram.

4. AnyCriteria is based on AnyContext, which determines AnyArchitecture. Architecture is used to create a new product.
   TC: What does AnyCriteria do? What controls AnyArchitecture? How and who decides on AnyCriteria?
5. AnyType specifies AnyEntity that is created by AnyArchitecture.
   TC: AnyType does what and what specifies AnyEntity?
6. AnyArchitecture determines AnyType. AnyArchitecture is done to create a product because of which sale of the product is created.
   TC: What does AnyArchitecture result? AnyArchitecture is done for what?
7. AnyMedia is used by AnyEntity for achieving creativity for a product. AnyMedia creates one or more AnyLog.
   TC: What is the purpose of AnyMedia? How AnyMedia is related to AnyLog?
8. AnyLog is related to AnyMedia. AnyLog is used to store the records of media.
   TC: AnyLog is related to what? What is the purpose of AnyLog?
9. AnyLog is related to AnyContext. AnyLog stores all the records of the entire context for AnyArchitecture. AnyLog is shown by company records.
   TC: AnyLog is related to what? AnyLog stores records of what?

17.2.6.3 Alternatives

Creativity is performed with success in mind and thus it gives us architecture in some form. They also provide creativity by participating to create with the other party. They can also create the architecture by forming an alliance with other party.

Case Study Sequence Diagram as shown in Figure 17.3.

17.3 SUMMARY

Architecture can be of many forms and types, varying from domain to domain. Architecture is a very important part of business, market, and government. In addition, the type of architecture varies depending upon the scenario in which it is used. Some organizations create software architecture, while others create electronic and electrical architecture. It is an effective way to interact with public and private enterprises. Architecture can be stored on different media available today, for example, computer server, Internet, etc. If we use stable AnyArchitecture design pattern, we can create any architecture in any context, with any criteria, with any rules.

17.4   REVIEW QUESTIONS

1. What is "architecture"? Can it be used in any other situation than what has already been mentioned in this chapter?
2. Find out all terms that are similar to "architecture" and can you use them interchangeably?
3. What are the inherent capabilities to achieve architecture? Describe each one of them.
4. Draw and explain the class diagram for stable architecture pattern? Identify all participants and explain their existence.
5. Create two situations other than those given in this chapter. Fit them with the architecture pattern.
6. Create a use case and an interaction diagram for all of the scenarios that appeared in the above question.
7. Draw a class diagram for the application of architecture pattern in providing architecture for a computer hardware designer. Describe the entire process. Create a use case and provide a brief description.
8. Draw a class diagram for the application of architecture pattern for a garden. Describe it, create use case for it, and provide its description. Draw a sequence diagram for the use case and explain it.
9. List some major differences between the architecture patterns described here and the traditional pattern developed elsewhere.
10. List some design, process, and implementation issues faced when you were implementing this pattern. Provide details.
11. Give some other applications where you can use architecture pattern.
12. Did you learn any lessons and details by studying this pattern?
13. Define and elaborate architecture stable analysis pattern.
14. Can you apply and extend this pattern to any other domain?
15. Can you document a CRC card for this pattern?
16. Can you identify some tradeoff points by using this pattern?
17. Create a sample pattern in the domain of space engineering (deep space probes).
18. Architecture stable pattern could be extended to any number of applications. (True/False)
19. Can you use architecture pattern in industrial applications like e-commerce, banking services, military, and governance?
20. List important pitfalls of this pattern.

17.5   EXERCISES

1. Architecture versus Design Stable Patterns
   a. What are the differences between Architecture Stable Pattern and Design Pattern?
   b. Create the Design Stable Pattern.
   c. Name and describe three contexts for each of them.
   d. Show how to apply the two patterns, include with one context of your choice and include a use case, the application of each pattern with your select context.
2. Architecture versus Analysis Stable Patterns
   a. What are the differences between Architecture Stable Pattern and Analysis Pattern?
   b. Create the Analysis Stable Pattern.
   c. Name and describe three contexts for each of them.
   d. Show how to apply the two patterns, include with one context of your choice and include a use case, the application of each pattern with your select context.
3. Create five applications of AnyArchitecture Pattern in a table as shown in Table 23.1 (Chapter 23)

4. Architecture versus *contemporary*: Stable Patterns
   a. What the differences between Architecture Stable Pattern and *contemporary*: Pattern?
   b. Create the *contemporary*: Stable Pattern.
   c. Name and describe three contexts for each of them.
   d. Show how to apply the two patterns, include with one context of your choice and include a use case, the application of each pattern with your select context.
5. Architecture versus Extensability Stable Patterns
   a. What are the differences between Architecture Stable Pattern and Extensability Pattern?
   b. Create the Extensability Stable Pattern.
   c. Name and describe three contexts for each of them.
   d. Show how to apply the two patterns, include with one context of your choice and include a use case, the application of each pattern with your select context.
6. Architecture versus System Stable Patterns
   a. What are the differences between Architecture Stable Pattern and System Pattern?
   b. Create the System Stable Pattern.
   c. Name and describe three contexts for each of them.
   d. Show how to apply the two patterns, include with one context of your choice and include a use case, the application of each pattern with your select context.
7. Architecture versus Structure Stable Patterns
   a. What are the differences between Architecture Stable Pattern and Structure Pattern?
   b. Create the Structure and Stable Pattern.
   c. Name and describe three contexts for each of them.
   d. Show how to apply the two patterns, include with one context of your choice and include a use case, the application of each pattern with your select context.

17.6   PROJECTS

Use Architecture in one of the following and create:

a. Architectural Style
b. Adaptive Architecture
c. Architecture Illustration
d. Architecture Critera
e. Blueprint
f. Product-Line Arcitecture
g. Service-Oriented Archiecture
h. Document Architecture
i. System Arcitecture
Three contexts for each one of them
Stable architecture pattern for each of them
An application of each one of them

REFERENCES

1. Winston Churchill (November 30, 1874–January 24, 1965), last modified on June 21, 2017, https://en.wikipedia.org/wiki/Winston_Churchill
2. D. Rowland and T.N. Howe. *Vitruvius: Ten Books on Architecture*, Cambridge University Press, Cambridge, 1999, ISBN 0-521-00292-3.
3. Le Corbusier. *Towards a New Architecture*, Dover Publications, Mineola, New York, 1985, ISBN 0-486-25023-7.
4. C. Alexander. *The Timeless Way of Building*, Oxford, UK: Oxford University Press, 1979, ISBN 978-0-19-502402-9.

Section IV

SDPs' Short-Size Documentation Template

This part consists of 15 chapters (14 short-size templates), summary chapters, and one sidebar.

Each chapter presents the detailed documentation pattern that consists of the following sections: (1) the name and type of the pattern, (2) context, (3) the functional and nonfunctional requirements, (4) solution (5).

1. AnyDecision Stable Design Pattern (Chapter 18)
2. AnyConflict Stable Design Pattern (Chapter 19)
3. AnyView Stable Design Pattern (Chapter 20)
4. AnyModel Stable Design Pattern (Chapter 21)
5. AnyReason Stable Design Pattern (Chapter 22)
6. AnyConsequence Stable Design Pattern (Chapter 23)
7. AnyImpact Stable Design Pattern (Chapter 24)
8. AnyHype Stable Design Pattern (Chapter 25)
9. AnyCause Stable Design Pattern (Chapter 26)
10. AnyCriteria Stable Design Pattern (Chapter 27)
11. AnyMechanism Stable Design Pattern (Chapter 28)
12. AnyMedia Stable Design Pattern (Chapter 29)
13. AnyLog Stable Design Pattern (Chapter 30)
14. AnyEvent Stable Design Pattern (Chapter 31)
15. Summary (Chapter 32) and a Sidebar Context in Different Domains (SB 29.1)

18 AnyDecision Stable Design Pattern

Sometimes it's the smallest decisions that can change your life forever.

Keri Russell [1]

Decision is an active English word when someone involves in the act or process of deciding. They may also conduct an act for making their mind to make a judgment. This chapter elaborates on designing a stable software pattern by taking into account various usages of the term *Decision*. The major advantage of creating such a pattern is seizing an opportunity to gain immense advantage in terms of robustness, extendibility, applicability, and stability. In addition, developers may also hope to apply the resulting pattern to all contexts and scenarios that involve making a decision. Working on a pattern involving Decision domain will also ensure savings in time and effort that are needed to create a traditional software pattern; in effect, developers need not work repeatedly on a given pattern whenever the situation demands a change in the architecture of the pattern.

Decision is a general English term that can be used in a specific context and its usage stretches across a number of usage scenarios that eventually pose a serious challenge to developers while creating a meaningful pattern. The AnyDecision stable pattern seeks to deal with finding out innumerable problems and find out a suitable solution for intention that is incidentally the influence of Decision.

18.1 INTRODUCTION

This chapter focuses on creating a stable pattern by using the English word *Decision*. Decision refers to making some judgment through the act of deciding or it is a process determining something as of a question by making a judgment. Decision could be both positive and negative depending on the error of judgment. Factually, no domain exists in the world that does not use the term Decision. The importance of making a decision is extremely relevant to almost all occupations. A corporate manager must decide to formulate an action plan to motivate his/her project team members to perform to the best of their abilities. The court of law will be compelled to make a decision to sentence a criminal or set him free for lack of enough evidence. A student may make a poor decision to drop out of school because of some silly and strange reasons; in fact, she/he may repent that decision later in the life. Sometimes, couples who are planning to divorce may find making a decision to end the relationship extremely disturbing considering the young age of their children. Considering these wide-ranging uses, the applicability of making a decision is rather multidimensional and it is applicable to many domains and different contexts.

This unique character and attribute is enough to motivate pattern developers to design a model for Decision; calling the model AnyDecision could be practical idea as it portrays all domains and areas where someone makes a decision. One of the major advantages of designing AnyDecision pattern is an opportunity to detect innumerable numbers of business objects (BOs) and a stable essential business theme (EBT). The ultimate goal of any pattern is to find the most appropriate EBT. In this chapter, the ultimate goal of AnyDecision intends to make a decision. A sturdy and robust model, AnyDecision stable pattern could be a universal application that outperforms any other traditional pattern that is made of the term *Decision*. As the core concept of Decision is almost similar to all scenarios of uses, it may be advantageous to create a general pattern model that can solve all recurring problems that usually arise while making a decision. The basic principles of stable software stability will also enable developers to create stable architectures that can work in all contexts irrespective of different usage scenarios [2–4].

18.2 PATTERN NAME: DECISION STABLE DESIGN PATTERN

18.2.1 DEFINITION

Decision can be defined as a conclusion or resolution reached after consideration of different factors. It is the action or process of deciding something or of resolving a question. Decision is chosen as a BO and is not limited to only one idea.

Decision is a general concept of thought process to select one logical option from other available options. A person must consider all positive and negative sides of each option while making a good decision. He/she must be able to predict the impact of each option and based on it decide the best possible option for the situation.

18.2.2 CONTEXT

Decision has innumerable applications in different domains, such as criminal laws, entrepreneurship, business studies, philosophy, academics, finance and economics, medical profession, and other areas that need decision-making. The following are some of the contexts where we can apply AnyDecision stable design pattern.

1. *Decision in Economics and Finance*: Economic decision is the process of an organization (AnyParty) taking business decisions involving revenue generation (AnyReason). The main purpose of making a decision is generally to create strategies and action plans (AnyConsequence) that help one another to make the company more valuable or to increase the owner's revenue (AnyReason). Those involved in the decision-making process must have access to the company's detailed financial reports and they should have a good understanding of the company's economic climate (AnyFactor).

   Success in the world of business depends on a company leader's (AnyParty) ability to make wise economic decisions. Since economic decisions have to do with the amount of money a business earns or is valued at, it is essential that the decision-making process relies heavily on the company's financial reports (AnyCriteria). The outcome of the decisions being made should involve some type of monetary reward, and the decision-makers generally expect the reward to be much greater than the cost or sacrifice that's required to obtain it [5].

2. *Decision in Medical*: Decision-making in medical field is a collaborative process that allows patients (AnyActor) and medical assistance providers (AnyActor) to make healthcare decisions together by taking into account the best scientific evidence available. It also involves taking into account the patient's values and preferences (AnyCriteria). Decision-making allows both the provider's expert knowledge and the patient's right to be fully informed of all care options and the potential harms and benefits (AnyFactor). This process provides patients with the support they need to make the most appropriate and individualized care decisions while allowing providers to feel confident in the care they prescribe (AnyReason). Decision-making helps providers suggest various treatment options in an unbiased and balanced way to patients so they can make an informed choice (AnyType).

3. *Judgment (Law)*: Judgment is a decision of a court (AnyParty) regarding the rights and liabilities of parties in a legal action or proceeding. Judgments also generally provide the court's explanation of why it has chosen to make a particular court order (AnyCriteria).

   A judgment may be provided either in written or oral form depending on the circumstances (AnyConsequece). Types of judgments (AnyType) can be distinguished on a number of grounds, including the procedures the parties must follow to obtain the judgment (AnyFactor), the issues the court will consider before rendering the judgment, and the effect of the judgment [6].

18.2.3 Problem

Decision can be applied to various situations in life. However, the major problem is how to make decision that are generalized enough, so that it can be applied to different domains. AnyParty and AnyActor should be able to make decision in any domain by using this solution. The problem also concentrates on providing enduring business techniques, which can be relevant to any given scenario and AnyParty or AnyActor can use these patterns to resolve their pending problems.

Different IOs show many changeable scenarios where EBT and BO could be used in an appropriate manner. Here, we will not have to make any changes to the system while changes will only be made to applications that one uses. EBT represents the core theme while the BO that is extremely stable would be internally adaptable to these IOs.

Every problem has following requirements.

18.2.3.1 Functional Requirements

1. *Intension*: The ultimate goal of intension is to know the internal content of a concept. It is the purpose of the effect of one's conduct for making the decision. Intension has at least one instance of application, which substitutes coextensive expressions into it and will not store the logical value. Intention is the EBT for the proposed pattern.
2. *AnyDecision* is a BO, AnyDecision emanates through someone's intention to make a decision. Decision leads to some consequences that could be good or bad. AnyCriteria influences AnyDecision while AnyParty and AnyActor should set the criteria.
3. *AnyParty*: Legal users using the system for intension for making the decision can use this concept. AnyParty has an intension and makes AnyDecision to implement it. AnyParty should follow some criteria to make a decision.
4. *AnyActor*: Actor is not part of the system. It acts on the objects present in the system. There are four types of actors namely—Software, Hardware, People, and Creature. Even AnyActor should set some well-defined criteria to make a decision.
5. *AnyConsequence*: Any decision leads to any consequence. This plays important role as it defines the result of the intension.
6. *AnyFactor*: Intension leads to factors that are to be considered to predict the consequences of decision. Factor also helps reaching the destination or walk on its path with a clear intension.
7. *AnyEvent*: AnyDecision is based on any event, which describes types that are going to make decision for well-defined intension. AnyEvent refers to the events that contain any activity that needs to be held according to the decision.
8. *AnyEntity*: Entity is a participant of the complete system that the pattern represents.
9. *AnyCriteria*: AnyParty or AnyActor defines the criteria through which the propagation of waves is going to happen. An entity (AnyActor/AnyParty) should have predefined criteria to make any type of decision.
10. *AnyType*: AnyDecision can be of any type depending upon the reason for intension. The decision can be related to any type.
11. *AnyReason*: It is a statement presented in justification or explanation for any intension.

18.2.3.2 Nonfunctional Requirements

1. *Reliability*: Decision made should be reliable and trustable. It should have competency of success encompassing the intention. Reliability exhibits a sense of solidity to decision that is made by a party.
2. *Scalability*: This is the nonfunctional requirement of decision. A dynamic representation is required to characterize constant changes and progress. It will be able to handle a growing amount of work in a capable manner. It has an ability to enlarge to accommodate that growth. It is the inherent ability not only to perform well in the rescaled situation, but also to actually take full advantage of the already made decision.

3. *Performance*: Performance is the action or process of carrying out or accomplishing an action, task, or function. A decision should ensure fair degree of accuracy and it should be able to compare itself with known standards of accuracy, completeness, cost, and speed.

18.2.4 SOLUTION

The below listed diagram provides a solution for the challenges and constraints mentioned in the previous section. The EBTs and BOs written in the class diagram can be applied to any domain.

18.2.4.1 Class Diagram for Decision Stable Design Pattern

Class Diagram Description (as shown in Figure 18.1):

1. AnyActor or AnyParty can have intension.
2. AnyParty/AnyAction follows AnyCriteria.
3. Intension is implemented through AnyDecision.
4. AnyDecision leads to AnyConsequence.
5. Intension is because of AnyReason.
6. AnyReason is composed of AnyType.
7. AnyDecision is based on AnyEvent.
8. AnyEvents depends on AnyFactor.
9. AnyCriteria influences AnyCriteria.
10. AnyDecision determines AnyEntity.

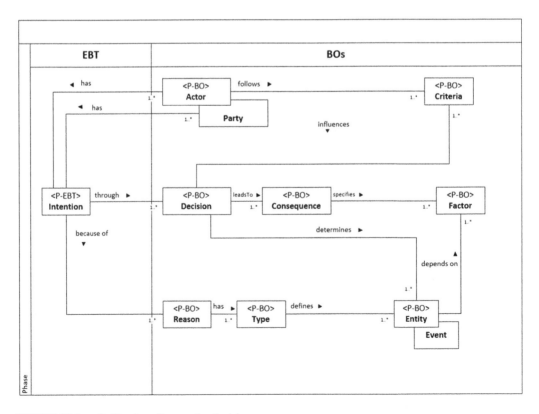

FIGURE 18.1   Stable class diagram for decision.

18.3   SUMMARY

This chapter proposes a stable design pattern for decision. This stable pattern has been developed to represent any type of decision by using software stability principles as proposed by many authors. The core knowledge of decision has been extracted from different domains and presented as EBTs and BOs. Intension is the EBT of the system. The core model can be extended with numerous IOs to design applications that are suitable to specific domains. This is illustrated with an application that uses the core model to implement any type of decision that eventually demonstrates the applicability and usefulness of the application.

18.4   REVIEW QUESTIONS

1. What is *Decision* and can you list different uses for this term?
2. Why decision is an important word for this chapter?
3. Is the term *Decision* a general one that propagates almost similar meanings for all versions and the word?
4. How does this term drive a pattern developer to design and create a stable pattern?
5. Does it solve the problem of reusability and adaptability that are likely to emerge in pattern making processes?
6. What is the EBT for Decision?
7. What is AnyDecision stable design pattern and list some of the important features of this pattern?
8. What are the different BOs for this pattern?
9. Is stable pattern for Decision (AnyDecision) better than a traditional model? If yes, provide proof and answers. Highlight why traditional pattern-making process is riddled with innumerable problems.
10. Recreate and design similar software patterns for at least five scenarios that are not covered in this chapter.
11. Write down use case diagrams for each one of these scenarios.
12. Write use case descriptions for each one of these examples.
13. Write down different contexts of applications where Decision works in combination with other factors.
14. What are some of the functional requirements of this pattern?
15. List some of the nonfunctional requirements and think of some more that are not covered in this chapter.
16. What are the major advantages and benefits of this pattern?
17. What are some of the pitfalls and disadvantages of creating this pattern? Could you list some of the potential challenges, which are likely to emerge while designing the pattern?
18. Is the AnyDecision pattern stable, robust, and reusable over different domains?
19. If so, give reasons why it is stable.
20. Draw your inferences and a brief summary for this chapter. Provide a bird's eye view of the entire chapter.

18.5   EXERCISES

1. Decision versus Judgment stable patterns
   a. What are the differences between stable decision pattern and judgment pattern?
   b. Create the judgment stable pattern.
   c. Name and describe three contexts for each of them.
   d. Show how to apply the two patterns including one context of your choice and including a use case, the application of each pattern with your select context.

2. Decision versus opinion stable patterns
   a. What are the differences between stable decision pattern and opinion pattern?
   b. Create the opinion stable pattern.
   c. Name and describe three contexts for each of them.
   d. Show how to apply the two patterns including one context of your choice and including a use case, the application of each pattern with your select context.
3. Create five applications of AnyDecision patterns in a table as shown in Table 23.1 (Chapter 23).
4. Create two applications of two contexts of AnyDecision stable design pattern and provide the following for each one of the applications
   a. Context
   b. Application diagram using the hype stable design patterns
   c. Detailed use case
   d. A sequence diagram for the application

18.6   PROJECTS

Use Decision in one of the following and create:

a. Making Decision
b. Decision Tree
c. Decision Game
Three contexts for each one of them
Stable architecture pattern for each of them
An application of each one of them

REFERENCES

1. Keri Lynn Russell (born March 23, 1976), last modified on May 31, 2017, https://en.wikipedia.org/wiki/Keri_Russell
2. M.E. Fayad and A. Altman. Introduction to software stability, *Communications of the ACM*, 44(9), 2001, 95–98.
3. M.E. Fayad. Accomplishing software stability, *Communications of the ACM*, 45(1), 2002, 95–98.
4. M.E. Fayad. How to deal with software stability, *Communications of the ACM*, 45(4), 2002, 109–112.
5. P. Ghirardato. *Decisions in Economics and Finance*. Online Journal no. 10203, Springer, September 2014.
6. B.A. Garner. *Black's Law Dictionary*, Thomson Reuters, Eagan, MN, Tenth Edition, May 9, 2014.

19 AnyConflict Stable Design Pattern

For good ideas and true innovation, you need human interaction, conflict, argument, debate.

Margaret Heffernan [1]

Conflict denotes different meanings under different contexts and circumstances. In simplest terms, it is a struggle for property, position, power, and command. However, in a definitive explanation, it means a strong disagreement between people, nationalities, and groups that eventually results in a violent or angry argument. Conflict could also be a visible barrier that prevents forming a consensus and common agreement between two groups of people. A literary meaning of conflict means the opposition of persons or forces that gives rise to the dramatic action in a drama or fiction. A conflict is a specific word that involves some sort of dissent and division between two parties. In other words, a conflict is a specific word that indicates the existence of a mutual discord.

This chapter focuses on creating a stable pattern by considering various usages of the English word conflict. This pattern also looks forward to decipher the central theme of conflict and provide a practical solution to merge all aspects of conflict into one application that can eventually work in different contexts and uses. Some of the major advantages of using this pattern are its stability, robustness, applicability, and extendibility across a wide range of domains. In addition, pattern developers are also expected to save considerable quantum of money and time that are needed to complete the process of pattern making. The pattern conceived here (AnyConflict) would also help them solve a myriad variety of problems that are expected to crop while designing the stable pattern.

19.1 INTRODUCTION

This chapter intends to design and create a stable pattern by using an English term called conflict. Conflict means a dissenting situation when angry and violent scenarios might occur because of serious disagreement. A conflict usually leads to a bitter and an acrimonious end. However, conflicts could be solved by mutual discussion and agreements; incidentally, people who are involved in conflicts may put an end to it by clarifying reasons that led to a conflict. Conflicts occur everywhere and in all domains of life. For example, conflicts exist between nations and states because of varying reasons. Conflicts may also occur in human relationship like in the case of husband–wife, father–son, and among persons who might be working anywhere. Conflicts also arise in politics and decision-making groups, corporate firms, and any other areas where conflicts are likely to emerge because of differences in opinions.

Conflicts are unique in their nature and attribute. It is universal and present in many scenarios of life. Its special nature may be just enough to motivate pattern developers to design stable model for conflict. Named, AnyConflict, this model is expected to represent all domains and realms where one expects a conflict to arise. With this pattern, developers may also be able to extract the main business theme behind conflict and different business objects (BOs) that are likely to drive the pattern forward with their technical contributions. The ultimate goal of any pattern is to find the most appropriate essential business theme; the EBT for conflict is to provide a clarification so that a pending conflict is solved successfully. Compared to a pattern that is developed by using a traditional method, a stable model will be universal and extremely beneficial. It will also assist in solving all pending problems that usually arise when finding a resolution for conflicts by providing clarifications. Ultimately, finding a stable architecture for a software pattern is the main goal for any pattern developer and this chapter intends to achieve this goal by developing a pattern model for conflict.

19.2 PATTERN NAME: ANYCONFLICT STABLE DESIGN PATTERN

The stable design pattern AnyConflict is based on the term conflict, which generally refers to a situation in which it is difficult for two things to exist together or be true at the same time. Conflict is a BO. A conflict might arise because of different reasons and it might occur between two individuals, two groups of people, or between two nations. The factor of generality is the main reason for choosing this term as it is appropriate and fit for all the possible scenarios of conflict. This will eventually lead to the process of creating a stable design pattern by using clarification as a conflict's enduring business theme (EBT).

19.2.1 CONTEXT

The AnyConflict stable design pattern can be applied to any domain or a situation where we have any kind of disagreement and when clarification is demanded to solve and find a resolution for conflict. Conflicts may also occur when people (or other parties) perceive that, because of a disagreement, there is a threat to their needs, interests, or concerns. It has applications in various domains such as interpersonal relationship, politics, business, sports, groups, and among animals too, if one of them encroach another's territory. The following are some of the contexts where we can apply AnyConflict stable design pattern.

Scenario 1: Conflict with customers. A car sales representative (AnyParty) sells a used car (AnyEntity) without a performance guarantee or warranty (AnyEvent). After sometime, the car breaks down (AnyReason) and the buyer (AnyParty) may angrily confront the seller (AnyType) who is the salesperson (AnyDisagreement) and demand a refund (AnyOutcome). To solve this conflict (AnyConflict), one will need to involve a manager (AnyParty) who has the authority to offer refunds, discounts, or other facilities.

Scenario 2: Conflict at workplace because of discrimination issues. A minority employee (AnyParty) in a team (AnyEntity) feels that he is consistently assigned (AnyEvent) the most menial work (AnyReason) tasks in the group. This employee may begin to harbor resentment (AnyDisagreement) against team members and managers (AnyParty) eventually lashing out through decreased productivity (AnyOutcome) or outright verbal conflict with others (AnyConflict). To resolve this issue, the manager sits down with the team and discusses several ways in which job tasks are assigned, and in making changes as necessary to ensure that tasks are divided equitably.

Scenario 3: Performance review Conflict. A company's employees (AnyParty) are extremely angry (AnyType) over not receiving expected pay raises, promotions, or other performance-related incentives (AnyReason). Consequently, they are showing their discontent (AnyType) through gossip and a negative attitude at work (AnyOutcome). The management (AnyParty) feels that they gave good work appraisals last year (AnyDisagreement). However, they are refusing to give more pay rises and benefits due to decreased business performance this year (AnyEvent). To resolve the conflict arising out of a negative performance review, the managers (AnyParty) are working directly with the employees to improve work performance and tie the completion of these goals with guaranteed incentives for the next year.

19.2.2 PROBLEM

The main problem is to create a stable pattern around conflict that can be applied to all types of contexts and situations. In addition, another perceived problem is to design it in such a way that it is stable and resource managing unlike a traditional pattern that is difficult to apply to differing scenarios that usually occur whenever situations of conflicts occur.

19.2.2.1 Functional Requirements

1. *AnyParty*: This refers to a legal user who acts on his or her own will. This can be the person or team or any organization involved in the conflict. An actor represents someone or something which is involved in conflicts.
2. *AnyDisagreement*. This refers to a situation, when two parties do not agree on something or if their opinions differ. This could be over the matter of wages, performance appraisal, or over a discrimination issue. Disagreement always leads to a conflict that could be either argumentative or confrontational.
3. *AnyReason*: This refers to the activity that leads to the conflict. This can be anything that leads to a disagreement over any matter. AnyReason always leads to AnyType of argument and eventual conflicts.
4. *AnyType*: This refers to the type of conflict. It can be in any form, either negative work environment or in the form of strikes or even domestic quarrels.
5. *AnyEntity*: This refers to anything that are involved in conflict. This class represents the entity, which is involved in any conflict and is named by type of any conflict.
6. *AnyEvent*: This refers to the event at the time of conflict or which leads to conflict. This class also represents the event itself, which is involved in AnyConflict and is named by type of AnyConflict.
7. *AnyMedia*: This refers to the environment in and around conflict. This class represents the medium through which conflicts take place. AnyEntity or AnyEvent resides on AnyMedia.
8. *AnyConflict*: This refers to the actual disagreement, which happens in the form of conflict. Conflicts always occur due to differences in opinions and they are marked by violent arguments and debates.
9. *AnyOutcome*: This refers to the eventual outcome of conflict, either a strike or dissent displayed by the car owner or negative work environment. Outcome could be positive or negative, result yielding or it could be a failure.

19.2.2.2 Nonfunctional Requirements

1. *Understanding*: This means the sympathy that comes from knowing how other people feel and why they do things or an agreement made in an informal way or not expressed in words. To understand others, a party should know that a conflict would result in agony and pain to others. Successful resolutions to any conflicts are possible only when there is a deep understanding between all parties concerned.
2. *Change*: This usually means to become different or to make someone or something different in terms of conflict. Change usually occurs after successful resolution of conflicts. Change will also occur after providing meaningful clarifications by parties involved in conflicts. Changes that occur after successful resolution are either positive or negative. For example, failed resolutions may lead to wars, divorces, or even acts of crime.
3. *Acknowledgment*: This refers to acceptance of the truth or existence of something or the action of expressing or displaying gratitude or appreciation for something. Unless one acknowledge that there is a conflict, resolutions and arbitrations are never possible.
4. *Struggling*: Conflicts always mean struggling and suffering. Conflicts that lead to quarrel in families and among nations usually end up in suffering and struggling. For example, a divorce will lead to mental agony as well as separation of children from either one of the parents. A conflict leading to a bitter war may result in economic recession, shortage of food, and other essential needs and impoverishment.

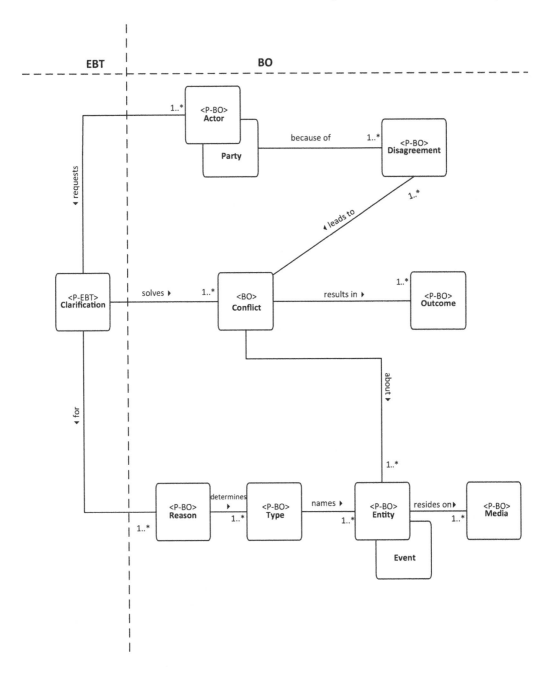

FIGURE 19.1 AnyConflict stable design pattern.

5. *Engagement*: Conflicts always involve engagement of two parties in negative connotations of mutual participation like arguments, quarrels, debates, crimes, and wars. However, positive engagements like mutual talks and resolution measures will always yield positive results.

19.2.3 SOLUTION

Class Diagram Description (as shown in Figure 19.1):

1. AnyParty requests clarification at the time of conflict.

2. AnyParty, because of AnyDisagreement, leads to AnyConflict.
3. AnyConflict can be solved by using AnyClarification.
4. AnyConflict produces AnyOutcome, which is either good or bad.
5. AnyConflict can occur because of AnyReason.
6. AnyConflict can be of any AnyType.
7. AnyType names AnyEntity and AnyEvent involved in AnyConflict.
8. AnyConflict can be in any type of environment, that is, AnyMedia.

19.3 SUMMARY

Software stability model (SSM) offers better stability and scalability than the ones provided by traditional modeling, since it ensures a holistic way of modeling, where the pattern defined can be applied to any type of applications irrespective of the domain. While modeling a system in traditional approach, the system becomes more specific to a particular application for which it is being modeled. Nevertheless, the SSM usually focuses the ultimate goal of the application and models the system accordingly. This makes SSM more flexible and maintainable. SSM points out many flaws of traditional modeling and it helps build better software models with its patterns [2–4].

19.4 REVIEW QUESTIONS

1. What do you understand by the word *conflict* and list different uses for this term?
2. Why conflict is an important word in the context of this chapter?
3. Conflict has a global significance—do you agree?
4. Is the term *conflict* a general word that assumes similar meanings for all forms of the word? Compare three scenarios where the essence of conflict is perceived—divorce, war, and a game of football.
5. How does this term motivate a software pattern developer to design and create a pattern?
6. Does it solve the problem of reusability and adaptability?
7. What is the EBT for conflict? Why do you list it as the main theme?
8. What is AnyConflict stable design pattern? Can you define it in detail?
9. What are the different BOs for conflict?
10. Is stable pattern for conflict (AnyConflict) better than the one that was created by using a traditional model? If yes, provide proof and answers.
11. Recreate similar software patterns for at least two scenarios that are not covered in this chapter.
12. Write down use case diagrams for each one of these scenarios. Highlight each step of the diagram with brief notes.
13. Write use case descriptions for each one of these examples.
14. Write down different contexts and realms where conflict works in combination with other factors.
15. What are some of the functional requirements?
16. List some of the nonfunctional requirements and think of some that are not covered in this chapter.
17. What are the major advantages of this pattern?
18. What are some of the disadvantages of creating this pattern? Could you list some of the potential challenges and problems that are likely to emerge while designing the pattern?
19. Is the AnyConflict pattern stable, robust, and reusable?
20. If so, give reasons why it is stable. Be elaborate in your answers.
21. Draw you inferences and a brief summary for this chapter.

19.5   EXERCISES

1. Conflict versus agreement stable patterns
   a. What are the differences between stable conflict pattern and agreement pattern?
   b. Create the agreement stable pattern.
   c. Name and describe three contexts for each of them.
   d. Show how to apply the two patterns, including one context of your choice and including a use case, the application of each pattern with your select context.
2. Conflict versus harmony stable patterns
   a. What are the differences between stable conflict pattern and harmony pattern?
   b. Create the opinion stable pattern.
   c. Name and describe three contexts for each of them.
   d. Show how to apply the two patterns, including one context of your choice and including a use case, the application of each pattern with your select context.
3. Create five applications of AnyConflict patterns in a table as shown in Table 23.1 (Chapter 23).
4. Create two applications of two contexts of AnyConflict stable design pattern and provide the following for each one of the applications:
   a. Context.
   b. Application diagram using the hype stable design patterns.
   c. Detailed use case.
   d. A sequence diagram for the application.

19.6   PROJECTS

Use conflict in one of the following and create:

a. Conflict Analysis
b. Conflict Resolution
c. Conflict Management
Three contexts for each one of them
Stable architecture pattern for each of them
An application of each one of them

REFERENCES

1. Margaret Heffernan (born 1955), last modified on June 13, 2017, https://en.wikipedia.org/wiki/Margaret_Heffern
2. M. Fayad and A. Altman. Introduction to software stability, *Communications of the ACM*, 44(9), 2001, 95–98.
3. M. Fayad. Accomplishing software stability, *Communications of the ACM*, 45(1), 2002, 95–98.
4. M.E. Fayad. How to deal with software stability, *Communications of the ACM*, 45(4), 2002, 109–112.

20 AnyView Stable Design Pattern

It's easier to go down a hill than up it but the view is much better at the top.

Henry Ward Beecher [1]

The applications for any view concept exist in different contexts of pattern modeling, but the traditional model represents only a single application. Nevertheless, all the applications related to any view are handled in almost similar manner. Since the concept of any view is similar in all of the applications, an analysis of any view concept to extract its core knowledge would be a valuable contribution to software developers and software domain alike. The any view stable analysis pattern is a sturdy model, which would provide the core knowledge of any view and it is based on the principles of software stability model (SSM). Findings: Flexibility and adaptability—the proposed design pattern is not specific to a single domain. It is relevant in numerous areas that need viewing; hence, it is very flexible and adaptable. To achieve the best goals of modeling, the proposed pattern will maintain a high level of extensibility. The outline of this any view stable analysis pattern is exceptionally interoperable because it can be identified with other stable patterns.

20.1 INTRODUCTION

In English language, *view* is an ability to see something or to be seen from a specific place. A person can view a beautiful landscape in a mountainous area, or he or she could sight something that is in the field of vision. View may also mean to look at something or inspect an object or a thing. In its verbal meaning, view could also mean to look at a person with a specific attitude. View could also be an opinion, belief, or idea. In the context of this chapter, view could be termed as something that could be seen and believed.

In common, a software pattern scenario like traditional modeling, any view provides just a single instance of solution. However, we propose a stable model of any view that is easy to extend and made applicable to different scenarios and concepts. In fact, there is need for such a stable model that could be used in many applications. The basic concepts of SSMs will help us create [2–4] AnyModel patterns and it is possible to hook other concrete objects with this theme to create a variety of applications. The AnyModel stable patterns will result in the culling of core knowledge of the concept, which will have different requirements, while enforcing it on other scenarios. The proposed design pattern is highly flexible and it is not tied to any single domain. It is quite relevant areas of computing that needs differing views. This approach will help us reduce the degree of complexity and reduced adaptability.

Section 20.2 discusses the pattern documentation that illustrates the context of the pattern along with some scenarios, discusses the functional and nonfunctional requirements, and illustrates the solution with the pattern diagram. Section 20.3 provides summary. Section 20.4 provides review questions. Section 20.5 lists a number of exercises. Section 20.6 presents projects. The chapter ends with a list of references.

20.2 PATTERN DOCUMENTATION

20.2.1 Pattern Name: AnyView Stable Design Pattern

View is business object (BO). A view is an opinion about something.

A view can be something that is seen, believed, outlined, or just spectacle. Other names for view are aspect, glimpse, outlook, perspective, or picture. Generality is the main reason for choosing this term as this term is appropriate for all the possible scenarios of AnyView. This will lead to a stable design pattern by using viewing as its enduring business theme (EBT).

20.2.2 Context

View has numerous applications in various domains such as entertainment, magazines, software engineering, and so on. Some types of views are graphical view, SQL view, model-view-controller design pattern, point of view, and many more. The following are some of the contexts, where we can apply AnyView stable design pattern.

1. *Computer View*: Computer view commonly known as computer vision is a process where a computer (AnyActor) can recognize images like human beings. It deals not only with recognizing (AnyView) but also with processing, analyzing, and understanding (Viewing) the images and making decisions. This type of object recognition techniques can be of great use in many areas. One example would be that it could help a blind person (AnyActor) in recognizing the high dimensional objects without touching (AnyCriteria) them. A common way of recognizing the objects for blind people is by touching and feeling the objects, but computer view or computer vision eliminates the necessity of touching every object, but can recognize objects just by processing images. A newspaper can be read aloud by image processing technology, which is a subset of computer view, just by putting the newspaper in front of the camera. Similar technology is also being used to produce unmanned vehicles, which are run by processing the view, and which is captured by them helping the vehicle to run without a human. A computer will capture images (AnyEntity) by using a camera (AnyMedia). Camera has different modes of operations such as automatic (AnyMode) or manual or night mode.
2. *Bird's Eye View*: Bird's eye view (AnyView) is a view from a top altitude, as viewed by a bird when it is flying. The observer is regarded as a flying animal in this case. Bird's eye view has various modes such as a photograph, video (AnyMode), or a drawing. A view from the top of a high mountain or a tower is also considered as bird's eye view. One uses this kind of view while preparing a map, layout, filmmaking, or a blueprint. In filmmaking, the cameraman (AnyParty) uses a bird's eye shot, say to capture a battle scene, (AnyEvent) to view the actors (AnyParty) from above and move near or away from the subjects. Generally for such shots, the camera needs to be taken to a higher altitude (AnyCriteria) than the usual. For taking cameras to a high altitude, the director and cinematographers mostly use a crane (AnyMedia) to achieve specific shots.

20.2.2.1 Functional Requirements

1. *Viewing*: This class represents the inspection of something. Viewing has an *elementOfInspection, type, and description*. It can have operations such as *survey(), inspect(), explore(), and observe()*.
2. *AnyView*: This class represents an instance of vision. A view can also be a picture of something or even an opinion depending on the context. View class can have attributes such as *context, mode, and range*. It can have operations such as *examine(), judge(), forSee(), and represent()*.
3. *AnyParty*: AnyParty class represents a country, political party, government, and persons belonging to an organization. AnyParty will generally have a *name, location,* and *phoneNumber*. It will have operations such as *monitor(), analyze(), and check()*.
4. *AnyActor*: AnyActor participates in various activities. An actor has a *name, id, and category*. It has operations such as *interact(), look(), plan(), and conclude()*.

5. *AnyCriteria*: AnyCriteria represent a class that is used to take decisions. It has attributes such as *standard, principle, proof, and judgment*. It has operations such as *test(), confirm(), verify(), and assess()*.

6. *AnyMedia*: This class represents the media through which AnyView takes place. Data and information reside inside this medium. It has operations attributes such as *name, availability, and is Available*. It has operations such as *display(), store(), capture(), broadcast(), and connect()*.

7. *AnyEntity*: This class represents the entity that is viewed. An entity can be any visible object. It has attributes such as *name, type, and position*. It has operations such as *status(), performFunction(), and update()*.

8. *AnyEvent*: This class represents any viewable occurrence. An event is something that happens. It has attributes such as *name, occasion, type, and outcome*. It has operations such as *appear(), happen(), and arrange()*.

9. *AnyMode*: AnyMode class represents the manner of doing something. It has attributes such as *method, type, condition, style, and name*. It has operations such as *provideChoice(), governQuality(), and regulateFashion()*.

20.2.2.2 Nonfunctional Requirements

1. *Quality*: A view should be of high quality. It should have the essential characteristics. Because of the increasing demand in consumer electronics, 3D display systems have come into existence. According to high-quality view synthesis algorithm [1] frames with virtual views can be corrected. The algorithm for high-quality view synthesis helps in capturing good human view perception.

2. *Effective*: A view should be able to accomplish a purpose. It should give the expected result. For example, in the case of human computer interaction which uses computer view or computer vision technology, effectiveness would mean that a system should be able to complete the tasks given to it in an accurate manner. In software engineering, when we are building system architecture, the view model needs to be effective else the framework will lack coherence.

3. *Accessible*: A view should be easily approachable and obtainable. We should be able to reach it easily. Say, there is an automated vehicle which can view, process, and analyze its surrounding in order to drive but is such a vehicle accessible to majority of car owners is a challenge. A computer view should be accessible by the unmanned vehicle in order to function. A cameraman cannot take a point of view shot when it is not accessible to him.

4. *Complete*: A view should be complete. Without completeness, a view may not make any sense. For example, a landscape view that is blocked by an object is never complete to the viewer or the camera. There should not be any hindrance to a view so that it looks complete in all respects. A view made about a person or an entity should be complete and it should not be half absorbed or it should not be biased.

5. *Normalized*: A view should be normal to the eyes of the viewer. A normal view can be of two types—a psychological mode of view as seen by in a human eye. Many times, such views may not be normal. For example, views made about others and crystallized in the mind. Such views could be highly skewed and biased. Another type of view is materialistic and external. For example, a view made on a painting should provide normal perceptions and it should not look abnormal.

6. *Illustrative or representable*: A view should represent something that is easy to represent. For example, a view of a great person like Mahatma Gandhi automatically draws the mind of a viewer to model himself to the ideals of that person. In other words, a view should be illustrative and identifiable to a person's mind.

7. *Perspective*: Views should provide a complete perspective of what is going on and how it has happened. For example, a person can be viewed with many possible scenarios. A view could inform us many things: A person can be good or bad depending on the perception

we generate. However, our views could be very biased and bad. A complete perspective of any view should be available to anyone who wants to get a view.

8. *Visibility*: A view should be visible completely before a viewer could make a perception. When a view is not visible, the viewer may not be able to understand or comprehend the real meaning of the view. Scenery should be completely visible, while a model of clay should be completely visible so that it could be evaluated properly.

9. *True Abstraction*: A view should denote true abstraction. Abstract is derived from a Latin word "pulled away or detached," and the basic idea is of something that is detached from physical or reality. Views should provide general ideas rather than specific people, objects, or actions. In other words, a viewer should express ideas and expressions without trying to create a realistic picture.

10. *Remarkable*: The view should be remarkable and significant. A view of natural scenery will remain for a long time in the minds of a viewer. A view made of a great personality will be remarkable and it will remain the minds for longer duration. Remarkable views are something unusual, extraordinary, or worthy of attention.

20.2.3 Solution

The following class diagram will provide details about views, AnyView, and Viewing.
Class diagram description (as shown in Figure 20.1):

1. One or more party/actor does viewing based on one or more any criteria.
2. Viewing utilizes one or more any view.

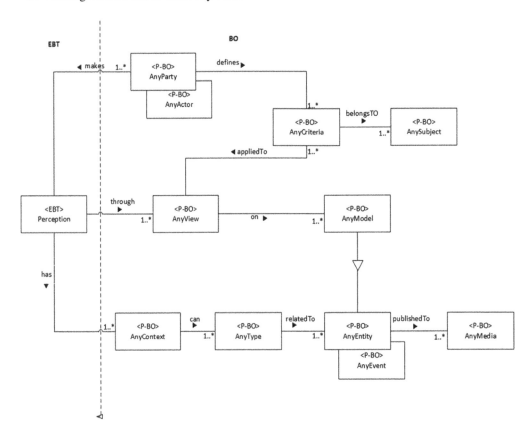

FIGURE 20.1 AnyView stable design pattern.

3. AnyView resides on one or more media and is controlled by any criteria.
4. AnyView is of zero or more entity/event.
5. AnyEntity/event has one or more any mode.

20.3 SUMMARY

The midsize template provided in this chapter demonstrates how one can identify core knowledge of AnyViews and in what manner it can be used to create applications for many different scenarios [5]. Identifying the main theme of pattern is quite difficult, and we have succeeded in deriving the main theme along with its BOs and IOs. This model can also be extended to identical contexts. Stability, reusability, and robustness of the pattern itself are a very big contributions apart from reduced time, money, and effort to create required patterns.

20.4 REVIEW QUESTIONS

1. Define and discuss the AnyView stability design pattern.
2. Do you consider view as the most appropriate BO in the pattern?
3. List a few application domains for the AnyView design pattern.
4. List four challenges that you might face while formulating the AnyView design pattern.
5. Is AnyView stable design pattern domain and application specific—True or False?
6. List any two constraints or challenges while formulating the AnyView design pattern.
7. List all the participants in the AnyView design pattern.
8. Illustrate the class diagram for the AnyView design pattern other than the one given in this chapter.
9. List one tradeoff while using this pattern.
10. Why the word view is too generic and general for use in making a stable pattern?
11. The word view connotes many meanings. Pick the one that best fits for this chapter.
12. List two main advantages or benefits of using the AnyView stable design pattern.
13. List two applications of the AnyView design pattern.
14. List some research issues for AnyView pattern chapter.
15. Is AnyView design pattern stable, interoperable, and reusable over time?
16. Why AnyView design pattern provides extensibility over a period?
17. List some of the related design patterns used in formulating the AnyView stable design.
18. List three test cases to test the participants of the AnyView design pattern.
19. Provide a brief summary of this chapter other than the one highlighted.
20. List three other contexts where view can be discussed as a topic for creating a stable pattern.

20.5 EXERCISES

1. View versus Model Stable Patterns
    a. What are the differences between stable view pattern and model pattern?
    b. Create the agreement stable pattern.
    c. Name and describe three contexts for each of them.
    d. Show how to apply the two patterns, include with one context of your choice and include a use case, the application of each pattern with your select context.
2. View versus observation stable patterns
    a. What are the differences between view stable pattern and observation pattern?
    b. Create the observation stable pattern.
    c. Name and describe three contexts for each of them.
    d. Show how to apply the two patterns, include with one context of your choice and include a use case, the application of each pattern with your select context.

3. Create five applications of AnyView patterns in a table as shown in Table 23.1 (Chapter 23).
4. Create two applications of two contexts of AnyView stable design pattern and provide the following for each one of the applications:
   a. Context
   b. Application diagram using the hype stable design patterns
   c. Detailed use case
   d. A sequence diagram for the application

20.6 PROJECTS

Use View in one of the following and create:

a. Map View
b. Relation View
c. Policy View
Three contexts for each one of them
Stable architecture pattern for each of them
An application of each one of them

REFERENCES

1. Henry Ward Beecher (June 24, 1813–March 8, 1887), last modified on June 15, 2017, https://en.wikipedia.org/wiki/Henry_Ward_Beecher
2. M.E. Fayad and A. Altman. Introduction to software stability, *Communications of the ACM*, 44(9), 2001, 95–98.
3. M.E. Fayad. Accomplishing software stability, *Communications of the ACM*, 45(1), 2002, 111–115.
4. M.E. Fayad. How to deal with software stability, *Communications of the ACM*, 45(3), 2002, 109–112.
5. M.E. Fayad, H.A. Sanchez, S.G.K. Hegde, A. Basia, and A. Vakil. *Software Patterns, Knowledge Maps, and Domain Analysis*, Auerbach Publications, Boca Raton, FL, Taylor & Francis Catalog #: K16540, December 2014, ISBN-13: 978-1466571433.

21 AnyModel Stable Design Pattern

We will not simply rely on a special fixed model for reform. All practical models are welcome.

Long Xinmin [1]

Model is a popular term in English language and it is used in a wide variety of contexts and scenarios depending on the situation. In essence, a model is an object, that is usually built to scale and it represents in detail another object that is larger than the model itself. In creative arts and sculpturing, a model is a preliminary work or prototype that serves as a plan from which a final product is made. Model could also be a human that serves as an object for an artist for a painter or photographer. Alternatively, it could also be a person that serves as the promoter for a fashion design show. With its dynamic range of uses in our daily life, *model* could also be an apt candidate in the domain of software pattern making.

Model is a very important issue that tries to represent an original object in its complete image through the depiction of a prototype or archetype. In the domain of stable software designing, model with its innumerable practical applications, could act as a bridge to link different areas, where the concept of modeling emerges as the central theme. The main motivation of this chapter is to create a valid solution to AnyModel to prevent working on finding similar solutions repeatedly from scratch whenever a developer attempts to create an application that demands the inclusion of modeling. The stable software pattern so created will find an array of important applications in many areas of life like fashion modeling, civil engineering, mechanical and electrical engineering, software domains, fine arts, math, space science, biological sciences, and others. This pattern will also help developers to find the real essence of the term Model that eventually helps in introducing stability, reusability, adaptability, and extensibility to the developed pattern application.

21.1 INTRODUCTION

Model is an English word that has many online dictionary meanings. It means a three dimensional representation of a thing or a structure. It could also mean a replica, copy, or a mock of a specific thing or object. In another context, it could mean a dummy, an imitation, a double, a duplicate, a look-alike, or reproduced object. Specifically speaking, model may also mean a thing or a person that used to follow or imitate as an example. Modeling is a verbal form that denotes a thing, an object, or a person for making comparisons.

The traditional software pattern model of any model provides a solution only to a single context of any model application. Hence, we propose a pattern model for the term model, which can be applied in various contexts and under different scenarios. Using the concept of the software stability model (SSM) [2–4]. AnyModel stable design model is derived with which other applications can be modeled just by linking it up with the required concrete objects.

The AnyModel stable design pattern creates core knowledge that applies to different applications, which will have different requirements for enforcing any model. The goal of this pattern is to discover the core knowledge behind the AnyModel concept, to build any type of application that is not specific to a certain domain. Once the core knowledge of the AnyModel concept is discovered, it can be used as a foundation to analyze the support requirements in every domain and any application.

SSM would make it easy to reuse any models because of the level and higher degree of abstraction it provides to the user. With this approach, we can reduce complexity and solve the problem of adaptability issues. Stable design patterns can be adapted to the user's level of expertise and organized based on users' any models. Other excellent principles of SSM such as scalability, extensibility, interoperability, and flexibility will be discussed further in the succeeding sections of this document [2–4].

This chapter starts with Section 21.2, which discusses the pattern documentation, illustrates the context of the pattern along with some scenarios, discusses the functional and nonfunctional requirements, and illustrates the solution with the pattern diagram. Section 21.3 provides the chapter summary. Section 21.4 provides 20 review questions. Section 21.5 lists 4 exercises. Section 21.6 provides projects.

21.2 PATTERN DOCUMENTATION

21.2.1 Pattern Name: AnyModel Stable Design Pattern

Model is business object (BO). A model could be a standard for making comparisons or a replica to show how something would look like.

A model can be an example, picture, 3D object, duplicate, pattern, instance, sample, and representation. Other names for model are illustration, miniature, exemplar, copy, dummy, and imitation. Generality is the main reason for choosing this term, as this term is appropriate for all the possible scenarios of model. This will lead to a stable design pattern by using modeling as its enduring business theme (EBT).

21.2.2 Context

Model has applications in various domains such as mathematics, computer sciences, music, business, psychology, arts, and entertainment. Some types of models are scientific models, 3D models, computer models, system models, conceptual models, mathematical models, and role models. The following are some of the contexts, where we can apply AnyModel stable design pattern.

1. *Mathematical Model*: A mathematical model is an illustration (Model) of a system, to represent how something works by using mathematical concepts. This process is known as mathematical modeling (Modeling). Mathematical equations are used to describe the behavior of the future in mathematical modeling. For example, a student (AnyParty) is given a math (AnyContext) problem as an assignment (AnyTest) by his professor (AnyParty) to find out the surface area of a cuboid, without using calculator (AnyCriteria). Solving such problem would be a lot easier if the student breaks down the problem by drawing a sample figure (AnyView) of a cuboid on a paper (AnyMedia) and then solving it. Generally, mathematical models can be divided in black box models and white box models. Consider white box (AnyType) model, where information about the working of a medicine (AnyEntity) is fully provided, such as how the medicine decays and how much medicine should be taken by a person.

2. *Fashion Model*: A fashion model (AnyModel) is a person who promotes clothes and accessories (AnyEntity) made by fashion (AnyContext) designers (AnyParty) in order to display their work generally at any fashion show (AnyMedia). A fashion model may also model items in photographs (AnyView) for advertisement purposes. Celebrities (AnyParty) are common in the field of fashion modeling (Modeling), who promote brands apart from their regular work. Supermodels (AnyType) are fashion models that are very expensive (AnyCriteria) to hire and are the face of a particular brand. All fashion models are live models, so a consumer is more encouraged to buy clothing promoted by their favorite celebrity, than by seeing it on a mannequin. It is also easy for fashion experts to analyze

(AnyTest) designer clothes at a fashion event, where several models would display clothes on themselves.

21.2.3 FUNCTIONAL REQUIREMENTS

1. *Modeling*: This class refers to making small copies of something. It has attributes such as *name, make, and date*. It has operations such as *represent(), provideCopy(), and visualize()*.
2. *AnyActor*: An actor represents someone or something, which is involved in modeling. It has attributes such as *name, id, and type*. It has operations such as *endorse(), design(), and illustrate()*.
3. *AnyParty*: This class represents party which is involved in violation. A party can be a country, political party, organization, or a person affiliated to an organization. AnyParty has attributes such as *name, id, and location*. It has operations such as *copy(), giveSample(), and imagine()*.
4. *AnyModel*: This class represents a small copy of something. It has attributes such as *type, version, and name*. It has operations such as *compare(), imitate(), and envision()*.
5. *AnyCriteria*: This class represents something that is utilized as a purpose behind settling on a judgment or choice. AnyCriteria defines a standard based on which one can judge. It has attributes such as *description, name, and id*. The operations of criteria can be *judge(), prove(), and measure()*.
6. *AnyTest*: This class represents the means by which one can measure the quality of something. It has attributes such as *name, numberOfQuestions, and procedure*. It has operations such as *check(), examine(), and verify()*.
7. *AnyView*: This class represents an instance of vision. A view can also be a picture of something or even an opinion depending on the context. View class can have attributes such as *context, mode, and range*. It can have operations such as *examine(), judge(), forSee(), and represent()*.
8. *AnyContext*: This class represents the instructions that tell the actor or party how something needs to be accomplished. It has attributes such as *description, scope, and category*. It has operations such as *validateMethodology(), analyzeCost(), and recommend()*.
9. *AnyType*: This class represents the type of model. AnyType has attributes such as *id, name, interfaceList, methodList, and property*. The operations are *change(), categorize(), and subtype()*.
10. *AnyEntity*: This class represents the entity, which is involved in any modeling and is named by type of any model. The attributes can be *id, name, and status*. Its operations can be to *update(), relationship(), and type()*.
11. *AnyEvent*: This class also represents the event which is involved in any modeling and is named by type of any model. It has attributes such as *name, occasion, type, and outcome*. It has operations such as *appear(), happen(), and arrange()*.
12. *AnyMedia*: This class represents the medium through which modeling takes place. AnyEntity or any event resides on any media. It has attributes such as *name, type, status, and capability*. It has operations such as *broadcast(), capture(), display(), select(), navigate(), and remove()*.

21.2.4 NONFUNCTIONAL REQUIREMENTS

1. *Helpful*: A model should make a job easier to do by giving assistance. Consider a situation, where a civil engineer draws the model of a house to show the clients how it would be constructed. The model drawn would visually represent the house and help the client to understand the layout in a better manner. Thus, a model should be of help in some manner.

2. *Acceptable*: A model should be fairly good and efficient of being accepted. Consider a supermodel in the context of fashion that has lots of experience in modeling. However, if a particular set of designer clothes do not fit so well on that supermodel, then it would be wise to choose a model on whom the clothes fit well. So, a model should be acceptable for the purpose it is being used for.

3. *Relevant*: A model should be relevant, such that it relates to a something in an appropriate manner. A 2D model is a great way to represent mathematical models, when solving problems related to 2D geometry, but when representations of 3D models is necessary like in games, a 2D model would be irrelevant. Hence, any model needs to be relevant.

4. *Completeness*: A model should be complete in all respects. A model that is incomplete may not make any sense and a person, who looks it finds it to be very worthless and valueless. In the domain of sculpturing, a sculptor should create a live statue by providing complete details.

5. *Consistency*: A model should be consistent in his or her approach and mode of working. For example, a live model that advertises for clothing should be consistent in her approach to catwalking so that the audience can watch every detail of her clothing, its style and substance. A model that mimics a live object should also be consistent in its designing and construction.

6. *Perfection*: A model created should be perfect and spotless in all respect in the areas of design, reproduction, and creation. A model painting of Mona Lisa should look exactly like the original Mona Lisa and not as a witch. A model stone carving object should mimic the original as far as possible too.

21.2.5 SOLUTION

Class Diagram Description (as shown in Figure 21.1):

1. One or more any party or any actor does modeling.
2. Modeling is done through one or more any model and within one or more any context.
3. AnyModel uses one or more any entity or any event with one or more any test.
4. One-to-many any evidence influences any model.

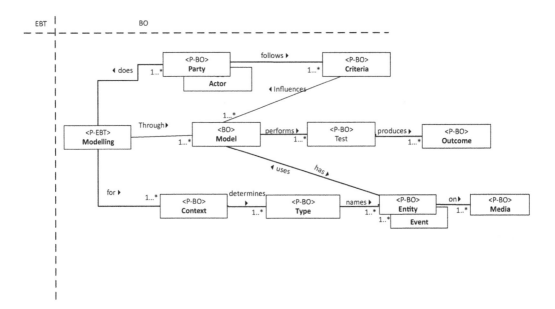

FIGURE 21.1 AnyModel stable design patterns.

5. AnyContext determines one to many any type.
6. AnyType names one-to-many any entity or any event.
7. AnyEntity or any event is about any model on one-to-many any media.
8. AnyTest produces one to many any view.

21.3   SUMMARY

The midsized template generated here exhibits how core knowledge of any model can be used for numerous applications [5]. With proper identification and detection, it is possible to cull out the main theme for the pattern along with its associated BOs and IOs. This is an important aspect of creating a pattern based on AnyModel. It also interprets stable design pattern for any model in an effective manner. This model can also be extended to identical contexts. Stability, reusability, and robustness of the pattern itself are a very big contribution apart from reduced time, money, and effort to create required patterns.

21.4   REVIEW QUESTIONS

1. Describe the term *Model* and list different uses for this term. Do you agree that the central theme of this term applies to all definitions of the word?
2. Why model is an important word in the context of this chapter?
3. Is the term *Model* a general one that assumes similar meanings for all forms of the word?
4. How does this term motivate a software pattern developer to design and create a pattern?
5. Does it solve the problem of reusability and adaptability?
6. What is the EBT for Model?
7. What is AnyModel stable design pattern?
8. What are the different BOs for Model?
9. Is stable pattern for Model (AnyModel) better than a traditional model? If yes, provide proof and answers.
10. Recreate similar software patterns for at least four scenarios that are not covered in this chapter.
11. Write down use case diagrams for each one of these scenarios.
12. Write use case descriptions for each one of these examples.
13. Write down different contexts where Model works in combination with other factors.
14. What are some of the functional requirements?
15. List some of the nonfunctional requirements and think of some that are not covered in this chapter.
16. What are the major advantages of this pattern?
17. What are some of the pitfalls of creating this pattern? Could you list some of the potential challenges that are likely to emerge while designing the pattern.
18. Is the AnyModel pattern stable and reusable?
19. If so, give reasons why it is stable.
20. Draw you inferences and a brief summary for this chapter.

21.5   EXERCISES

1. Model versus representation stable patterns
   a. What are the differences between stable model pattern and representation pattern?
   b. Create the representation stable pattern.
   c. Name and describe three contexts for each of them.
   d. Show how to apply the two patterns including one context of your choice and including a use case, the application of each pattern with your select context.

2. Model versus illustration stable patterns
   a. What are the differences between model stable pattern and illustration pattern?
   b. Create the illustration stable pattern.
   c. Name and describe three contexts for each of them.
   d. Show how to apply the two patterns including one context of your choice and including a use case, the application of each pattern with your select context.
3. Create five applications of AnyModel Patterns in a table as shown in Table 23.1 (Chapter 23).
4. Create two applications of two contexts of AnyModel stable design pattern and provide the following for each one of the applications:
   a. Context
   b. Application diagram using the AnyModel stable design patterns
   c. Detailed use case
   d. A sequence diagram for the application

21.6   PROJECTS

Use Model in one of the following and create:

a. Scale model
b. Business model
c. System model
d. Conceptual model
e. Economical model
f. Role model
g. Model-view controller
Three contexts for each one of them
Stable architecture pattern for each of them
An application of each one of them

REFERENCES

1. Long (Ren) Xinmin (December 5, 1915–February 12, 2017), last modified on February 25, 2017, https://en.wikipedia.org/wiki/Ren_Xinmin
2. M.E. Fayad and A. Altman. Introduction to software stability, *Communications of the ACM*, 44(9), 2001, 95–98.
3. M.E. Fayad. Accomplishing software stability, *Communications of the ACM*, 45(1), 2002, 111–115.
4. M.E. Fayad. How to deal with software stability, *Communications of the ACM*, 45(4), 2002, 109–112.
5. M.E. Fayad, H.A. Sanchez, S.G.K. Hegde, A. Basia, and A. Vakil. *Software Patterns, Knowledge Maps, and Domain Analysis*, Auerbach Publications, Boca Raton, FL, Taylor & Francis Catalog #: K16540, December 2014, ISBN-13: 978-1466571433.

22 AnyReason Stable Design Pattern

Thinking is the hardest work there is, which is probably the reason why so few engage in it.

Henry Ford [1]

Another candidate that is appropriate for designing a stable pattern is the English word *reason*. A frequently used word, reason is linked to many aspects of our life. The word reason could be used multidimensionally in different contexts and realms of English use. It is mainly associated with deep thinking, human cognition, and intellect. Reason is an ability that allows someone to think about one idea and link it to another idea that is as important as the first one. A simple definition of reason is a statement that is presented in justification or explanation of a belief, action, or an action. Reason is also used in philosophy and debating a certain topic, when a person is forced to give reasons for the arguments forwarded to the audience.

Given its wider applicability in English language, it may be possible to create a stable pattern around reason. Understanding deeper meaning of reason would help pattern developers to decode the inner meaning of the word and eventually provide a practical solution to combine all aspects of reason into a single application that can work in various contexts and spheres. A stable pattern developed by using reason would be truly stable, robust, universally applicable, and extendable over swathe of different fields and areas. This chapter seeks to develop a stable pattern named AnyReason and it is expected to help developers solve a number of problems that are likely to emerge while designing a stable pattern.

22.1 INTRODUCTION

With its varied English usage and wider application contexts, the word reason could be a potentially good candidate for creating a stable pattern called AnyReason. According to free online dictionaries, reason means the capacity for consciously making sense of things, applying logic, establishing, verifying and evaluating facts, and justifying conventions and practices. It is widely related to such areas of human activities as philosophy, science, language, math, and art; in fact, it is a significant attribute of human nature. A person, who gets internally motivated, is well equipped to provide reasons for actions taken or explanation provided. He/she may also be successful in providing valid reasons for developing internal motivation. Humans give any kind of reasons to defend their action or activities. Without the power of reasoning, humans will never achieve the desired results they want in their life.

Hence, it is presumed that designing a stable pattern for reason seems to be pertinent and timely. Named, AnyReason, this stable pattern is likely to represent all scenarios and sphere of life where one would expect people to provide reasons to develop motivation. Understanding the inner meaning of reasons is quite challenging and this chapter intends to extract the main business of theme of reason along with its several business objects (BOs) that are known to propel the stable pattern forward to realize innumerable technical contributions. The ultimate goal of this chapter is to design a stable pattern that is much more sophisticated and superior than the pattern created by using a conventional method. The pattern—AnyReason will also solve all technical problems that might arise when finding a practical solution for reason.

22.2 PATTERN DOCUMENTATION

22.2.1 Name: Reason Stable Design Pattern

The stable design pattern for reason describes the term "Reason" that AnyParty or AnyActor uses to get motivation. Usually, we give several reasons to get motivated and hence the enduring business theme (EBT) for "Reason" is said to be "Motivation." AnyActor or AnyParty can get motivated within AnyContext and a reason can determine the type of motivation (AnyType). AnyParty or AnyActor collects AnyEvidence. Reason can influence AnyEvidence. A reason is about AnyEntity or AnyEvent. Similarly, AnyEntity or AnyEvent can use a reason. A reason can have AnyConsequence. The reason that AnyEnity or AnyEvent gives can be recorded on AnyMedia.

22.2.2 Context

AnyActor or AnyParty in any domain to get motivated can use AnyReason Design Pattern. Reason has innumerable applications in different areas of human life like philosophy, logic, science, art, personal relationship, mathematics, sociology, and others. Some types of reasons are scientific reasons, philosophical reasons, mathematical reasons, personal reasons, business reasons, logical reasons, and technical reasons. Some of the scenarios where AnyReason are listed below:

22.2.2.1 Education

The main driving force to reach to achieve excellence (AnyReason) in the field of education is internal motivation. A student (AnyActor) can get motivated to continue education (AnyContext) by enrolling to a university (AnyParty) for gaining knowledge. To do so, the student has to complete the application process (AnyEvent). It is a kind of a mental motivation (AnyType) that gets one's mind ready to reach academic excellence by obtaining more knowledge by attending numerous courses (AnyEntity) in the university. The reason to continue education can get one a better job soon after the degree is awarded (AnyConsequence). The degree is awarded by issuing a Degree Certificate (AnyMedia). As part of the admission process, any person is required to produce their marks sheets and few other documents as requested by the university (AnyEvidence) to get the admission.

22.2.2.2 Physical Fitness

As the saying goes, "A Healthy Body is a Healthy Mind," the main reason to stay fit (AnyReason) is to develop motivation to develop a healthy mind and fit body. In the context of Physical Fitness (AnyContext), a person (AnyActor) can get motivated to stay fit by enrolling himself or herself to a gymnasium. By regularly exercising (AnyEvent) by lifting weights (AnyEntity), the person can continue to by physically and mentally fit (AnyType). Because of fitness, a person can be successful in leading a good life (AnyConsequence). To enroll to a gymnasium may require some evidence to prove that you are not a person with physical or mental disability (AnyEvidence). If any person competes in a challenge, they can be recognized with a certificate (AnyMedia).

22.2.2.3 Workplace

The main reason to get promotion (AnyReason) in an office is the motivation to perform better and to get better pay. In the context of office work life (AnyContext), a person (AnyActor) can choose to work in a company (AnyParty) and get promotions quicker. In the process of getting a promotion (AnyEvent), the reporting manager (AnyParty) can use the achievements of the person (AnyEvidence) in order to qualify the person to promotion. Superior work performance, when compared to the other members, can be fruitful in getting a promotion to the person (AnyConsequence). The promotion decision is notified by email (AnyMedia).

22.2.3 Problem

The main problem is to create a stable design pattern for "AnyReason," and show that it can be applied across several contexts in various domains by using different events and prove that they all can have different consequences.

22.2.3.1 Functional Requirements

The following are the functional requirements exhibited in any given scenario:

1. *Behavior*: The behavior of AnyActor or AnyParty can be understood by using one or more scenarios.
2. *AnyActor*: An actor is the actual user who can be motivated because of one or more reasons within a given context. An actor represents someone or something, which is involved in reasoning.
3. *AnyParty*: A party is an organization or something that has a legal compliance. The actor can be part of the party or can use the services provided by the party in a given context of the scenario. Even a party can be motivated because of one or more reason. This class represents the party, which is involved in giving reasons. A party can be a country, political party, organization, or a person affiliated to an organization.
4. *AnyContext*: It can be defined as a general situation in which AnyActor or AnyParty can be motivated to give one or more reasons. This class represents the instructions that tell the actor or party how reasoning could be performed.
5. *AnyType*: It is used to identify in what way AnyActor or AnyParty is motivated. This class represents the type of reason.
6. *AnyEvent*: An event is a sequence of steps that motivates the person to give reasons in the given context. This class also represents the event, which is involved in reasoning and is named by the type of reason.
7. *AnyEntity*: An entity is an individual unit, who gives reason, because he/she motivated in the given context. This class represents the entity, which is involved in reasoning and is named by type of reasoning.
8. *AnyEvidence*: AnyActor or AnyParty collects a set of evidences to get motivated and give reasons. This type of reason always influences evidence and the quality of evidence is always governed by reasons.
9. *AnyMedia*: All the documents that need to be stored for future use and reference that ensures AnyActor or AnyParty is motivated is done using AnyMedia. This class represents the medium through which reasoning takes place. AnyEntity or any event resides on any media.
10. *AnyConsequence*: The motivation that the person gets from various reasons can have one or more consequences on AnyActor or AnyParty. Reason always has a consequence as an aftermath which could be positive or negative, successful or a failure.

22.2.3.2 Nonfunctional Requirements

1. *Convincing*: A reason that AnyPerson or AnyActor gives should be convincing enough to defend and produce positive consequences. Reasons should be convincing to make others agree to points put forward by an entity. Reason should also be capable of causing someone to believe that something is true or real.
2. *Acceptable*: The reasons that AnyActor or AnyParty gives to get motivated must be true and acceptable. A party, who hears them, should agree reasons provided by an entity. In other words, they should be suitable. In other words, reasons should be reasonable, beyond doubt and clear.
3. *Justifiable*: The reasons given to get motivated must also justify that some good can happen to the Actor or the Party. Justifiable also means something that can be shown to be right or reasonable. It should be defensible and arguable. Any reasons given by an entity should be having sufficient grounds for justification.

22.2.4 SOLUTION: PATTERN CLASS DIAGRAM

Class Diagram Description (as shown in Figure 22.1):

1. One or more AnyActor or AnyParty can be motivated by one or more reasons.
2. AnyActor or AnyParty collects one or more evidences.
3. Motivation is within one or more context.
4. Motivation uses one or more reasons.
5. A reason can determine one or more types of motivations.
6. AnyContext can have one or more types.
7. A reason influences one or more evidences.
8. A reason can have one or more consequences.
9. A reason is about one or more entities/events.
10. AnyEntity or AnyEvent uses one or more reasons.
11. AnyType can name one or more entity or event.
12. The results of AnyEntity or AnyEvent can be stored on one or more media.

22.3 SUMMARY

The smaller pattern template generated here shows how inner knowledge of any reason can be used for developing numerous applications that are practical and reusable [2]. With proper identification, evaluation, and detection, a pattern developer can easily extract the main theme of AnyReason pattern along with innumerable business objects called BOs. This chapter also explored the possibility and options to create a stable pattern that is easy to use and apply to any kind of usage domain. In other words, AnyReason pattern is sophisticated and practical that it realizes many benefits and advantages that can be applied to almost all realms of life.

22.4 REVIEW QUESTIONS

1. Why the term reason is important in English language? Can you list different uses of this term? Describe the term *Model* and list different uses for this term.
2. Why reason is a critical word in the context of this chapter?
3. Is the term *reason* a general word that assumes similar meanings in different contexts of usages?
4. Do you believe that this word can motivate a software pattern developer to design and create a pattern?
5. Does it solve the problem of reusability, robustness, and adaptability?
6. What is the EBT for reason?
7. What is AnyReason stable design pattern?
8. What are the different BOs for reason? Can you think of more for this chapter?
9. Does a traditional model create stable pattern for reason (AnyReason) better than stable method? If yes, provide proof and answers.
10. Recreate similar software patterns for at least six scenarios that are not covered in this chapter.
11. Write down use case diagrams for each one of these scenarios. Highlight them.
12. Write use case descriptions for each one of these examples. Provide explanations for each one of them.
13. Write down different contexts and realms where reason works in combination with other factors.
14. What are some of the functional requirements?

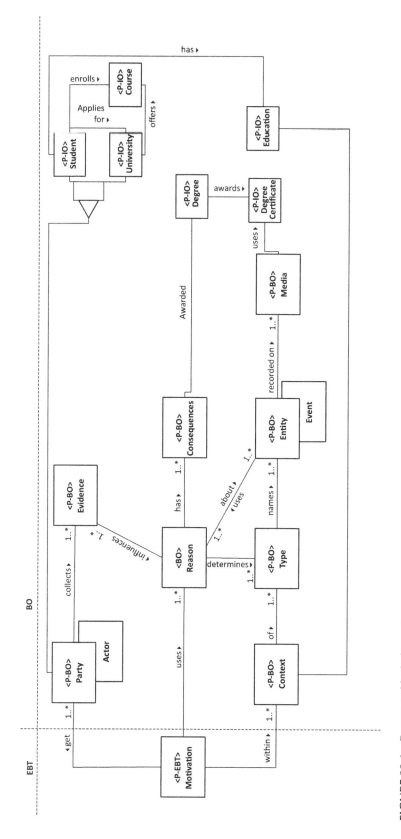

FIGURE 22.1 Reason stable design pattern.

15. List some of the nonfunctional requirements and think of some that are not covered in this chapter.
16. What are the major advantages and benefits of using this pattern?
17. What are some of the pitfalls of creating this pattern? Could you list some of the potential challenges which are likely to emerge while designing the pattern? Provide reasons why they are challenging.
18. Is the AnyReason pattern stable, rugged, flexible, and reusable?
19. If so, give reasons why it is stable. Link your answers with the motivational angle, which is incidentally the main theme of this chapter.
20. Draw you inferences and provide a brief summary for this chapter.

22.5 EXERCISES

1. Reason versus cause stable patterns
   a. What are the differences between reason stable pattern and cause pattern?
   b. Create the cause stable pattern.
   c. Name and describe three contexts for each of them.
   d. Show how to apply the two patterns including one context of your choice and including a use case, the application of each pattern with your select context.
2. Reason versus excuse stable patterns
   a. What the differences between reason stable pattern and excuse pattern?
   b. Create the excuse stable pattern.
   c. Name and describe three contexts for each of them.
   d. Show how to apply the two patterns including one context of your choice and including a use case, the application of each pattern with your select context.
3. Create five applications of AnyReason patterns in a table as shown in Table 23.1 (Chapter 23).
4. Create two applications of two contexts of AnyReason stable design pattern and provide the following for each one of the applications:
   a. Context
   b. Application diagram using the hype stable design patterns
   c. Detailed use case
   d. A sequence diagram for the application

22.6 PROJECTS

Use reason in one of the following and create:

a. Informative Reason
b. Right Reason
c. Intuitive Reason
Three contexts for each one of them
Stable architecture pattern for each of them
An application of each one of them

REFERENCES

1. Henry Ford (July 30, 1863–April 7, 1947), last modified on June 22, 2017, https://en.wikipedia.org/wiki/Henry_Ford
2. M.E. Fayad, H.A. Sanchez, S.G.K. Hegde, A. Basia, and A. Vakil. *Software Patterns, Knowledge Maps, and Domain Analysis*, Auerbach Publications, Boca Raton, FL, Taylor & Francis Catalog #: K16540, December 2014, ISBN-13: 978-1466571433.

23 AnyConsequence Stable Design Pattern

Knowledge is the consequence of time, and multitude of days are fittest to teach wisdom.

Jeremy Collier [1]

Consequence is a commonly used word that is used almost every day. Commonly, the word consequence means the effect, result, or the outcome of something that occurred earlier. In other words, it is the aftereffect of an action that occurs earlier. In another definition, consequence is an act or instance of following something as an aftereffect or result or outcome. It is also something that occurs as a result of a specific action or set of conditions.

Consequence is a critical aspect that relates to the basic issue of finding the impact of consequence for any given situation. Building a stable software pattern on consequence would provide unlimited opportunities to build software systems that are robust, extensible, applicable, and extendable over different domains and contexts. In addition, it will also provide a lasting solution to any problem under consequence domain; in essence, a developer need not work on any given pattern repeatedly which seems to the most common problem with traditional models that are so common today. In addition, a developer needs to work on the same problem again and again whenever he or she needs to either upgrade software system or rework on the core of the problem. Consequence is too a general English word to be used within a given context and its usage stretches across innumerable usage scenarios. The most common ones are interpersonal relationship, logic, science, arts, software programming, politics, and theology. The pattern created around consequence could be a stand-alone system or a part of other patterns.

The AnyConsequence pattern deals with the core of problem in depth to find a solution for the aftermath of consequence which is incidentally, the impact of consequence. The pattern designed after the term consequence is reusable, extendable, and robust under any context and usage scenario. Once this pattern is created, a developer need not work again and repeatedly to recreate and rebuild it from the scratch.

23.1 INTRODUCTION

Consequence is an important term in English that is used almost every day in our lives as an aftereffect of something that might have occurred before. A consequence could be positive or negative depending on the situation. Its impact could be very mild or it could be very serious. No domain exists that does not use the word consequence. There is no domain in the word without an impact occurring due to Consequence. All domains need a stable pattern model for AnyConsequence; in fact, this model helps a developer in saving precious time and money to recreate the pattern repeatedly from its scratch. This chapter offers a unique case of creating a stable pattern around the word consequence and it also provides an opportunity to find out innumerable business objects (BOs) and a stable essential business concept (EBT).

The final goal of any pattern is to find the most appropriate EBT. Here, the ultimate goal of AnyConsequence is the impact. The AnyConsequence pattern is an extremely stable model that consists of one EBT and several BOs. This pattern helps us in solving any type of pattern that investigates any domains like interpersonal relationship, logic, science, arts, software programming, politics, and theology.

The word consequence can be applied in many scenarios and usage systems. The consequence of personal rivalry and animosity could be serious and its impact could be both negative and positive. A negative impact could be a serious quarrel or a positive one could be a gain in trust and belief. Consequence is generally used in legal domains. For example, the bitter consequence of committing a felony is the serious impact of facing strict prison sentence. Alternatively, a case of forgery may result in a negative impact that could result in a jail sentence. System and drug abuse will lead to serious consequences of lengthy medical, social, and mental problem; in fact, the impact of substance abuse is physical and mental degradation among the users. In a college scenario, the consequence of plagiarism and copying is academic punishment that could be summary dismissal from the college or debar from a semester. Hence, the usage scenarios for the word consequence are many and every instance of consequence always leads to an impact.

The core concept of consequence is same for all scenarios. This necessitates the need for creating a general model of pattern to solve these commonly occurring problems. This chapter will propose a suitable design for a stable design pattern for Consequence. The name of the design pattern is AnyConsequence stable design pattern. The motivation for this chapter is to list down all requirements and design a stable architecture for consequence using the concepts of the software stability model. The main goal is to practice utilizing the concepts of the software stability model to create stable software architectures.

23.2 PATTERN DOCUMENTATION

23.2.1 AnyConsequence Stable Design Pattern

The name of the pattern is AnyConsequence stable design pattern. A consequence is a result of a course of action (or of a decision) taken by the decision maker. In an analysis, the consequences of a course of action are determined (predicted) by the use of models.

The software stability model is a model for designing stable software that encompasses three main entities: EBTs, BOs, and IOs. An EBT is an enduring business theme, which represents the ultimate quality factor of a concept. BOs are business objects which remain externally stable overtime. IOs are industrial objects which do not remain externally stable overtime.

In this model, *consequence* is a BO and its EBT is *impact* since the "consequences" of an action are everything the action brings about, including the action itself.

23.2.2 Context

Usually, we go through decision-making processes more than once on a daily basis and we also make choices, which bear certain responsibilities and consequences, whether positive or negative. This section will describe three different scenarios in which strategies are important and provide the IOs for each scenario.

23.2.3 Legal Consequence

In a typical running the Red Light (AnyContext), classified as a serious traffic violation (AnyType) is one of the leading causes of injuries among the drivers in the state of California. A driver (AnyActor) in a pursuit of seeking temporary thrill exceeds the safe speed limit of the vehicle (AnyEntity) eventually and jumps the red signal. Habitual wrongdoers can be sentenced to jail 9AnyConsequence0, by law enforcement agency (AnyParty), based on the recorded footage/past breaches of law (AnyMedia) from the DMV records (AnyEvidence).

23.2.4 Consequence of Substance Abuse

Regardless of which side of the argument you find yourself, a lot of teenagers (AnyActor) experiment with illicit drugs/alcohol (AnyEntity). Passing time (AnyAction) in this manner often proves to be a slippery slope to addiction (AnyContext). The impact of substance abuse (AnyType) can be long lasting and users show signs of neurological problems (AnyEvidence). A psycho analysis test report (AnyMedia) from a doctor (AnyActor) can be used to assess degree of judgment impairment (AnyConsequence) among the long-term users.

23.2.5   CONSEQUENCE OF FAILURE

To pass out GRE/GMAT with higher marks (AnyEvent), a competitive exam (AnyType) for securing admission (AnyContext), a student should not have an apathetic attitude. If the student (AnyActor) postpones all his or her academic work (AnyEntity) and miss lectures and review sessions, the understanding of the subject will be inhibited and he or she will be unable to prove necessary competence (AnyEvidence). Limited and shoddy understanding of subject matter would ultimately result in a bad score (AnyConsequence) and wasted opportunity (AnyConsequence).

23.2.6   PROBLEM

This chapter expects to generate a solution to the stable design pattern for "Consequence." It also expects to show that this stable design pattern can be applied across several domains and under different scenarios.

23.2.6.1   Functional Requirements

1. *AnyActor*: AnyActor may represent roles played by a person or creature. The actor signifies a person who has outlined AnyEvidence. Actor is an external entity and not the part of the subject of the review. AnyActor is affected by impact which is the EBT for the pattern system.
2. *AnyParty*: AnyParty refers to any legal users such as group of like-minded people or business that makes a unanimous choice to aid an action. AnyParty also outlines AnyConsequence.
3. *AnyEvidence*: Evidence is the event, which a person reports against the entity to report an issue. AnyEvidence is the result of AnyConsequence.
4. *AnyConsequence*: Based on the implied action dictated by the choice, it will have a corresponding AnyConsequence. This nature of result, however, depends on the user choice in the model and almost always carries some kind of a suggestion with it. AnyConsequence has AnyContext and it involves AnyEntity.
5. *AnyContext*: Context is the set of situations that inscribe an event. Evaluation involves different context for different AnyEntity based on which AnyConsequence is demanded.
6. *AnyType*: The entity will belong to a particular type. So, the types should be properly distinguished. AnyContext determines AnyType.
7. *AnyEntity*: AnyEntity is a something that exists in itself whether can be in real or hypothetically. Entity need not be compulsorily tangible. AnyEntity refers to the system which involves AnyConsequence.
8. *AnyEvent*: AnyEvent is an event that occurs during the course of the strategy.
9. *AnyMedia*: AnyMedia is the object on which the data resides and can be used to store logs.

23.2.6.2   Nonfunctional Requirements

1. *Desirability*: Desirability assesses how well a combination of variables satisfies the goals one has defined for the responses. The final effect of the consequence model can be a positive or a negative effect. AnyConsequence that occurs should be desirable for all parties. If the consequence is not desirable, then one may not expect a positive impact.
2. *Causality*: Causality is a phenomenon whereby one cause is the effect of another. Their practical importance is that they lead one to produce or to prevent causally related events by direct or indirect intervention. A consequence may have a serious causality for the other. The impact of this causality could be highly negative too. A consequence should not lead to an impact that becomes causality for a party.

3. *Predictability*: Prediction is a mere extrapolation of given data into the future. They make use of the conclusions drawn from the premises of available data using theories and models. The consequence and its subsequent impact should be predictable. AnyConsequence that occurs in future should be predicated well in advance and sufficient care should be taken to avoid unpredictability.
4. *Impactfulness*: Impactfulness can be defined as the tangible or intangible effects of the entity actions. The degree of impact that arises out of consequence is also very important factor while we study the core issue that relates to the quality of impact.

23.2.7 Solution

23.2.7.1 Class Diagram

Class Description (as shown in Figure 23.1):

1. An impact can have multiple Consequences.
2. AnyImpact is about AnyContext.
3. AnyParty/AnyActor outlines AnyEvidence.
4. AnyEvidence impels AnyConsequence.
5. AnyConsequence involves AnyEntity.
6. AnyEntity used by AnyConsequence.
7. AnyConsequence has AnyContext.
8. AnyContext determines AnyType.
9. AnyType has an entity.
10. AnyEvent stored on AnyMedia.

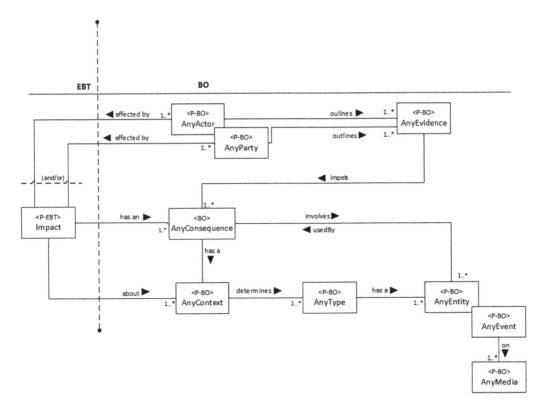

FIGURE 23.1 AnyConsequence stable design pattern.

TABLE 23.1
Contexts of Applications

EBT	BO	Legal Consequence—IOs	Substance Abuse Consequence—IOs	Failure Consequence—IOs
Impact	AnyActor/AnyParty	Driver, law enforcement	Teenagers, doctor	Student, college
	AnyEvidence	DMV records	Neurological problems	Competence
	AnyConsequence	Jail sentence	Impaired judgment	Course repetition
	AnyContext	Running the Red Light	Addiction	Securing admission
	AnyType	Traffic violation	Substance abuse	Competitive exam
	AnyEntity/AnyEvent	Automobile	Illicit drugs/alcohol	Academic work
	AnyMedia	Recorded footage/past citations	Medical reports	Transcripts

23.2.8 APPLICATIONS

Applications as shown in Table 23.1.

23.3 SUMMARY

Consequence is an excellent topic for creating a stable pattern. The consequence of any action leads to an impact whose effect could be either positive or negative. However, the word consequence is too general for anyone to understand and comprehend its core meaning. Although, it is quite challenging to understand core problem of extracting the EBT for the pattern, the entire process becomes easy when a developer studies different nonfunctional requirements set for the pattern [2–4]. Once the core issue is understood and its principle is extended over, anyone can create a stable pattern for consequence. The resulting pattern is robust, reusable, extendable, and applicable to all domains and contexts.

23.4 REVIEW QUESTIONS

1. Define the basics of AnyConsequence stability design pattern.
2. Why do you consider AnyConsequence as the most important pattern?
3. List a few application domains for the AnyConsequence design pattern.
4. List four challenges or constraints, while formulating the AnyConsequence design pattern.
5. AnyConsequence as a stable design pattern is domain and application specific—say True or False.
6. List the participants in the AnyConsequence design pattern.
7. Illustrate the class diagram for the AnyConsequence design pattern and provide notes.
8. List one tradeoff while using this pattern.
9. Does AnyConsequence pattern involves use of other patterns too? Explain.
10. List two main advantages of using the AnyConsequence stable design pattern.
11. List two applications of the AnyConsequence design pattern that is applicable to our daily lives.
12. Is this pattern stable and robust over time?
13. Highlight some research issues of this pattern.
14. Is this pattern extensible over time?
15. List two or three modeling challenges that arise when making a stable pattern for AnyConsequence.
16. List two scenarios that will not be covered by AnyConsequence design pattern.
17. List three test cases to test all participants of the AnyConsequence design pattern.

23.5 EXERCISES

1. Consequence versus reaction stable patterns
   a. What are the differences between consequence stable pattern and reaction pattern?
   b. Create the reaction stable pattern.
   c. Name and describe three contexts for each of them.
   d. Show how to apply the two patterns including one context of your choice and including a use case, the application of each pattern with your select context.
2. Consequence versus impact stable patterns
   a. What are the differences between consequence stable pattern and impact pattern?
   b. Create the impact stable pattern.
   c. Name and describe three contexts for each of them.
   d. Show how to apply the two patterns including with one context of your choice and including a use case, the application of each pattern with your select context.
3. Create five new applications of AnyConsequence patterns in a table as shown in Table 23.1.
4. Create two applications of two contexts of AnyConsequence stable design pattern and provide the following for each one of the applications:
   a. Context
   b. Application diagram using the hype stable design patterns
   c. Detailed use case
   d. A sequence diagram for the application

23.6 PROJECTS

Use consequence in one of the following and create:

a. Logical consequence
b. Behavior consequence
Three contexts for each one of them
Stable architecture pattern for each of them
An application of each one of them

REFERENCES

1. Jeremy Collier (September 23, 1650–April 26, 1726), last modified on October 19, 2016, https://en.wikipedia.org/wiki/Jeremy_Collier
2. M.E. Fayad and A. Altman. Introduction to software stability, *Communications of the ACM*, 44(9), 2001, 95–98.
3. M.E. Fayad. Accomplishing software stability, *Communications of the ACM*, 45(1), 2002, 95–98.
4. M.E. Fayad. How to deal with software stability, *Communications of the ACM*, 45(4), 2002, 109–112.

24 AnyImpact Stable Design Pattern

The only limit to your impact is your imagination and commitment.

Tony Robbins [1]

According to online dictionaries, impact is a noun word that is used commonly in almost all walks of daily life. It can also be used as a verb when used with and without objects. Used commonly, impact could mean to make an impression. Sometimes, linguists also refer to this word as to encroach or infringe something. When used followed by on or upon, impact could mean striking, dashing, or colliding. This chapter proposes to use the term *impact* to create a stable pattern model that can apply all domains that use this term. One of the main goals of developing this pattern is to avoid working from scratch whenever someone wants use this term in an applicable domain; in fact, this unique pattern helps developers avoid repetition and monotony. A stable pattern built on impact may help developers work on special issues concerning business, technology, climatology, industry, and manufacturing and many other areas.

One may also use this pattern system as a stand-alone entity or also as an essential part of any other system that seeks to address issues regarding impact of something. The AnyImpact stable pattern is a stable solution that addresses the core of the problem in a holistic manner to derive proof for any type of impact. By creating this pattern, we can reuse it in any domain and avoid developing the model from scratch.

24.1 INTRODUCTION

Impact is an active noun that indicates action and protectiveness. In essence, one cannot find a phase in life without something affecting everyday activities. Words that are synonymous with the term impact are affect, impress, influence, move, reach, strike, sway, and touch. However, the real meaning of this word may change according to circumstances that occur during a specific situation. In mechanics, an impact is a high force or shock that is applied over a short duration when two bodies collide with each other. Impact factor (IF) is an academic term widely used in measuring average numbers of citations to journal articles published in an academic journal.

In the world of social media, the term *impact* is used to measure the impact on a wide swath of social media who use portals like Facebook and Twitter. In climatology, scientists assess, evaluate, and discuss the overall impact of weather and pollution on the world weather. Likewise, scientists also discuss and evaluate the impact of greenhouse gas on the ozone layer and global warming. In business, managers and CEOs assess the impact of advertisement, marketing, and promotion on consumer response and eventual turnover. In public speaking, an emotionally surcharged and forceful speech will have a propound impact on the audience. In medical field, health professionals will try to find out the impact of a specific drug on the overall health and well-being of an individual. In sociology, demographers evaluate the impact of increasing population on the overall demographic health of a country. In space science, the overall impact of a meteor on the surface of the earth will be very useful while studying mineral properties of a particular soil. In studies involving culture and tradition, one may study the impact of culture and tradition on a particular nationality. This demonstrates that the word *impact* has a diverse use in different fields of daily life.

To summarize, the core concept of impact remains similar in the examples given above. However, this compels us to create a generalized and unique model to tag the word *impact* to a unique model to solve existing problems in the process of creating a stable pattern on AnyImpact. Three specific purposes are served here after creating a stable model for AnyImpact: reusability, repeatability, and robustness. Representing this pattern by tagging with an enduring business concept (EBT) will reveal a series of business objects (BOs) that act as subordinates to the process of creating a stable pattern. The ultimate goal of AnyImpact is profitability as it is the most enduring term when compared to other usages for the word *impact*.

24.2 PATTERN DOCUMENTATION

24.2.1 PATTERN NAME: ANYIMPACT STABLE DESIGN PATTERN

The concept *impact* suggests that any change in a given situation when an impact is applied. Impact can be positive or negative. The AnyImpact design pattern is used in many ways for different situations and is applied to different scenarios. Impact can also be stated as the striking of one thing to other, a forceful action, or collision. For example, social media keeps the youth updated with what is happening around the world, helps them stay connected with friends, and interact with their family members even if they are distance apart. This will strengthen relationship among them, even if they work and live in different locations and this is considered to be a positive impact.

In addition, social media sites have provided a robust and easy to use platform, whereby the youth can create groups and pages based on their common interests, start building connections, and find newer career opportunities just by updating their present status.

24.2.2 KNOWN AS

AnyImpact denotes change or transformation in something. Proper impact applied at the right time may lead to change in an environment. For example, social media changes in the minds of the youth are the most significant. On the other hand, climatologically occurring changes like global warming, breach in ozone layer, and development of greenhouse emission may lead to transformation in the earth's identity. Although, the word *impact* has many other synonymous words like affect, impress, influence, move, reach, strike, sway, and touch; they cannot reflect the real meaning as *change* does so successfully and meaningfully. Given below are some words that are synonymous to the word *change*:

Affect: Affect means to have an effect upon or make believe with the intent to deceive. Although, this word sounds similar to *change*, it may not completely impart its real meaning and depth. For example, change could be slow and gradual, but it would be permanent. On the contrary, affecting someone with something that could be related to subjective aspect of feeling or emotion.

Impress: Impress as a verb can mean, "Living a mark." One might impress someone with a show of strength or intellect. For example, a person may impress his or her superior with superior managing skills and talent. However, impress is not a permanent and it could fizzle away very soon without causing a definitive change.

Influence: It is the power to induce a significant effect on someone or something. Influence could be positive or negative, long lasting or short duration, permanent or temporary, and it could be simply changing a person's thinking. This may not be synonymous to the one we are discussing in this chapter. Change is often permanent and negative or positive.

24.2.3 CONTEXT

The pattern of AnyImpact can occur in many areas of like as explained in detail in the earlier sections of this chapter. Let us consider the example of a life coach who is proficient in changing other peoples' life with his thought provoking speeches and lectures. When a person hears these lectures and speeches over and again, his or her mind may change in a remarkable manner suggesting a drastic change in the way they think and behave. Armed with an air of positivity, the listener may transform his personality completely. Let us assume another similar example. In a business organization, employee motivation is an important issue that might result in increased turnover and profits. When employees are motivated with better work culture and better incentives, one can expect a significant change in the way the entire organization works and perform their duties. Hence, AnyImpact pattern could be used in a number of scenarios and situations. The following two examples provide the practical representation of AnyImpact: impact of social media on youth and impact on global climate patterns.

1. *Impact of social media on youth*: Change being the ultimate goal of the AnyImpact could be used in the domain of social media. AnyMedia plays an important role in youth's (AnyActor) lives. AnyActor could be the youth, who will have a deep impact exerted by the social media. AnyMedia (social media) can change youths in a drastic way. AnyReason causes AnyImpact that is related to different situations. AnyImpact of the social media on youth can be of Anylevel based on AnyConsequences that occur in a specific situation where AnyMedia is involved. AnyMedia provides AnyPlatform for AnyActor, where youth (AnyActor) can create groups and pages based on their common likes and preferences, start building connections and opportunities to promote their career and personality by discussing on various issues and topics.

2. *Impact in the case of climate*: Global warming is the most important issue that can alter earth's climate patterns forever. AnyParty is responsible for negative changes in climate patterns that lead to global warming. AnyOutcome is the result of permanent changes in the global climate patterns. AnyMedia is a part of this change that is happening in the environment due the impact of climate disturbances. Serious changes in the world climate and ultimately in the environmental patterns may lead to AnyConsequences. A change in climate drives AnyRisk that might be extremely negative and harmful for the environment.

24.2.4 PROBLEM

24.2.4.1 Detailed Requirements

AnyImpact consists of both functional and nonfunctional requirements and here are some examples.

24.2.4.2 Functional Requirements

1. *Change*: The ultimate goal of AnyImpact would be the *change* which is essentially the EBT for this pattern. In any field, whenever impact is applied in any situation, change invariably happens. Impact can be on youth or old because of social media and other similar tools. An impact on climate may cause global warming and an irreversible change in environment.

2. *AnyImpact*: AnyImpact is a BO for change that can occur in any given situation. AnyImpact always leads to the change. AnyImpact describes AnyLevel and generates consequences and it may relate to AnyType.

3. *AnyParty*: AnyParty undergoes slow but gradual change. AnyParty is responsible for AnyConsequences. AnyImpact affects AnyPary that is participating in AnyEvent. The party represents people who are participating in a given scenario, where impact might happen. AnyParty will experience AnyImpact and is responsible for AnyConsequences.

4. *AnyConsequences*: AnyParty has AnyConsequences when AnyImpact occurs. There can be a number of consequences irrespective of a given situation. In addition, consequences may lead to changes occurring in a particular situation. AnyMedia participates in AnyConsequences that the latter can be of AnyType.

5. *AnyType*: AnyImpact can be of AnyType. AnyType drives the change in a given situation because of the impact. AnyType is related to any event that may cause a change. It is applicable to any type of application until and unless the rules are followed.
6. *AnyMedia*: AnyMedia participates AnyConsequences that might arise in any situation because of AnyImpact. AnyMedia defines AnyEvent that happens when changes appear.
7. *AnyEvent*: Each change triggers an event that needs to be synchronized with the subsequent impact. AnyEvent relates to AnyMedia and AnyType causes AnyEvent.
8. *AnyLevel*: Impact may occur on different levels that lead to consequences. AnyImpact describes AnyLevel which leads to AnyConsequences.

24.2.4.3 Nonfunctional Requirements

1. *Strong*: Impact happening in any situation should be strong in order to make visible changes. Impact should lead to a definite change as in the case of social media impact on youth. Any change due to strong impact is permanent and it always shows perceptible results. Without a strong impact, it is almost difficult to see changes and this applies to the pattern under discussion here. In the example of employee motivation, significant changes can be seen when the employee internal motivation is the strongest.
2. *Categorical*: Impact should be categorical, that is, it should be absolute. It should be without exceptions and conditions. Changes that might occur due to an impact should be definite and certain. In other words, it should be irreversible and permanent. Otherwise, it is almost impossible to tag the word *change* with EBT.
3. *Relevant*: Reasons behind any impact of any system should be relevant. In the example, reasons for enhancing employee motivation are to increase annual turnover and to assure a significant increase in the employee remuneration. Before creating a stable pattern on impact, one should also make sure that the EBT makes visible changes in something. It should also drive someone toward changes and transformation. EBT should also drive someone to change as in the case of social media.

24.2.4.4 AnyImpact Stable Design Pattern

Pattern Class Diagram Description (as shown in Figure 24.1):

1. AnyParty is responsible of AnyConsequences occurring because of the impact.
2. AnyImpact describes different levels.
3. AnyType is a part of impact that is happening in the system.
4. AnyEvent relates to the media that is participating in the system.
5. AnyLevel leads to the consequences based on the degree of impact.
6. AnyMedia participates in the system in order to make changes happen.
7. AnyType can be decided by segmentation on the basis of reason.
8. AnyParty undergoes different changes happening in the system.

24.3 SUMMARY

The term *change* is a dynamic phenomenon and any stable pattern designed and built on this term should be categorical and relevant because of the generality factor that is associated with the word *impact*. Because it is used widely in an array of situations and contexts, defining and finding the most appropriate EBT for the patterns assumes utmost importance. AnyImpact stable pattern is stable, robust, and reusable because it takes into consideration many aspects of the word *impact*. In other words, AnyImpact pattern is definite and assertive in nature with its unique architecture that is stable over period and repeatable over any number of usages.

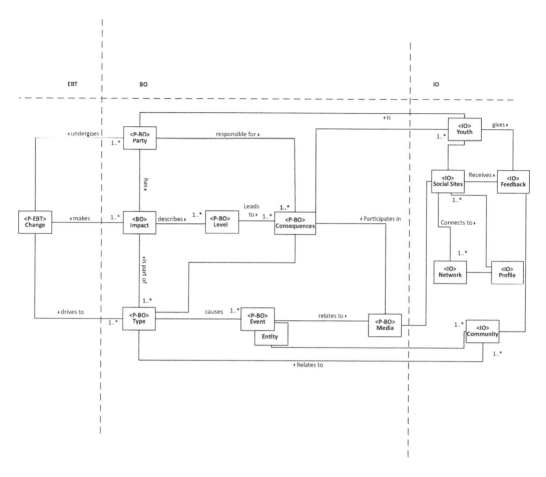

FIGURE 24.1 AnyImpact stable design pattern.

24.4 REVIEW QUESTIONS

1. Define the AnyImpact stability design pattern.
2. List a few numbers of applications for the pattern.
3. List four challenges while formulating the AnyImpact design pattern.
4. AnyImpact as a stable design pattern is domain and application specific—say True or False.
5. List any two constraints and challenges while formulating the AnyImpact design pattern.
6. List all participants in the AnyImpact design pattern.
7. Illustrate the class diagram for the AnyImpact design pattern.
8. The AnyImpact pattern involves the use of other patterns too. Briefly explain.
9. List two main advantages of using AnyImpact pattern.
10. List two disadvantages of using AnyImpact pattern.
11. Write down different applications of this pattern.
12. Explain why the word *change* has been used as the EBT for the pattern.
13. Is this pattern stable over time?
14. Write down three research issues pertaining to AnyImpact stable pattern.
15. Is this pattern extensible over time?
16. Write down two modeling issues that one might come across during the process of pattern making.

17. Are there any scenarios that are not covered in this chapter?
18. Write down pattern diagrams for three scenarios that are not covered in this chapter.
19. What are the core problems that influence the pattern under question?
20. Write down all the BOs given in this chapter and discuss why they have been tagged as BOs.

24.5　EXERCISES

1. Impact versus avoidance stable patterns
   a. What are the differences between impact stable pattern and avoidance pattern?
   b. Create the avoidance stable pattern.
   c. Name and describe three contexts for each of them.
   d. Show how to apply the two patterns including one context of your choice and including a use case, the application of each pattern with your select context.
2. Impact versus failure stable patterns
   a. What are the differences between failure stable pattern and impact pattern?
   b. Create the failure stable pattern.
   c. Name and describe three contexts for each of them.
   d. Show how to apply the two patterns including one context of your choice and including a use case, the application of each pattern with your select context.
3. Create five applications of AnyImpact patterns in a table as shown in Table 23.1 (Chapter 23).
4. Create two applications of two contexts and provide the following for each one of the applications:
   a. Context
   b. Application diagram using the impact stable design patterns
   c. Detailed use case
   d. A sequence diagram for the application

24.6　PROJECTS

Use Impact in one of the following and create:

   a. IF
   b. Impact Event
   c. Impact Evaluation
   d. Impact Record
   Three contexts for each one of them
   Stable architecture pattern for each of them
   An application of each one of them

REFERENCE

1. Tony Robbins (born February 29, 1960), last modified on June 18, 2017, https://en.wikipedia.org/wiki/Tony_Robbins

25 AnyHype Stable Design Pattern

Hype is a good thing, but it can also be a bad thing if you are not prepared to back it up.

Chet Faker [1]

25.1 INTRODUCTION

Hype is a basic dictionary word that seeks to empower, promote, edify, and elevate a specific event, a particular person, or a marketable product. In fact, hype is a grandiose plan to add a touch of halo or aura to an event person or a product to add an air of invincibility and extravagancy. The main goal behind developing this pattern is to create a stable model that can apply to all possible domains. The main objective of this pattern is also to provide a lasting solution to prevent efforts to find other solutions repetitively from scratch every time when developers need hype in an applicable domain. In fact, by creating a stable AnyHype stable pattern, one can use its principles in special areas like business, marketing, and personal edification. In other words, this pattern could be used as a stand-alone system or as a part of any other system that seeks to create hype and promotion around an event, person, or product. The AnyHype pattern is a viable notion that dissects the core in a comprehensive manner to cull out proof for any kind of proposition. The basic idea behind developing this stable pattern is twofold—to reuse the pattern in any application and domain, and to avoid developing the model again from its scratch.

Hype is a hyperbolic dictionary word that plays a vital role in many aspects of our life. Hype is a kind of promotion that consists of exaggerated claims or boasting. On a negative connotation, hype is something that is deliberately and deceptively misleading. Finding a domain without this aspect is almost difficult and the term "popularity" is almost difficult to prove without discussing "hype." In a general sense, "hype" refers to any overt promotion done to establish the popularity of an event, person, or a product. In other words, to create hype, one may need to set in motion a series of exaggerated claims that is sometimes not true in life. This word indicates specialized meanings and understanding in different contexts. Hype is used excessively in the area of marketing and promotion of a product or event. It also means that someone would carry out an exaggerated public relation event to add color and charisma to an event or product.

Hype may also be used in providing excessive publicity to an event. For example, a murder and its subsequent trial may be given excessive coverage to create a sense of suspense and thrill in the minds of public. Similarly, courtroom proceedings of murder trial may also provide an exaggerated coverage to create hype in the mind of people. For example, O.J. Simpson's trial went on for many months and TV and newspaper coverage ensured that it remained in the focal point of people's attention for many months.

A person may be lifted to the position of a god or hero by creating hype around him or her. For example, a pop singer may be given excessive coverage to create hype around the star. This hype will eventually help promoters in two ways: selling of their brand through events or concerts, and ensuring that the pop star's charisma remains the same even after a year or two. A big corporation or company may falsely mislead public by claiming outrageous things like profits or turnover. For example, a corporate manager may claim that there is no energy crisis in the country, which is actually hype and an exaggeration. In selling and marketing, an advertising company may make an outrageous claim to market their products and services. Unfortunately, people from all over the world fall prey to hype created by companies, businesses, and firms. Thus, hype is an extensively used technique to lend an air of halo to an event, person, or a product.

To summarize, the core concept of hype remains similar in the entire examples highlighted above; in fact, this necessitates the need for creating a generalized and unique model to solve any problem that exists while creating a stable pattern for hype. AnyHype pattern created here will ensure reusability as well as repeatability while developing any application. It will be clearer to some extent when we represent the pattern by using the concept of EBT. This helps us to reach one eventual goal for the pattern. The ultimate goal of AnyHype is popularity. It can also be a presentation of impact or influence by using exaggerated claims.

25.2 PATTERN DOCUMENTATION

25.2.1 Name: AnyHype Stable Design Pattern

The stable design pattern for AnyHype describes the term hype that is used by AnyParty or AnyActor. Hype involves extravagant promotion of a particular event, product, or an individual. The EBT that describes the ultimate goal for "AnyHype" is popularity. Popularity refers to a social phenomenon which is indicative of what is being liked or patronized at a point in time along with its extent. Popularity is what helps many business organizations and individuals to achieve desired results. The ultimate goal of AnyHype is to increase popularity and to leverage the benefits of it.

25.2.2 Contexts

AnyHype stable design pattern finds varied applications ranging from business organizations to individuals. The following are some of the areas where hype helps to increase popularity.

1. *Promotion of a product or service:* Every business organization (AnyParty) can create hype to increase the popularity of its products or services and consequently ensure better results. Hype can take place through advertisements, billboards, articles, and newspapers (AnyMedia). Hype has an impact (AnyImpact) on the popularity of the product or service.
2. *Individual popularity:* Public figures such as politicians, actors, and sportsmen (AnyActor) can make use of hype to improve their image. This helps to achieve their objectives. It can take place through various forms of media including posters, news, and articles (AnyMedia).
3. *Works of Art and Entertainment:* The organization or individuals behind works of art and entertainment like books, songs, paintings, and movies can create a hype to increase the popularity of their work. This in turn helps to generate more revenues (AnyImpact).

25.2.3 Problem

Popularity is important for all organizations and individuals. It can help them achieve the goals and objectives set by them. The problem requires the stable design pattern that caters to the diverse set of problems and application areas along with the nuances involved. The pattern concentrates on solving the problem such that any organization or individual would be able to apply this pattern to achieve their objectives.

25.2.3.1 Functional Requirements

1. *Popularity:* It refers to a social phenomenon that is indicative of what is being liked or in vogue. Popularity represents the level of individual and mass liking for an individual, product, service, or an event. A game of Super Bowl is as popular as ever, while the World Cup football matches gain immense popularity lent by excessive hype created around them. AnyHype created by AnyParty through AnyMedia achieves popularity. Likewise, AnyActor uses AnyEvent by using AnyMedia to gain popularity. AnyActor also uses AnyReason to create AnyImpact for AnyDuration.

2. *AnyHype*: It refers to extravagant promotion of a product or an individual. AnyHype last as long as AnyImpact exists on the ground. AnyImpact could be sustained by consistent use of AnyMedia and AnyEvent as long as momentum is sustainable.

3. *AnyParty*: It refers to a business organization promoting their products and services. They define AnyConstraint that needs to be executed. AnyParty could also be a government, association, firm, or an entity that seeks to gain popularity.

4. *AnyEvent*: AnyEvent refers to the events that are being carried out for popularity. AnyEvent could be a function, community gathering, press conference, promotion, or a workshop.

5. *AnyMedia*: It refers to social media, articles, newspapers, television where AnyEvent is published. AnyMedia will be effective as long as it creates a lasting AnyImpact.

6. *AnyActor*: It refers to any individual like a politician, sportsman, actors who use AnyHype to gain popularity. AnyActor is the eventual beneficiary of AnyHype. In other words, AnyActor is the product of AnyHype.

7. *AnyReason*: It refers to the reason and factors that contribute to popularity. A politician may want to gain immediate popularity by creating sustained AnyHype. Similarly, a business firm and its products may become immensely popular by creating AnyHype by using a series of techniques.

8. *AnyImpact*: It refers to the impact of the hype on the popularity. AnyImpact lasts as long as it retains its effectiveness. To sustain it, AnyActor may create fresh AnyHype events to create more AnyImpact.

9. *AnyDuration*: It refers to the time for which the hype runs. AnyDuration depends on the effectiveness of AnyImpact; a dwindling effectiveness may demand fresh efforts to create another round of AnyHype.

10. *AnyConstraint*: It refers to the constraints set forth by AnyParty or AnyActor. Constraints could be anything that either prevents or impedes the propagation of AnyHype.

11. *AnyType*: AnyReason determines Anytype that names AnyEvent. Reasons set forth for AnyHype will determine AnyEvent that is the most effective for propagating AnyHype.

25.2.3.2 Nonfunctional Requirements

1. *Effective*: It should be effective to ensure that the desired results are achieved. Here, the desired results mean gaining popularity by using an effective campaign that could be through TV, print, and online promotional events. A promotional campaign will be effective only when it produces sufficient impact on the minds of public. An effective hype would stay on for a longer period than an ineffective one.

2. *Convincible*: It is supposed to be extravagant but it should be convincible. Creating hype is excessively pricey and expensive. However, if it fails to convince people, it will be doomed to fail in the end. AnyMedia and AnyEvent should convince people to accept the hype that something is popular.

3. *Acceptable*: It should be acceptable to ensure achievement of purpose. People should also accept that hype created by promotional events is true and real. Popularity can be gained only when people start to believe that the hype created for a person, event, or product is real and truthful.

25.2.4 SOLUTION

25.2.4.1 Class Diagram

Class Diagram Description (as shown in Figure 25.1):

1. AnyParty or AnyActor has popularity.
2. Popularity can be gained using AnyHype.
3. AnyHype has AnyDuration.

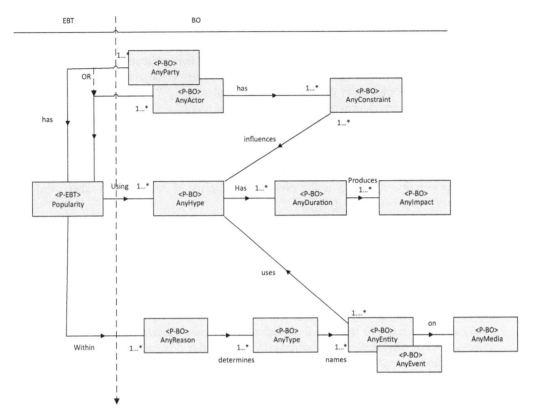

FIGURE 25.1 AnyHype stable design pattern.

4. AnyHype which has AnyDuration produces AnyImpact.
5. AnyParty has one or more AnyConstraint.
6. Any Entity uses AnyHype.
7. Popularity is within AnyReason.
8. AnyReason determines AnyType.
9. AnyType names AnyEvent.
10. AnyEvent is manifested on AnyMedia.

25.3 SUMMARY

The word "hype" is applicable more to advertisement and promotion. Although excessively exaggerative, hype created in a grand way could last longer and induce the much desired effect that a party wants to propagate. In a sense, the word "hype" is too general to be applicable to any one special domain. Hence, developers might need to find out the core meaning of this word before creating a stable pattern. In the context of this chapter, we have considered hype as leading to a gain in popularity to a person, event, or product. Although, hype is an abstract concept that can have different meanings, a careful examination of the word will help developers create and design a stable pattern that is extendable, reusable, and replicable in diverse situations and under different contexts.

25.4 REVIEW QUESTIONS

1. Define the word "hype" under different contexts.
2. Explain why this word is more applicable to marketing and promotional events.
3. List five areas where one can apply the concept of hype.

4. Popularity is the EBT of the term "hype." Explain why and give three reasons with your answer.
5. Discuss why media and the degree of impact created by promotional events lead to a gain in popularity.
6. List any two constraints while formulating the AnyHype design pattern.
7. List the participants in the AnyHype design pattern.
8. Illustrate the class diagram for the AnyHype design pattern.
9. Create two class diagrams other than the one given in this chapter.
10. Discuss class diagram discussion given in this chapter.
11. List two main advantages of using the AnyHype stable design pattern.
12. List additional functional requirements for this pattern.
13. List two applications of the AnyHype design pattern.
14. Can you list some of the research issues related to the AnyHype design pattern?

25.5   EXERCISES

1. Hype versus trend stable patterns
   a.   What the differences between hype stable pattern and trend pattern?
   b.   Create the trend stable pattern.
   c.   Name and describe three contexts for each of them.
   d.   Show how to apply the two patterns including one context of your choice and including a use case, the application of each pattern with your selected context.
2. Hype versus relevance stable patterns
   a.   What is the difference between relevance stable pattern and hype pattern?
   b.   Create the relevance stable pattern.
   c.   Name and describe three contexts for each of them.
   d.   Show how to apply the two patterns including one context of your choice and including a use case, the application of each pattern with your selected context.
3. Create five applications of AnyHype patterns in a table as shown in Table 23.1 (Chapter 23).
4. Create two applications of two contexts and provide the following for each one of the applications:
   a.   Context.
   b.   Application diagram using the hype stable design patterns.
   c.   Detailed use case.
   d.   A sequence diagram for the application.

25.6   PROJECTS

Use Hype in one of the following and create:

a.   Jargon Hype
b.   Hype Cycle
Three contexts for each one of them
Stable architecture pattern for each of them
An application of each one of them

REFERENCE

1. Nicholas James "Nick" Murphy (stage name is Chet Faker, born June 23, 1988), last modified on June 16, 2017, https://en.wikipedia.org/wiki/Nick_Murphy_(musician)

26 AnyCause Stable Design Pattern

Truth never damages a cause that is just.

Mahatma Gandhi [1]

One of the well-known words that could be a potential candidate to create a stable software pattern is Cause. A commonly used term, Cause relates to many aspects of daily life. Within the domain of English usage, this word is considered multidimensional with diverse application contexts. The word Cause is something that produces an effect or a result or a consequence. It may even result in an action or an aftereffect. Understanding the wider context of its usage in day-to-day world might help software pattern developers to design and conceive a stable pattern that could eventually provide a series of solutions to a wider domain of applications of Cause.

It is important to note here that software pattern developers may use Cause as a modeling concept to design and create an application-based and stable software pattern. Innumerable advantages and lasting benefits are possible with such an application, as it is perceived that it will provide a lasting solution to the recurring problem of working repeatedly on similar solutions from scratch. In the meanwhile, cost of money, effort and time involved in creating such solutions would be far lesser and significant. Henceforth, the stable pattern designed by using the word Cause would be referred to as AnyCause pattern, and this is likely to extend a series of benefits in different contexts and in diverse domains.

One of the significant advantages of using AnyCause pattern is its ability to ensure factors of stability, reusability, adaptability, and expendability. In addition, it is also likely to help software pattern developers to understand and comprehend the real meaning of the word Cause that is currently used in a wide array of usage contexts related to logic, science, engineering, business, math, law, and others. Furthermore, AnyCause stable pattern is thought to be multidimensional in a sense that it can represent all possible scenarios and contexts where one would likely come across numerous causes that eventually end in creating effects of diverse nature. However, understanding causes that result in some effects is truly challenging. To address this unique problem, this chapter attempts to extract the main theme that surround a Cause and pull out innumerable business objects that drive a stable pattern to ensure several technical contributions that are beneficial and effective for the domain of stable pattern development.

26.1 INTRODUCTION

The English word Cause has diverse uses in English language. With its innumerable application contexts, it also signifies something that eventually produces an effect, result, or consequence. A person, event, or conditions could cause a series of effects. A "cause and effect" is a well-known concept and an effect represents a basis for an action or response. In other words, it also symbolizes a valid reason to create an effect. According to yourdictionay.com, "cause and effect is a relationship between events or things, where one is the result of the other or others. This is a combination of action and reaction." Some common usages of this relationship suggest us that the word Cause could be used to denote definite effects that could be quantified and indexed with perfection. For example, heavy rains lashing for many hours may lead to a flooded street or an underpass. The result of unhygienic teeth is the development of cavities and tooth pain. Too many cigarettes may eventually lead to the onset of lung cancer. Many previous instances of ice ages might have resulted in the extinction of many plants and animals. A deadly oil spill will eventually lead to the death of thousands of animals that live in the water.

Eventually, the word Cause is a challenging concept that could be used to create a stable pattern called AnyReason. The decisive goal of this chapter is to design a stable pattern that is much more sophisticated, polished, and superior than a pattern that is created by using a conventional method. The pattern—AnyCause will also solve many known technical problems that likely arise when a developer is trying to find a practically feasible solution for Cause.

26.2 PATTERN DOCUMENTATION

26.2.1 NAME: ANYCAUSE STABLE DESIGN PATTERN

A cause is the reason or motive for some action. The word Cause also means an explanation or answer to a WHY question. The key idea is that a Cause helps to identify the reason for behaving in a particular way or for feeling a particular emotion. This will lead to a stable design pattern for *Cause* by using it as one of the *business objects* (*BOs*) and *justification* as its *enduring business theme* (*EBT*). If implemented in this way, this pattern can be used in all the areas where the concept of justification and cause are applied.

26.2.2 CONTEXT

A Cause would provide the justification for an action. Justification is an acceptable reason for doing something. Justification is also used in a number of domains, for example, legal laws and philosophical studies. In law-related matters, the cause of an action are a set of facts that are sufficient to justify a legal right to sue to demand money, property, or the enforcement of a right against any other party. The main goal of the Cause is to furnish a suitable explanation for the justification. Some of the most common, day-to-day scenarios where Cause is encountered are philosophy, logic, law, court proceedings, scientific investigations, engineering, business, and family-related matters. The following are the *scenarios* where Cause can be applied:

26.2.2.1 Hospital

In a hospital, many patients enroll for seeking treatment. These patients (AnyActor) arrive with visible pain or anguish that generally suggests that something is wrong with their health. This visible agony is the cause to (AnyCause) believe that the patient is not in perfect health. The doctor checks up patient's condition and suggests a therapy or gives a prescription to bring down the level of pain and anguish (AnyEntity). A different number of contexts are responsible (AnyContext) for patient's illness and the physician's special evaluation of patient is the testimony to this fact. Eventually, physician's justification of existence of illness derives from the causative factors (AnyCause). In other words, justification is the central theme of the pattern and it is also referred to as an EBT.

26.2.2.2 High Court

In the high court, the main players that dominate the proceedings are the judge, lawyers, and victims. It is also a tribunal with the vested authority to adjudicate and finalize legal, civil, and criminal disputes between different parties (AnyParty). It also carries out different court-related activities (AnyActivities) that dispenses justice and adjudications. The judge arrives at a legally tenable judgment after listening to all parties in dispute and it could be a guilty or not guilty verdict. The judge gives justification for the verdict after considering different contexts (AnyContext) and different criteria (AnyCriteria). In the end, a cause will lead to the justification of crime committed and a legally valid verdict based on the evidences and witnesses presented before the bench.

26.2.2.3 Police Arrest

In the example of a police arrest of a suspect, the most probable cause for the crime will become the key issue. The police invariably needs a cause or motive for crime before seeking an arrest warrant from the court. A police officer (AnyActor) will reply on various criteria (AnyCriteria) before deciding to make an arrest. To establish the cause for the crime, the police should be able to point out objective circumstances (AnyContext) and mechanisms (AnyMechanism) used to commit a crime; they are the most important legal requirements to establish that someone has committed some sort of crime. Eventually, the police will need to establish the existence of a solid cause to justify police actions like arrest or seizure.

26.2.3 PROBLEM

26.2.3.1 Functional Requirement

1. *AnyActor*: AnyActor is the actor who justifies a subject of Cause. An actor is a catalyst that seeks a cause to create an effect after justifying the use of Cause. The main attributes of AnyActor are an ID, actorName, type, role, member, affair, activity, and category. The different operations of AnyActor are playRole(), selectCriteria(), analyzeContexts(), and provideReason().
   playRole(): To justify the Cause.
   selectCriteria(): To select the subject criteria for justification.
   anaylzeContexts(): To analyze the contexts of the Cause.
   studyCause(): To study the main Cause.
2. *AnyCriteria*: AnyCriteria is a set of rules or restrictions. AnyActor follows AnyCriteria while justifying the explanation for the cause through different contexts (AnyContext). The different operations of AnyCriteria are analyzeRequirement(), followConstraint(), and satisfyCriteria(). The attributes would be limitations, rules, laws, description, and consequence.
3. *AnyEntity*: AnyEntity is the part of a Cause. AnyEntity is an individual unit who finds a cause because he/she wants to justify some action in a given context or scenario. AnyEntity is used by AnyMechanism, which has to be accessible and interfaceable with the system so that it can be used in performing the required tasks. The main attributes would be ID, entityName, entityType, position, state, and type. Some of the operations of AnyEntity in our case would be type(), relationship(), and performFunction().
4. *AnyMechanism*: AnyMechanism is the method that would be used by AnyActor to justify the Cause of an action. Different attributes of AnyMecahnism would be ID, mechanismName, status, application, usefulness, and description. It has various operations like identify(), analyze (), and specifyMethod().
   identify(): To identify what actually the cause is about.
   analyze(): To analyze the reason behind the cause.
   specifyMethod(): To specify the method of the mechanism.
5. *AnyType*: It is used to identify in what way AnyActor or AnyParty uses a cause. This class represents the type of cause. AnyType is the type of Cause and it varies from one context to the other based on the agenda of justification. Some of the attributes are ID, typeName, properties, methodList, attributeList, and clientList. Different opeartions of AnyType are change(), explain(), label(), and agenda().
6. *AnyCause*: AnyActor uses *AnyCause* to justify any actions based on AnyCriteria. Some of the attributes of AnyCause are ID, causeName, causeProperties, category, and purpose. Different operations of AnyCause would be asses(), examine(), specify(), and justify().
7. *AnyOutcome*: An outcome is something that happens because of an activity or process. The outcome could be either positive or negative depending on the type of Cause. AnyActor uses AnyMechanism to use the cause to create AnyOutcome. Some attributes of

AnyOutcome are ID, properties, cause, conditions, and outcome. Some of the operations of AnyOutcome are inspect(), scrutinize(), determine(), and decide().

8. *AnyContext*: It can be defined as a situation in which AnyActor or AnyParty can be motivated to use an existing cause and justify it to create some specified outcomes. This class represents the instructions that tell the actor or party how cause could be exploited to create a result. Some attributes of AnyContext are ID, ContextName, ContextProperties, ContextCategory, limitations, and constrains. Many operations of AnyContext would be list(), understandConstraints(), satisfyConditions(), and giveJustification().

9. *AnyLaw*: AnyParty should observe specific laws (AnyLaw) when demanding justification for a cause. Some attributes of AnyLaw would be ID, lawName, lawProperties, conditions, and impacts. Different operations of AnyLaw would be specifyConditions(), indexRules(), validateChecklist(), and checkConditions().

26.2.3.2   Nonfunctional Requirement

1. *Acceptable*: Any justification attributed should be acceptable to everyone. Hence, it should be appropriate and acceptable. For example, in the court of law, justification for any result resulting out of a specific cause should be acceptable and appropriate by the jury as well as judge.

2. *Relevant*: Both the context and criteria used by the actor should be relevant enough to examine and evaluate the explanation for the cause used. In other words, they must be applicable to the cause domain for proper justification. An irrelevant cause may lead to improper justification and possible ill effects that could be detrimental to the actor involved.

3. *Convincible*: Any proposals proposed by one actor by using an entity must be convincible to another actor who has to give justification of the cause. An actor should be able to provide proper justification after causing some actions. If the actor cannot convince any actions without any proper justification, the entire exercise will become useless.

4. *Complete*: The justification of the cause must be complete and satisfy the requirements of cause. Incomplete or unsatisfactory justification of any cause may not be beneficial to the actor involved. Additionally, an actor should be able to defend the cause by giving complete justifications for the actions and scenarios caused.

26.2.4   Solution: Pattern Class Diagram

Class Diagram Description (as shown in Figure 26.1):

1. AnyParty is involved in the justification process.
2. AnyActor follows AnyCriteria to perform justification.
3. AnyActor uses AnyMechanism.
4. AnyMechanism produces AnyOutcome from AnyCause.
5. AnyCause is of AnyType that specifies any AnyContext.
6. Justifications are based on AnyCriteria.
7. AnyContext is analyzed by AnyCause to produce justification.
8. AnyType determines AnyEvent and AnyEntity.
9. AnyActor provides justification using AnyMechanism.
10. AnyLaw influences AnyActor to provide justification.

26.3   SUMMARY

The word Cause is a suitable candidate for designing a stable pattern as this chapter successfully demonstrated with a classic diagram of pattern. The resulting pattern is an encompassing entity

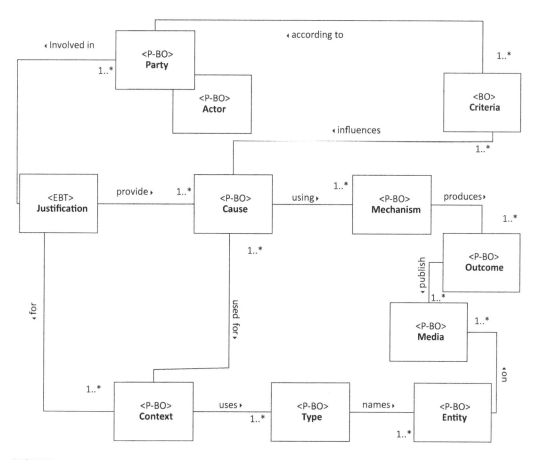

FIGURE 26.1   AnyCause pattern class diagram.

that provides the developer with unlimited extensibility, extendibility, robustness, and stability. The pattern demonstrated in this chapter also helps developers overcome many pitfalls that are most likely to emerge while designing the pattern. Furthermore, it also assists them in ensuring several technical contributions that are beneficial and effective for the domain of stable pattern development. Apart from its multidimensional nature, this pattern is also multifunctional with its diverse application capabilities and technical contributions to the domain of pattern making.

26.4   REVIEW QUESTIONS

1. What is the real meaning of Cause?
2. Can you interpret different meanings of the word Cause?
3. How does it relate to the creation of a software pattern that is stable and robust?
4. What is the Essential Business Theme for the word Cause?
5. Name its business objects.
6. List some contexts where you can use AnyCause stable pattern. Are they different from each other?
7. What are some of the known functional requirements?
8. What are some of the identified nonfunctional requirements?

Is it possible to draw a pattern diagram for Cause? If yes, write a detailed diagram highlighting all BOs.

26.5   EXERCISES

1. Cause versus effect stable patterns
    a.   What are the differences between Cause stable pattern and effect pattern?
    b.   Create the effect stable pattern.
    c.   Name and describe three contexts for each of them.
    d.   Show how to apply the two patterns including one context of your choice and including a use case, the application of each pattern with your select context.
2. Cause versus mission stable patterns
    a.   What are the differences between cause stable pattern and mission pattern?
    b.   Create the mission stable pattern.
    c.   Name and describe three contexts for each of them.
    d.   Show how to apply the two patterns including one context of your choice and including a use case, the application of each pattern with your select context.
3. Create five applications of AnyCause patterns in a table as shown in Table 23.1 (Chapter 23).
4. Create two applications of two contexts and provide the following for each one of the applications:
    a.   Context
    b.   Application diagram using the cause stable design patterns
    c.   Detailed use case
    d.   A sequence diagram for the application

26.6   PROJECTS

Use cause in one of the following and create:

    a. Common Cause
    b. Legal Cause
    Three contexts for each one of them
    Stable architecture pattern for each of them
    An application of each one of them

REFERENCE

1. Mahatma Gandhi (October 2, 1869–January 30, 1948), last modified on June 24, 2017, https://en.wikipedia.org/wiki/Mahatma_Gandhi

27 AnyCriteria Stable Design Pattern

An inner process stands in need of outward criteria.

Ludwig Wittgenstein [1]

The word criteria in English is a precise term that is used as a standard, rule, or test over which a decision, evaluation, inference, or judgment can be based. Some of the most common synonyms to the word criteria are basis, reference, standard, norm, yardstick, benchmark, touchstone, test, formula, measure, gauge, scale, barometer, indicator, and litmus test. A specific word, criteria is a highly targeted word that is used specifically under certain circumstances where a standard is used for judging something. With its broad usages in everyday English, the word criteria could act as a tool to evaluate and assess something in a wide variety of contexts and applications. The important goal of this chapter is to use criteria as a model concept and central theme for designing an application that can help developers create a stable pattern.

It is also envisaged that this chapter will try to find a lasting solution by creating a stable software pattern called AnyCriteria; in effect, this will also help developers avoid the costly mistake of finding similar solutions from scratch and by spending huge amounts of money and time. The stable software pattern will find a large range of uses in many areas of life like academics, business, logic, science, arts, engineering, math, and many other areas where someone is likely to use criteria. This pattern will also assist everyone involved in software engineering to find the real meaning of the word criteria and to ensure introduction of stability, reusability, adaptability, and extensibility to the developed application.

27.1 INTRODUCTION

Successfully used as a qualifying word, criteria is a universal term that finds hundreds of daily uses in varying contexts. In the area of linguistics, internal motivation could be the most important criteria to achieve success in learning English as a second language. In human resources development domain of a business firm, hiring managers will be looking for different criterion like education, skills, and experience when hiring a new candidate. In an immigration department that looks after granting citizenship to foreigners, candidates are evaluated based on some criteria like age, health, skills, education, and experience. In genetics, researchers evaluate several criteria like chromosomal contents, DNA mapping, hereditary factors, and other parameters to predict a person's personal appearance and disposition.

These examples demonstrate that criteria are a multidimensional word that can find several usages in different occupations. Applying criteria to the domain of stable software making is rather challenging. A traditional software pattern model provides its solution to a single context of any criteria model. However, a software stability model (SSM) is known to help developers to create a universal application that can satisfy all types of contexts and scenarios. One of the other goals of this chapter is to discover the core knowledge that acts behind AnyCriteria concept, which eventually helps in building an application that is not specific to a given domain. SSM is reusable, it has a higher degree of abstraction, less complex, and truly organized; these qualities work in a perfect combination to extend the stable software model with several generic qualities like scalability, extensibility, interoperability, and flexibility.

27.2 PATTERN DOCUMENTATION

27.2.1 NAME: ANYCRITERIA STABLE DESIGN PATTERN

The stable design pattern for AnyCriteria describes the term criteria. AnyParty or AnyActor who wants to evaluate AnyEntity uses this term. The enduring business theme (EBT) that describes the ultimate goal of "AnyCriteria" is evaluation. If AnyParty wants to evaluate an entity, it will define criteria under which it has to be evaluated. The party or actor can have AnyReason to evaluate an entity. The ultimate goal of AnyCriteria is to evaluate an entity by using AnyMechanism. Once the entity is evaluated, AnyOutcome is published in AnyMedia like e-mail or letters to communicate with others. Once AnyOutcome is ensured, a checklist or log (AnyLog) is generated to verify whether the criteria are applied properly for evaluation of an entity or not.

27.2.2 CONTEXT

AnyCriteria stable design pattern can be applied to any domain where an AnyEntity needs to be evaluated under AnyCriteria for AnyReasons. Criteria serve as a principle or a standard by which something can be judged or evaluated. The AnyCriteria pattern can exist in many domains of daily life and some of the instances were explained in detail in the introductory section of this chapter. To recap different contexts that relate to the use of criteria in daily life, here is an example of a baseball coach who is trying to select the best team for the final match. Setting a list of criterion for final selection, the coach starts interviewing prospective candidates to check their ability to give their best performance in the last but important game. Eventually, he evaluates all players based on predefined criteria for each player. Incidentally, the coach sets himself a frame within which winning criteria are defined and each player is evaluated based on those qualifying factors. Similarly, a number of scenarios and contexts come under the purview of this pattern like the selection of a candidate for a job posting, choosing a student for a PhD slot, and many others. Here are some more examples of representation of AnyCriteria pattern that occur in our society.

1. *Loan Evaluation Criteria*: To avail a loan in any bank, the applicants must meet certain eligibility criteria. The criteria will be set by the banks for AnyReason. The applicant's applications will be evaluated by AnyMechanism which will be influenced by AnyCriteria. The evaluation outcome will then be tracked as log file (AnyLog) for future references and the decision will be published and sent to the applicants, via e-mail (AnyMedia), informing whether they are qualified for the loan or not.

2. *Admission Criteria in Graduate Schools*: To be considered for admission in any graduate school, the applicants must meet certain eligibility criteria that is set by the academic board. The applicant's applications will be evaluated by AnyMechanism which will be influenced by AnyCriteria. The evaluation result will then be tracked as a log file (AnyLog) for future references and the result will be published and dispatched to the applicants through e-mail (AnyMedia) informing whether they are qualified for the admission or not.

3. *Criteria for Choosing a Perfect Job*: Once a student finishes his/her degree in a graduate school, the next obvious step is to find a lucrative job with a decent pay package and perquisites. However, choosing a perfect job is not too easy. The person who wants a job may need to make an informed decision to choose the best job that is in the market. One will need to consider several criteria like annual pay package, available benefits, job security, and other factors. Using AnyMechanism, a prospective student may find the best job by filtering out those parameters that do not suit the candidate. The job seeker will have AnyReason to

carry out this evaluation. Finally, once a proper evaluation is carried out, the candidate can appear for the interview and wait for AnyOutcome from the employer to know whether the appointment letter has been dispatched through e-mail or letter (AnyMedia).

27.2.3 PROBLEM

The main problem that this chapter envisages is to find out a solution to different contexts where the term criteria occurs and to design a stable pattern for it. One of the main problems that one comes across while creating AnyCriteria pattern is its applicability across different domains where the issue of criteria is involved. Forthwith, this chapter seeks to create a pattern that can work in all domains and under all contexts.

27.2.3.1 Functional Requirement

1. *AnyActor*: AnyActor refers to the person, who prepares the AnyAgenda, such as event manager or the person, who participates in the AnyActivity that is listed down in AnyAgenda. The event manager has to define AnyConstraint that needs to be applied while listing the AnyActivity in AnyAgenda. The participant must follow different constraints defined in agenda.
2. *AnyParty*: AnyParty refers to the any legal users such as an event committee or a business that consists of an event organizer (AnyActor) that plans AnyAgenda to organize the workshop or business meeting. They also define the AnyDuration and the AnyActivity that are needed to be executed.
3. *AnyMedia*: AnyMedia refers to Internet, e-mail, and newspaper where the AnyAgenda is published so that the participants (AnyActor) can come to know about AnyEvent and AnyDuration. This is important as the participant will prepare himself/herself based on information displayed in AnyAgenda.
4. *AnyLog*: AnyLog refers to the checklist or any logging details that contains all the AnyActivity details that are executed as per the AnyAgenda and the status of the AnyActivity. The AnyLog can be used to verify if all the AnyActivities are completed and any pending activities are still need to be executed.
5. *AnyCriteria*: This refers to any standard by which something can be evaluated. AnyActor, when he/she wants to evaluate something, considers some criteria like safety, cutoff, and benefits. To evaluate AnyEntity, criteria are considered as a benchmark.
6. *AnyMechanism*: This refers to a technique by which evaluation of AnyEntity occurs. AnyCriteria influences mechanism. Mechanism can be any technique like priority-based, first-come first-serve basis. AnyCriteria will influence AnyMechanism. AnyMechanism will evaluate AnyEntity.
7. *AnyEvent*: Series of events carried throughout the process of creating the pattern.

27.2.3.2 Nonfunctional Requirement

1. *Availability*: AnyCriteria defined by AnyActor or AnyParty must be published by any means so that it is available to everyone, and it will be clear to all the applicants stating their eligibility whether they have met the benchmark for admission or not.
2. *Executable*: AnyCriteria must be executable. The ultimate goal of AnyCriteria is to evaluate all the entities based on the criteria. AnyCriteria on which evaluation is not possible is of little use. So while defining criteria, AnyActor must ensure that they are applicable on a larger context.
3. *Performance*: AnyCriteria must always be set right because the AnyOutcome should be efficient and it performs better. If the criteria are not set properly, then evaluation of AnyEntity might fail and thus AnyActor or AnyParty will not get AnyOutcome as expected.

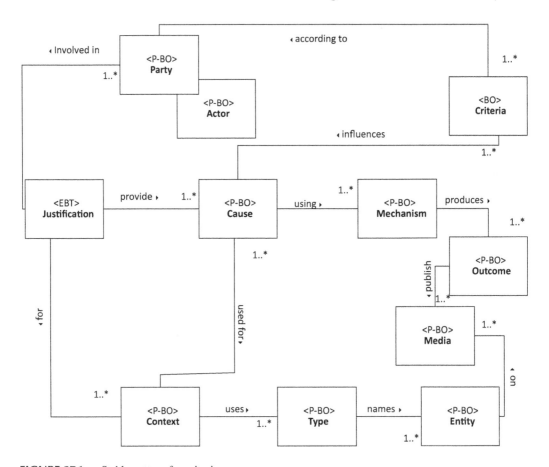

FIGURE 27.1   Stable pattern for criteria.

27.2.4 SOLUTION

27.2.4.1   Class Diagram of AnyCriteria Stable Design Pattern

Class Diagram Description (as shown in Figure 27.1):

1. One or more AnyParty or AnyActor does evaluation of AnyCriteria.
2. Evaluation of entities can be done through one or more AnyCriteria.
3. Evaluation must be done through applying AnyMechanism.
4. AnyMechanism is influenced by AnyCriteria.
5. AnyMechanism gives one or more outcomes.
6. Evaluation is done on AnyReason.
7. AnyCriteria is published in one or more AnyMedia.
8. AnyParty defines one or more AnyCriteria for evaluation the entity.
9. AnyOutcome is logged on AnyLog for future verification.

27.3 SUMMARY

SSM offers better stability and scalability than the one that is generated by using traditional modeling because it provides a holistic way of modeling, where the pattern defined in SSM can be applied in every application irrespective of the domain. While modeling a system in traditional approach, the system becomes more specific to a particular application for which it is being modeled. Nevertheless, the SSM focuses on the ultimate goal of the application and models the system accordingly. This approach makes SSM more flexible and maintainable. SSM also detects any flaws in traditional modeling and thereby builds better software models with its extremely useful patterns.

27.4   REVIEW QUESTIONS

1. Define criteria.
2. List some scenarios where this term will play its role.
3. Can you use the word criteria interchangeably with other words?
4. What are some of the most common synonyms of criteria?
5. Can you make a brief note to explain why this word is fit enough to be used as a base for creating a stable pattern?
6. What are some of the motivations that might compel you to design a stable pattern around the word Criteria?
7. Can you explain AnyCriteria stable design pattern and its importance in the field of stable pattern making?
8. Why criteria are an important word in the context of this chapter?
9. Is the term *criteria* a general one that assumes similar meanings for all forms of the word?
10. What is the EBT for criteria?
11. What are the different BOs for criteria?
12. Is stable pattern for criteria (AnyCriteria) better than a traditional model? If yes, provide proof and answers.
13. Design similar software patterns for at least three scenarios that are not covered in this chapter.
14. Write down use case diagrams for each one of these scenarios.
15. Write use case descriptions for each one of these examples.
16. Write down different contexts where criteria works in combination with other factors.
17. What are some of the functional requirements for this pattern?
18. Write down some of the nonfunctional requirements and think of some more that are not covered in this chapter.
19. What are the major benefits of this pattern?
20. What are some of the pitfalls of creating this pattern?
21. Is the AnyCriteria pattern stable and reusable?
22. List some of the potential challenges that are likely to emerge while designing the pattern.

27.5   EXERCISES

1. Criteria versus benchmark stable patterns
   a.   What are the differences between criteria stable pattern and benchmark pattern?
   b.   Create the benchmark stable pattern.
   c.   Name and describe three contexts for each of them.
   d.   Show how to apply the two patterns including one context of your choice and including a use case, the application of each pattern with your select context.
2. Criteria versus opinion stable patterns
   a.   What the differences between criteria stable pattern and opinion pattern?
   b.   Create the opinion stable pattern.
   c.   Name and describe three contexts for each of them.
   d.   Show how to apply the two patterns including one context of your choice and including a use case, the application of each pattern with your select context.
3. Create five applications of AnyCriteria patterns in a table as shown in Table 23.1 (Chapter 23).

4. Create two applications of two contexts and provide the following for each one of the applications:
   a. Context.
   b. Application diagram using the hype stable design patterns.
   c. Detailed use case.
   d. A sequence diagram for the application.

27.6 PROJECTS

Use criteria in one of the following and create:

   a. Evaluation Criteria
   b. Comparison Criteria
   Three contexts for each one of them
   Stable architecture pattern for each of them
   An application of each one of them

REFERENCE

1. Ludwig Wittgenstein (April 26, 1889–April 29, 1951), last modified on June 15, 2017, https://en.wikipedia.org/wiki/Ludwig_Wittgenstein

28 AnyMechanism Stable Design Pattern

The talent for discovering the unique and marketable characteristics of a product and service is a designer's most valuable asset.

Primo Angeli [1]

AnyMechanism or AnyService design patterns deal with the concept of service, mechanism, assistance, and the interactions among various parties that deal with services. The concept of assistance or using a mechanism or a service to fulfill functionalities are demonstrated in multiple contexts, though each of these contexts usually use different services while the parties dealing with services are unique to the domain or the context. The AnyMechanism or AnyService design pattern makes it easier to capture underlying concept of service and related interactions that exist in different application domains in a generic way by using the concepts of software stability model [2–4]. The pattern presents a generic model for services that can be extended for different domains and contexts, which is performed, by capturing the core knowledge of services and assistance.

28.1 INTRODUCTION

The concept of service to seek assistance is used in diverse application domains. Each of these application domains may be using different types of services that are performed by different service providers and to be used by different types of consumers. Nevertheless, the concept of using a service for assistance is a common underlying theme in all of these application domains. For example, in a financial application scenario payment assistance is obtained by using the credit card service offered by credit card companies and their merchants. In a yellow pages service, the yellow pages company provides merchant listings for the subscribers who are trying to find a business address. The previous two examples demonstrate two services that are very different as different consumers use these two services. However, both of these examples use services to provide assistance while interactions among service consumers and providers are quite common. The AnyService design pattern captures the core knowledge of both assistance and service in all these application domains to develop a generic and stable model applicable across multiple domains.

Traditional design practices demand designing different solutions for each of the application scenarios from different domains individually even though they share the same underlying concepts of service and assistance. The software stability model enables designers to develop generic models that are domain independent and extendable for different application domains. This involves capturing the core knowledge of the underlying concepts into enduring business themes (EBTs) and business objects (BOs). The stable model thus developed can be further enhanced to suit any application domain by extending the BOs as industrial objects (IOs) that is specific to that application.

The objective of this chapter is to design a stable design pattern for service and assistance by using the software stability model [2,3] and by capturing the core knowledge of service. The *AnyService* stable design pattern [2] so developed can be used as a basis to model different applications by using services and assistance.

28.2 PATTERN DOCUMENTATION

28.2.1 PATTERN NAME: ANYMECHANISM DESIGN PATTERN

Service is a term used to describe any useful function provided by one party to be used by other parties. Examples of services can be found in diverse systems like credit card processing or weather information.

28.2.2 KNOWN AS

1. *Identical Terms*: Assistance, aid, duty, favor, use, providing, avail, courtesy, maintenance, and supply are some of the common and identical terms related to service. All these terms are very similar and they could be used interchangeably. Any person can do favor to any other person who is in need of one. Favor can be found in diverse situations like lending money to anyone who is in need of it. Similarly, assistance is given to someone who needs help or guidance in completing the task. In general, it is similar to offer service.
2. *Identical but with some differences*: Terms like benefit, business, dispensation duty, kindness, value, worship, action, active duty, army, combat, duty, fighting, operation, and sting are some of the identical one although there are some sort of differences among them. These terms are very similar to service. The use of these terms varies under various contexts and scenarios. Hence, these terms cannot always be used interchangeably with the term AnyService.
3. *Seems identical but they are not*: Advantage, applicability, appropriateness, employ, employment, labor, use, utility, and usefulness are the terms which initially may seem identical in meaning to the term service. However, a deeper check suggests us that these terms are not the same; hence, we cannot use these terms interchangeably with the term AnyService. Service is to provide or offer something, whereas advantage is something we benefit after using any particular service.
4. *No similarity, just ignore them*: Account, check, fitness, indulgence, ministration, office, overhaul, relevance, work, ceremonial, ceremony, formality, function, liturgy, observance, ritual, and sermon are some of the terms which have no similarities.

28.2.3 CONTEXT

Service is a well-defined unit of work performed by a service provider to achieve desired results for the service consumer [5]. Service is a well-known concept with the increasing popularity of service-oriented architecture domains. The concept of service is applicable in various domains like information technology, financial systems, and health services. For instance, in a financial industry, some of the services may include credit card processing and portfolio maintenance. The consumers who are using any such service will benefit from it. In information systems, examples of services may include providing security services like user authentication, encryption, web site verification, or other computational services like scientific calculations, etc. Other examples of services in various fields include reservation services in travel and recreation industry, diagnosis and treatment services in healthcare industry, etc. Thus, service has always been a term and concept that is endeared by millions to receive. Service is provided to people who may or may not need it. It has been extended to people who worked and desired to get it and to those people who had no idea that it is being offered to them.

The pattern that is presented here is a global pattern and it can be applied to most scenarios.

The important idea is to reuse the pattern for any kind of service, be it service to a senior citizen or service to the poor. This pattern aims to solve and represent most situations involved with service.

1. *Credit Card Processing Service*: Processing payments (AnyService) by using credit cards is a very common consumer service used in merchant transactions that are both online and offline. The service is generally provided by the credit card companies themselves or third-party financial institutions (AnyParty). The consumers for this service could be anyone who gets payment from the credit card service provider (AnyActor). This application illustrates the *assistance* EBT to identify and use the *Credit Card Processing Service* by the merchant (AnyMechanism). The service provider will be identified once the service is chosen (AnyContext) and when a service contract is established between the service provider and service consumer.

2. *Directory Assistance Service*: Directory assistance is another commonly used service from a very different domain (AnyService). The same companies providing the phone service, whether it is a landline or cell phone (AnyParty), generally provide this service. It is also possible to procure services from a third party (AnyActor) in which case it is a different number (analogous to a different service end in our case). The consumers of this service are the phone service customers (AnyParty). This application illustrates the service consumer who is also the phone user using the assistance EBT (AnyMechanism) to identify the service and use it (AnyType), similar to the above application #1, although both these application are taken from two different domains. This example illustrates the usefulness of this pattern in multiple application domains.

28.2.4 PROBLEM

Due to wider applicability of services, it would be helpful to design a pattern that can be used across various domains for services. To achieve this goal, we will focus on capturing the core concept of a service. The problem at hand is to create a stable pattern that represents the common concerns behind services; incidentally, this is easily adaptable and extendable to any kind of service across multiple domains.

28.2.4.1 Functional Requirements

Computation: As an EBT for AnyMechanism

1. *AnyMechanism*: It represents the service itself. The *AnyService* pattern is associated with the EBT of *assistance*. It holds all the details of the service like identity, service description, and the methods it supports. It is provided by AnyServiceProvider and consumed by AnyParty.

2. *AnyParty*: It represents the main user of the pattern, the consumer of the assistance. It holds the details of the consumer like identity and credentials in case the service requires authentication. It uses the assistance to find the service that can perform the assistance needed by it and contracts with the AnyServiceProvider on the terms of use, and then uses the service to get the assistance.

3. *AnyActor*: The person or an entity, who is a part of the system and who interacts with the system, is known as the actor. It can be a creature, hardware, software, or any person. In this context, we can term a person, a start-up, or the organization which develops mechanism-based pattern as an actor.

4. *AnyRule*: A set of rules, regulations, and policies that are imposed or forced upon by a party or actor can be termed as any rule. For example, when a corporate customer demands that the industrial machinery purchased should provide a warranty for 10 years and that cost of spares must be free for the first 2 years, then this scenario could be termed as the one that follows a rigid rule.

5. *AnyDuration*: The period or the time that it takes to develop a mechanism-based stable pattern is known as duration.

6. *AnyOutcome*: AnyOutcome is defined as consequence or result of something. It is the result of an action which could be either positive or negative.

7. *AnyContext*: The context determines the situation where any type of interaction occurs. It restricts the interaction to a specified set of boundaries so that it may not be deviated. For example, in case of classroom interaction, the context of interaction will be the subject that the teacher will teach.

8. *AnyType*: It is a class or group which can be identified in mechanism-based stable pattern development is known as a type. In this case, the type used for pattern development forms a legitimate example of AnyType.

9. *AnyEntity*: The entities and attributes used in mechanism-based stable pattern development can be termed as an entity.

10. *AnyEvent*: A series of events that are carried throughout the process of creating the pattern.

11. *AnyMedia*: AnyMedia refers to Internet, e-mail, and newspaper where the AnyAgenda is published so that the participants (AnyActor) can come to know about AnyEvent and AnyDuration. This is important as the participant will prepare himself/herself based on information displayed in AnyAgenda.

28.2.4.2 Nonfunctional Requirements

1. *Convenience*: Any mechanism used should embody the benefits of convenience. A mechanism indirectly refers to a tool used to achieve an outcome. Therefore, the mechanism used should be convenient enough to provide the best result that should satisfy all the parties involved in the process.

2. *Systematic*: Any mechanism used in the process of creating the software should be systematic, orderly, methodical, and logical. Otherwise, it may be very difficult to expect the best possible outcome from the mechanism used. For example, industrial machinery should be specific and systematic so that it works according to a predesigned procedure. Such machines will provide a desired outcome in the form of a product that is easy and convenient to use.

3. *Working*: Any mechanism used should be in a working and deliverable condition. A non-working mechanism will never provide the desired result and the expected outcome. When a carpenter uses tools to carve a piece of wood, he/she should use only working ones that are in impeccable condition.

4. *Structural*: The word "structural" relates to or affects the manner in which something is built or created. In the context of mechanism, it means the way in which it is designed and created to provide the best possible outcome. The mechanism used to achieve something should be structurally strong and robust to work seamlessly in an efficient manner.

5. *Efficient*: Efficiency is a hallmark of anything that works toward achieving excellence and desirable results. A mechanism that one uses should not only be systematic and structural it should also be efficient and capable of assisting people get what they want in the form of satisfactory results.

6. *Useful*: The mechanism used to create a stable pattern should be useful and applicable to all possible scenarios. It should help a software pattern developer to reach a particular goal. Usefulness of a mechanism tried is indicated by the extent by which a developer achieves excellence and completeness in the process of creating a pattern.

28.3 EXERCISES

1. Mechanism versus action stable patterns
   a. What are the differences between mechanism stable pattern and action pattern?
   b. Create the action stable pattern.
   c. Name and describe three contexts for each of them.

    d. Show how to apply the two patterns including one context of your choice and including a use case, the application of each pattern with your select context.

2. Mechanism versus activity stable patterns
    a. What are the differences between mechanism stable pattern and activity pattern?
    b. Create the activity stable pattern.
    c. Name and describe three contexts for each of them.
    d. Show how to apply the two patterns including one context of your choice and including a use case, the application of each pattern with your select context.

3. Create five applications of AnyMechanism patterns in a table as shown in Table 23.1 (Chapter 23).

4. Use the functional requirements to draw the class diagram of AnyMechanism.

5. Create two applications of two contexts and provide the following for each one of the applications:
    a. Context.
    b. Application diagram using the hype stable design patterns.
    c. Detailed use case.
    d. A sequence diagram for the application.

28.4 PROJECTS

Use mechanism in one of the following and create:

a. Defense Mechanism
b. Reaction Mechanism
Three contexts for each one of them
Stable architecture pattern for each of them
An application of each one of them

REFERENCES

1. Primo Angeli (May 5, 1906–October 25, 2003), accessed on September 27, 2016, https://de.wikipedia.org/wiki/Primo_Angeli
2. M.E. Fayad and A. Altman. Introduction to software stability, *Communications of the ACM*, 44(9), 2001, 95–98.
3. M.E. Fayad. Accomplishing software stability, *Communications of the ACM*, 45(1), 2002, 111–115.
4. M.E. Fayad. How to deal with software stability, *Communications of the ACM*, 45(4), 2002, 109–112.
5. R. Wolski, N. Spring, and J. Hayes. The network weather service: A distributed resource performance forecasting service for metacomputing, *Future Generation Computer Systems*, 15(5–6), 1999, 757–768.

29 AnyMedia Stable Design Pattern

The media can be a really strong vehicle.

Cecilia Bartoli [1]

29.1 INTRODUCTION

In a common usage, AnyMedia is referred to a means of communication like radio and television, newspapers, and magazines that reach or influence people on a wide scale. In a term that is more general, AnyMedia also refers to an intervening agency, means, or instrument by which something is conveyed or accomplished.

AnyMedia is used by several applications and controls many diverse contexts. Media is used for assisting communication between many parties in various and diversely different formats. For example, recording media uses devices to store information. While in print media, communication is delivered via paper or canvas like newspapers, magazines, pamphlets, books, etc. On the other hand, in electronic media, communications are delivered via electronic or electromechanical energy as in television. Media can transmit audio, visual, or both audio and visual data. Depending on what it transmits, media can be classified into audio type or visual type. In addition, media can have various types like multimedia, hypermedia, and digital media. In multimedia communications, multiple forms of information and content processing are used. Hypermedia is media with hyperlinks, while digital media uses electronic media to store, transmit, and receive digitized information. Media has a vast scope of applicability like presentation, appealing, propaganda, advertisement, distribution, mobility, etc. (refer to SB29.1).

Since the function of AnyMedia depends on the context in which AnyMedia is being used, the traditional approaches to software design will not yield a stable and reusable model. However, by using software stability model (SSM), AnyMedia can be represented in any context by using a single model. The SSM requires creation of knowledge maps by identifying underlying enduring business themes (EBTs) and business objects (BOs). By hooking industrial objects (IOs), that are specific to each application, the model is applicable to any application domain [2–4]. The resulting AnyMedia pattern is stable, reusable, extendable, and adaptable. Thus, any number of applications can be built by using this common model. The AnyMedia design pattern tries to capture the core knowledge of AnyMedia that is common to all the application scenarios listed above to create a stable design pattern.

The goal of this chapter is to design a stability model for AnyMedia by creating a knowledge map of AnyMedia. This knowledge map or core knowledge can then serve as building block for modeling different applications in diverse domains. The rest of this chapter presents AnyMedia, a stable design pattern based on the SSM.

Media can be of various types depending on the system. For example, for selling productions such as beverages, print media such as newspapers, and visual media like television can be used. Media can be electronic or physical. Electronic media examples include the Internet, television, radio, etc. Physical media examples include magazines, newspaper, catalogues, billboards, and brochures. The diversity of media for the same business use case makes it challenging to design a system that can interact with all the different media. Therefore, AnyMedia design pattern, a stable design pattern, can help minimize effort of time and money.

Media plays a key role in the system design. With the ever-changing role of media and advent of new type of media, it is important to use media. For example, advertising was once limited to print media and television. Nevertheless, Internet plays a very important role in advertisement today. Media can be effective in changing a company's target demography. Media opens a completely new avenue for companies to bring the different products and services to the customers and partners. Politicians use media to publicize their propaganda and message. Media is a key part of the system for various organizations such as governments, companies, corporations, and nonprofit organizations. Media touches almost every sphere of modern life. Media plays an important role in the design of software and systems. For example, a software application for the Internet media has to be designed so that the software design incorporates different aspects of the Internet and leverages different technologies and protocols supported by the Internet. A system designed for using the television media needs to understand the different intricacies of the television media and interact with the media effectively. Design for AnyMedia must be compatible with the different BOs such as parties, domains. Design for AnyMedia should also make sure that it supports the various media types that the design it applies.

AnyMedia is a useful stable pattern. AnyMedia's EBT is applicability. Based on the EBT, one can easily integrate with different BOs. Then, one can add different IOs depending on the domain. Different business entities can use the same media within the same system design. AnyMedia design pattern is useful in designing a system that needs to support various media. AnyMedia helps designers add new media for integration, and at the same time, remove old media that are no longer profitable. AnyMedia pattern lets designers change the domain and the application type more easily. AnyMedia pattern is stable, reusable, extendable, and adaptable. Thus, any number of applications can be built by using this common model. The AnyMedia design pattern tries to capture the core knowledge of media that is common to all the application scenarios.

29.2 PATTERN DOCUMENTATION

29.2.1 Pattern Name: AnyMedia Stable Design Pattern

The AnyMedia design pattern, depending on the applicability chosen for a particular application, controls the context of usage of media and thereby achieves a general pattern that can be used across any application. This pattern is required to model the core knowledge of AnyMedia without tying the pattern to a specific application or domain; hence, the name AnyMedia. Again, since media is semitangible and externally stable, AnyMedia name is chosen and it is BO.

AnyMedia name is the most appropriate, as the media's role in the application is being generalized to be used in any domain and application. AnyMedia does not strongly bind to any specific application or domain. The interaction between parties, domain, and applications with media is generalized to work with any domain, and so the name AnyMedia is chosen for this stable design pattern.

29.2.2 Known As

This pattern is similar to AnyMedium pattern.

AnyMedia and AnyMedium sound very similar and are used interchangeably in common context. AnyMedium, in addition to being any physical material that records or holds recorded information, also means a middle state or condition (means). For example, air acts as a medium for transmitting electromagnetic waves. Since AnyMedia and AnyMedium do not represent the same things, they cannot be used interchangeably.

29.2.3 Context

AnyMedia design pattern is conceptually modeled to address the issue of applicability for any context in use. There are various situations where media plays a key role in our day-to-day life. The goal is to establish an understanding of how any media pattern can serve applicability in various contexts. This is achieved by specifying the generic link between media and applicability by AnyParty for any given criteria.

A system might have one or more types of medias participating in coordination to achieve the goal of the system. However, each media plays a specific role in any system. AnyMedia cannot exist alone in any application and requires assistance from AnyParty or AnyActor to achieve its goal. In most general context, AnyMedia refers to various means of communication. Audio, video, or both can be used as a means of communication. To accomplish the applicability in terms of Distribution, Advertisement, and Propaganda, AnyMedia supports systems like health awareness, road safety, Internet advertising, specific information sites like Google Maps and Google Weather, disaster management, event management, and so on.

In short, the AnyMedia design pattern can be used in any application or context where information needs to be disseminated to large audience. In the later section, application scenarios to illustrate the applicability of AnyMedia pattern are demonstrated. AnyMedia concept is illustrated in two distinct scenarios like road safety awareness and Internet advertising to illustrate how media can achieve its applicability in different applications.

1. *Use of AnyMedia for road safety awareness*: In this example, AnyMedia is evaluated in a road safety awareness scenario. Educating public on traffic rules usually happens through different types of media. This example models the educating process of traffic rules to the public via the help of media. Attached below are the sequence diagram and the stability model for educating public on traffic rules.

Concept	Description
Applicability (EBT)	Broadcast
AnyMedia	Television, radio, newspaper
AnyParty	Program director
AnyCriteria	Safety rule
AnyContext	Road safety
MediaType	Audio, visual

2. *Using Internet as a media to advertise food products*: AnyParty can do advertising for any product or service. AnyParty determines the context of the product domain and the reachable criteria. Once it is decided, AnyParty decides the type of media that suits the context and criteria. Internet is the most popular media, and if the selection criteria (could be based on various factors like viewership, cost, etc.) is met, as specified and required by AnyParty, it is chosen for application. The table below summarizes how AnyMedia design pattern fits in this application scenario.

Concept	Description
Applicability (EBT)	Advertising and propaganda
AnyMedia	Internet
AnyParty	Product manager
AnyCriteria	Online marketing
AnyContext	Food product promotion
MediaType	Multimedia

29.2.4 PROBLEM

The AnyMedia design pattern is applicable across diverse application domains that have different functionality, usage, and nature. Thus, modeling a generic design pattern that can be applied to all these domains makes good sense. However, each problem domain uses media for its own custom needs. For instance, education uses media for educating people. Similarly, broadcasting uses media to transmit information or entertain public. The challenge is to build the right model for media, which can be applied to any context. Thus, priority should be given to below listed points before modeling AnyMedia pattern.

- The model should capture the core knowledge of the media problem.
- The model can be reused to model the media problem in any application.
- The ultimate goal for AnyMedia must be established along with the other patterns used in the design pattern of AnyMedia.
- Choosing the right media type depending on the applicability of the application.

Since AnyMedia is used in different context in different domains, building a generic model is very challenging. Using SSM, this problem is solved and a generic model is modeled for different domains. This model is illustrated and described in the solution section.

1. *AnyMedia*: A medium in which objects reside like data and information. AnyMedia refers to Internet, e-mail, newspaper, where the AnyAgenda is published so that the participants (AnyActor) can come to know about AnyEvent.
    a. *Searching (context)*: Here, searching is performed and the result from the search would be displayed.
    b. *Attributes*: id, mediaName, mediaType, capability, entry, securityLevel, status, sector, security, communicationLink, communicationParameters, userDetails.
    c. *Operations*: connect(), broadcast(), capture(), store(), display(), access(), remove(), select(), navigate(), secure(), format(), identifyMedia(), defineMediaProperties(), identifyMediaCapacity(), loadInformation(), present(), filter(), broadcast(), locate().
    d. *access()*: Allows information to be stored or retrieved in the media.
    e. *select()*: Selects the media to be used.
    f. *remove()*: Removes the media or the information in the media.
    g. *navigates()*: Navigates to different parts of the media.
    h. *secure()*: Secures the media by providing a security level or encryption.
    i. *format()*: Formats the media, rearranges the sectors and content so it can be navigated.
2. *AnyContext*: This provides a context for various objects and process to take place. For example, a presidential election could be the context, when parties involved will use media like TV, Internet, newspaper, and audio interviews to pass on their messages to the public.
    a. *Attributes*: boundary, subject, visibility, contextName, domain, description, pattern, applicability
    b. *Operations*: access(), relate(), exercise(), mark(), scale()
    c. *relate()*: Relates context to each other.
    d. *scale()*: Scales the domain/context up and down by removing or adding different elements.
    e. *constraint()*: Restricts a context so it only applies to a certain area.
    f. *filter()*: Filters the context, remove specific domains.
    g. *capture()*: Captures a context from the user or various sources data mining sources.
3. *AnyCriteria or AnyCriterion*: Contains a set of criteria used for constraining or matching or verification. Provide a set of criteria defined by a party to be used in the given goal.
    a. *Attributes*: id, name, condition, property, priority, conditiontype, definedby, conditiondetails, purpose
    b. *Operations*: define(): defines the criteria, verify(): verifies the criteria to see if it is actually valid, apply(): selects the criteria to be used, prioritize(): sets a priority level for the criteria. This is relative to other criteria in the system, parse(): parses the conditions that are associated with the criteria, exhibit(), relate(), evaluate(), validate(), measure(), verify(), prioritize(),

4. *AnyParty*: A party that participates in various activities. AnyParty is a legal user of the system.
   a. *Attributes*: id, partyName, type, role, member, affair, activity, partiesInvolved, id, activity, category (or orientation).
   b. *Operations*: participate(), playRole(), interact(), group(), associate(), organize(), request(), setCriteria(), switchRole(), partake(), join(), monitor(), explore(), discover(), receive(), gatherData(), integrate(), agree(), disagree(), leave(), switchRole().
   c. *participates()*: Participates in an activity or group.
   d. *join()*: Join members to the party.
   e. *leave()*: Force members to leave the party.
   f. *group()*: Groups the various parties in the system.
   g. *organize()*: Organize the various parties in the system.
   h. Applicability such as desired or appeal, assessment, viewing, advertisement, mobility, etc. Determines and provides a method to see what objects are applicable within a certain context.
   i. *Context*: Determines which media is applicable for searching.
   j. *Attributes*: Level, observation, domain, handler
   k. *Operations*: scoping(), select(), apply(), match()
   l. *apply()*: See whether or not a context is applicable to the media.
   m. *relevance()*: Returns a level of relevance to the user. This might be related to the success of the search or identity, criteria, where the searching takes place.
   n. *regulate()*: Regulates whether or not the context is applicable.
   o. *demonstrate()*: Demonstrates how would various contexts be applicable.
   p. *prefer()*: Allows the user to determine what is applicable and what they prefer.

29.2.4.1 Functional Requirements

1. *Applicability*: This class specifies how effectively media can be used within a given context. Context could influence the mode of applicability. For instance, in the context of advertising applicability, we could use different types of media like television, newspaper, radio, Internet, magazines, etc. Some other applicable instances include distribution, advertisement, propaganda, and mobility.
2. *AnyMedia*: This class is used by applications to handle applicability. For instance, one can use media for advertising, publishing, broadcasting, etc. At the same time, some other parties could try to use media for education, e-commerce, etc.
3. *AnyParty*: This class represents the people who choose applicability depending on the criteria specified for the problem domain. AnyParty might be humans, organization, or a group with specific orientation. AnyParty should be capable of specifying the criteria for the context clearly.
4. *AnyActor*: It refers to the person who prepares the AnyCriteria such as election propaganda and campaigning, who participates in the AnyEvent that is listed down in AnyContext.
5. *AnyCriteria*: This class specifies the condition or rules which controls the context of the application domain. For example, in terms of traffic awareness, the Criteria could be the rules of traffic. AnyParty enforces the Criteria.
6. *AnyContext*: This class represents the scope of the problem domain that guides AnyParty to choose the media type. The context is affected by the criteria. For example, in the field of Internet advertising, the context of the advertisement could vary depending on the age group of the audience it is intended for.
7. *AnyType*: This class represents the type of media that is used by application. AnyMedia picks the type of media depending on the application context. It is used to identify in what way AnyActor or AnyParty uses media. This class represents the type of media too. AnyType is the type of media and it varies from one context to the other based on the type of advertisement or promotion.

8. AnyEntity: It is the part of a AnyMedia. AnyEntity is an individual unit who finds media because he/she wants to promote/advertise something in AnyContext.
9. AnyEvent: A series of events carried throughout the process of creating AnyMedia campaign. This class represents any event where media is used. An event is something that happens.
10. AnyImpact: AnyImpact is a BO for media that can occur in any given situation or context. AnyImpact always leads to a perceptible change or transformation. AnyImpact describes AnyEvent and generates consequences, and it may relate to AnySubject.
11. AnySubject: Media could relate to any type of topic. A future event may decide the type of subject for which media will be used to create an impact. AnySubject is chosen by AnyActor and AnyParty based on type of advertisement or promotion.

29.2.4.2   Nonfunctional Requirements

1. Effective: Media that is used for any context should be effective and efficient. Otherwise, the main goal set for any event would be lost and the level of impact expected by a party will be reduced drastically.
2. Influence: Media that is used should also be influential so that it can affect and influence the mind and behavior of the target group. In other words, any media format that is not influential would be a waste of money, time, and effort. For example, media used in an election should be able to influence the minds of electorate and it should convert into votes for the party who fights the election.
3. Advertisement: A media used should be in the format of an advertisement. For example, a product advertisement launched by a manufacturing company should conduct media campaign to advertise its product line.
4. Marketing: Use of media is essentially a marketing exercise and it is the heart and soul of any media campaign. Media should be used to market an entity or a product in an effective manner. Marketing is also a tool to build a brand around an entity.
5. Connectivity: An effective media campaign should result in improved and extended connectivity among different stakeholders involved in the exercise. If the use of media is less successful, then the entire campaign could be termed a wasteful exercise.
6. Manipulation: Using media should ensure targeted manipulation of a party. Media should be effective to the extent that it results in desired outcomes; in effect, media should lead a party to change or transform one's mind and opinions.

29.2.5   Solution

SSM satisfies the classes of the system in three layers—EBT, BO, and IO layers. Each class in the stability model is classified into one of the three layers. The proposed solution is to focus on the core concept of media that is common across applications by eliminating the other domain specific entities. Figure 29.1 represents the static model of the pattern.

29.3   SUMMARY

Though building a stable design pattern for AnyMedia that is reusable and reapplicable across diverse domains is difficult and requires thorough understanding of the problem, it is worth the effort and time for a developer who wants to design an effective pattern that is stable and robust. By modeling AnyMedia pattern by using SSM, one can ensure a pattern that is reusable, extensible, and stable pattern.

The correct identification of EBT and BOs for AnyMedia is a challenging task and it requires some experience. Once EBT and BOs are correctly identified, next challenge is to determine the relationship between EBT and BOs so that AnyMedia pattern can hold true in any context of usage

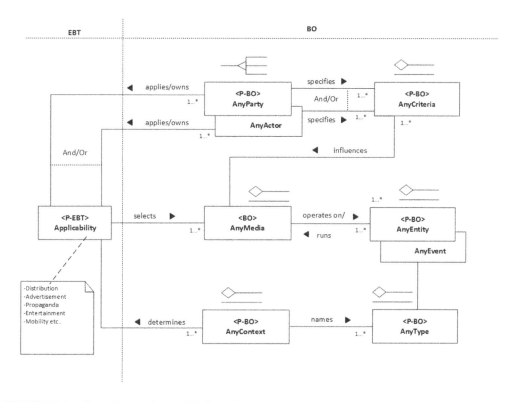

FIGURE 29.1   Class diagram for AnyMedia design pattern.

for media. Once this is ensured, depending on the application, the IOs are attached to the hooks so provided by BOs. Thus, by using AnyMedia pattern as a basis, infinite number of applications can be built by simply plugging in the application specific IOs to the pattern. This will eventually result in reduced cost, effort, and stable solution. Hence, AnyMedia design pattern is very useful and beneficial.

29.4   OPEN AND RESEARCH ISSUES

How can the AnyMedia pattern be applied to nondigital media formats. Examples, snail mail, carrier pigeon, etc.

In spite of several advantages offered by a AnyMedia stable pattern, a developer can identify certain pitfalls and lacunae while applying it to certain media formats like nondigital and traditional approaches. Nondigital media formats may include age-old postal communication systems, telegraphic messages, carrier pigeon communication system of conventional world, and Morse code systems and sign languages. Certain important questions could be posed while applying AnyMedia to the aforementioned media systems. First, is it possible to identify essential BOs from such a media system? Second, can a developer plug IOs to identified BOs? To create an effective and stable AnyMedia pattern with a nondigital media format, a pattern designer should choose maximum numbers of BOs although identifying all of them could be very challenging. For example, in the case of carrier pigeon media format, the media content that is delivered by a party through a carrier pigeon could be either lost or stolen by another party who could be a nonplayer in the entire system. Hence, some research issues may force a pattern developer to think over devising a system that takes into consideration such invariables and find a solution that is satisfactory for all domains and scenarios.

29.5   REVIEW QUESTIONS

1. (T/F) AnyMedia design pattern is domain specific.
2. (T/F) AnyMedia always involve some sort of communication.
3. (T/F) AnyMedia cannot be inanimate.
4. (T/F) The AnyMedia pattern describes a system.
5. (T/F) The AnyMedia design pattern can be applied and extended to any domain.
6. Define AnyMedia design pattern.
7. List two major applications of the AnyMedia design pattern.
8. Why does the AnyMedia pattern have "Any" as a prefix?
9. List three challenges in formulating the AnyMedia pattern.
10. List constraints that arose during the formulation of the AnyMedia pattern.
11. Draw the AnyMedia design pattern.
12. List the participants (EBTs and BOs) in the AnyMedia design pattern.
13. Draw the CRC cards for two of the BOs in the AnyMedia design pattern.
14. List four main benefits and usage notes for the AnyMedia design pattern.
15. List some of the patterns related to the AnyMedia design pattern.
16. List two scenarios that would not fit within the context of the AnyMedia design pattern.
17. List three business issues in the context of the AnyMedia design pattern.
18. Briefly explain how software stability concepts have been incorporated in the AnyMedia design pattern.
19. List a couple of research issues relevant to the AnyMedia design pattern.
20. How will the AnyMedia design pattern be stable over time?
21. Explain briefly how the AnyMedia design pattern provides a high level of extensibility.
22. What do the term "media" mean? Can you use it in any other context than what has already been mentioned in this chapter?
23. List the synonyms of "media." Can these terms be used interchangeably with respect to the AnyMedia pattern?
24. What are the requirements for a "media"? Describe each of them.
25. List differences between the AnyMedia pattern described here and the traditional methods.
26. List some design and implementation issues faced when implementing the AnyMedia pattern. Explain each issue.
27. Give some applications where the AnyMedia pattern is used.
28. What lessons have you learnt by studying the AnyMedia pattern?
29. List some of the domains in which the AnyMedia design pattern can be applied.
30. List some domains that you believe the AnyMedia pattern would not apply. Explain your reasons.
31. List three challenges in formulating the AnyMedia design pattern.
32. List three different constraints in the AnyMedia pattern.
33. Is the AnyMedia pattern incomplete without the use of other patterns? Explain briefly.
34. What do you think are the implementation issues for the AnyCriteria BO, when used in the AnyMedia design pattern?
35. What do you think are the implementation issues for the AnyContext BO when used with the AnyMedia design pattern?

29.6   EXERCISES

1. Apply the AnyMedia design pattern and apply it to the broadcast of a television show. The show can have any number of seasons with numerous episodes for each season. The AnyMedia pattern can cover multiple formats for consumption, broadcast TV, satellite TV, cable TV, Internet streaming services, DVDs, or Internet downloads.

 a. Draw the class diagram.
 b. Create the sequence diagram that has a user finding the television show based on their available media formats.
2. Apply the AnyMedia design pattern to magazine subscriptions. There are different subscription tiers that determine how subscribers are able to consume the content. The subscription tiers are described in the following table. Draw the class diagram.

Tier 3	All three types from Tier 1		
Tier 2	Choice of two types from Tier 1		
Tier 1	Online exclusives	Printed copy	Electronic copy

3. Create an AnyMedia pattern system for running a presidential election system that uses a media system including TV, Internet, and print media. Incidentally, this system is found to be extremely effective and it is known to influence voters' opinions on a specific candidate; a candidate may use this system to build a brand around him/her, influence voter's mind and steer the course of the election process in a positive manner.
 a. Draw a class diagram.
 b. Define the entire system with reference to its BOs and IOs.
 c. Highlight different factors (nonfunctional factors) that affect the media system.
 d. Find out problems that you would come across while implanting the stable pattern.
4. Media versus advertisement stable patterns
 a. What are the differences between media stable pattern and advertisement pattern?
 b. Create the advertisement stable pattern.
 c. Name and describe three contexts for each of them.
 d. Show how to apply the two patterns, including with one context of your choice and including a use case, the application of each pattern with your selected context.
5. Media versus marketing stable patterns
 a. What are the differences between media stable pattern and marketing pattern?
 b. Create the marketing stable pattern.
 c. Name and describe three contexts for each of them.
 d. Show how to apply the two patterns, including with one context of your choice and including a use case, the application of each pattern with your selected context
6. Create five applications of AnyMedia patterns in a table as shown in Table 23.1 (Chapter 23).
7. Create two applications of two contexts of AnyView stable design pattern and provide the following for each one of the applications:
 a. Context.
 b. Application diagram using the hype stable design patterns.
 c. Detailed use case.
 d. A sequence diagram for the application.

29.7 PROJECTS

1. Create a global delivery system that utilizes the sender and receiver's location, and the type of package sent to create a delivery route to any place in the world. For example, an email can be sent for transmitting simple data if the given parameters fit the usage of email. However, if someone desires to deliver a crate of items to a remote location in the Amazon, the system must:
 a. Take the starting coordinates and the parameters of the cargo.
 b. Take the ending coordinates.

   c. Generate a plan/manifest for making the delivery*
      i. Delivery to nearest port city
      ii. Cargo ship for delivery
      iii. Receipt of cargo at destination port
      iv. Final delivery

Assume that there are delivery services that go anywhere in the world and that cost is not a concern. Delivery across the globe can take place on land, air, or sea. Vehicles to use are plane, boat, cargo truck/van. Create size and weight constraints for each type of transportation. For example, cargo with a large footprint or extreme weight may only travel on land and sea.

1. Create class diagrams, sequence diagrams, and CRC cards.
2. Define BOs that are involved in the entire process.

Create an Internet promotion campaign for a certain product. This campaign may involve using e-mail, targeted message delivery, and use of advertisement dissemination by using online marketing systems like Google AdWords and AdChoices. Define a system that include issues like the number of individuals and businesses that are intended for message delivery, their locations, budget and time involved, number of advertisement channels to be used, and content to be distributed.

1. Create a class diagram.
2. Write a sequence diagram.
3. Draw CRC cards.
4. List different BOs and IOs for the system.
5. Highlight different problems that are known to exist in the system.

2. Use Media in one of the following:
   a. i. Advertising Media
      ii. Broadcast Media
      iii. Interactive Media
      iv. Recording Media
   b. Create three contexts for each one of them.
   c. Create stable architecture pattern for each of them.
   d. Create an application of each one of them.

REFERENCES

1. Cecilia Bartoli (born June 4, 1966), last modified on June 1, 2017, https://en.wikipedia.org/wiki/Cecilia_Bartoli
2. M.E. Fayad and A. Altman. Introduction to software stability, *Communications of the ACM*, 44(9), 2001, 95–98.
3. M.E. Fayad. Accomplishing software stability, *Communications of the ACM*, 45(1), 2002, 111–115.
4. M.E. Fayad. How to deal with software stability, *Communications of the ACM*, 45(4), 2002, 109–112.

SIDEBAR 29.1   CONTEXT IN DIFFERENT DOMAINS

Media refers to various channels of communication for dissemination of information and events. Media serves people diverse functions such as communicating news and information, advertising, publishing, entertainment, etc. Media can be physical (e.g., books, magazines, newspaper) or electronic (e.g., Internet, telephone, broadcasts). The press is also sometimes referred to as the media since the press was chronologically the first entity to communicate information to the public. However, today the Internet and television have a bigger share of mass communication market.

Unlike a channel that is limited to a contiguous physical medium between the sender and a receiver of communications, media includes the institutions that determine the nature, programming, and form of distribution. These days media is not only a channel for news and information but also a key marketing tool for businesses, which uses media to promote goods and services and communicate information to business partners. Businesses also use media to build public relations. Media is used in news, fine arts, engineering, marketing, education, commerce, industry, entertainment, military, nonprofit services, public services, transportation, science, and journalism to name a few. Media touches various realms of modern life and has come of age as a significant part of human culture.

Accessibility of media has undergone a sea change. Internet and television have now penetrated most of the populace around the world. Many are aware of different kinds of media and use it in some way or another. Media also conforms to the legal, cultural, and political rules and regulations of any given location. Some media are actually shrinking in terms of usage among the general population. For example, radio is not as popular as television and the Internet these days, as it used to be in the early 1900s. The types, market share, and dynamics of media are ever changing.

AnyMedia is a general concept with wide range of application in many different contexts. The AnyMedia stable design pattern aims at analyzing the general and important concept of AnyMedia. Since AnyMedia pattern is introduced based on the stable design pattern, it makes it easier to employ this pattern in many different applications. This is possible by just hooking the unstable industrial objects to the stable business objects according to the application under study. Here, we will introduce the AnyMedia design pattern based on stability model and introduce few scenarios where this pattern can be used.

30 AnyLog Stable Design Pattern

The *AnyLog* pattern model the core knowledge of any type of Log, in the form of a written record. The Log finds extensive use in the computing industry. The pattern makes it easy to mold different kinds of logs rather than thinking of the problem each time from scratch. This pattern can also be utilized to shape any kind of log in any kind of application and it can be reused as part of a new model.

30.1 INTRODUCTION

"Log" as a record implies any significant information stored in a media for future reference. Different applications for log use different media and store entirely different kind of information. The information in a log is stored as one or more entries. For example, a traveler's log can be a diary, whereas a web site traffic log can be a file on an electronic disk. The entries in a traveler's log may be his daily expenses and places visited all written in a free form (unformatted), whereas a web site traffic log may contain formatted data such as client IP address, time, date, and others. Since log finds use in diverse applications, it is important to understand the core knowledge that occurs behind the log. The AnyLog design pattern aims at capturing such core knowledge that is common in all applications that use log.

In the traditional way of modeling, we have created separate model of log for each application. This exercise involved enough time and effort to start and finish the entire modeling process from scratch every time. On the other hand, the stable design pattern is an excellent way to start and end the entire process. The AnyLog pattern describes the characteristics, behavior, and lists all the key players and their relationships that are involved in any kind of log application. This pattern can be easily extended to recreate the model for any type of log application.

A fundamental question is "Can we develop a pattern that captures the atomic log notion, and thus can serve as a base for modeling any kind of logs?" The main objective of this chapter is to offer an answer to this question by discussing and documenting the atomic pattern *AnyLog*. This pattern models the core knowledge of a log, thereby making it easy to reuse this pattern and extend it to model different kinds of logs, rather than reworking the same problem each time from scratch. AnyLog pattern is a stable design pattern [1–3] that is build based on the software stability concepts.

30.1.1 SOFTWARE STABILITY AND STABLE ANALYSIS PATTERNS: BRIEF BACKGROUND

The pattern proposed in this chapter is based on the concept of stable analysis patterns that were introduced in References 1–3. The main working idea behind stable analysis patterns is to analyze the problem under consideration in terms of software stability concepts. Software stability stratifies the classes of the system into three layers: the EBTs layer (contains classes that present the enduring and basic knowledge of the underlying industry or business, and hence, they are extremely stable), the BOs layer (contains classes that map the EBTs of the system into more concrete objects. BOs are tangible and externally stable, but they are internally adaptable), and the IOs layer (contains classes that map the BOs of the system into physical objects).

The three main layers of the software stability model are depicted in Figure SBF3.1. In generating a stable design pattern, we may need to recognize the concept that always follows the object, irrespective of the application. This chapter suggests a possible design pattern for any type of log.

30.2 PATTERN DOCUMENTATION

30.2.1 Pattern Name: AnyLog Stable Design Pattern

The term "log" implies maintaining a record of some kind. It could be record of performance or events. It could also be a record of the activity of a certain system stored for later reference. A log could be used in many ways and contexts. In aviation and sea voyage domains, a log is an official record of events that might occur during travelling. Usually, a log is a written account preserving knowledge of facts or events. Keeping a log or maintaining a log could encompass several areas of daily life such as making a log of events in a police station, keeping a log of records in a datacenter, creating a log of activities, or maintaining a personal memoir in the form of a logbook. Some of the synonymous words that represent log are record, journal, diary, chronicle, and record book.

30.2.2 Context

Log is a record that stores some information valuable to its creator. The log has applicability in various fields such as information technology, travel, art, etc. For example, a file that contains the information on how many hits or impressions a web page is receiving is known as a traffic log. Another example of a log is a record of a computer or application's activity, used for system information, backup, and recovery. Another case of use of a log is that a ship captain stores events in a voyage in a trip log.

This section will describe two different scenarios in which a log plays its vital role:

- *A captain who maintains a log of ship's activities*: Seafaring is an important aspect of a ship's life (AnyActivity). The captain of the ship (AnyActor) is duty bound to maintain a complete log of all ship's activities especially its safety and security (AnyContext). Avoiding accidents and unexpected events (AnyType) that might cause deaths and injuries to sailors (AnyParty) is the most important issue while the ship is sailing from one port to the other. The captain should maintain a log by storing it in a proper format like saving it (AnyRecording) in the hard drive of a computer or in a CD/DVD or by sending the data to a cloud-based platform (AnyMedia). The stored data may contain information such as ship's average speed, wind direction, weather patterns, sea surface temperature, engine room activities, turbine condition, engine room safety, fuel used, and communication exchanged between the ship and ports (AnyType).
- *A web master's log*: Monitoring a web site's online performance is a very important activity of a web portal (AnyActivity) and a web master (AnyActor) should oversee the overall performance to check whether it is doing well in certain aspects like page hits per day, the level of security, load testing, antivirus performance level, and others (AnyType). A web master's basic duty is to ensure that web site performs to the best possible levels (AnyContext) and establishes its online identity with thousands of visitors (AnyParty). The web master is also required to maintain a comprehensive web activity log and preserving it by storing it in a proper media format like a CD, DVD, or by sending the data to a cloud-based online platform (AnyMedia).

30.2.3 Problem

30.2.3.1 Functional Requirements

1. *Recording*: This class indicates the purpose of using a log. It is modeled as an EBT as it conveys the core goal of using any kind of log.
2. *UnformattedEntry*: It is an abstraction that is representative of all entries in the log that does not have a format or any kind of unformatted data. Example of an unformatted entry would be the numbers in a telephone diary not arranged according to alphabetical

order, whereas if they were arranged in alphabetical order, they would be formatted entries.

3. *FormattedEntry*: An abstraction for all entries that have a particular format and it is indicative of all the formatted data in the log. For example, if entries in a web site administrator's log are arranged according to a particular date and time and to their time of occurrence, they are termed as formatted entries.

4. *AnyLog*: Represents the logging process itself. The class contains the behaviors and attributes that regulate the actual logging process. The AnyLog pattern is associated with an EBT of Recording, where one's purpose is to preserve the information.

5. *AnyParty*: Represents the party that maintains the log. Party can be a customer, company, any organization, etc.

6. *AnyEntry*: Represents the information that should be maintained and stored in the log.

7. *AnyMedia*: It identifies and defines the media on which the log is created. It also represents the media by which the log is to be displayed.

30.2.3.2 Nonfunctional Requirements

1. *Sufficient*: A log should be complete in all basic aspects of maintenance and the data collected by it should be sufficient to draw a valid inference or summary of events that actually occur during log making. Nothing can be inferred or made out of an incomplete log created by any person.

2. *Easy to understand*: The log should be simple, comprehensible, and easy to understand. Only the log maker and not others understand complex logs. Hence, creating a simple log will enable anyone to read and understand the contents of the log.

3. *Clear*: A log should be clear, lucid, and precise to save time and effort needed to understand the content. It should also reveal and divulge the reason for making the log and parameters that define the framework within which it was created.

4. *Accessibility*: A log should be easily accessible to anyone who wants to see its content. The person who creates a log should store it in such a place that everyone can access it immediately. A log in an e-format should neither be protected with an unknown password nor should it be stored in a remote location; even if it is stored in this way, the log creator should inform others the precise location where it is stored.

5. *Stable*: A stable log provides a real-time picture of events that occurred previously and details that are recorded in the log could be considered as an authentic record for all future applications, corrective actions, and recommendations.

6. *Scalable*: Scalability is the ability of any system or process to manage an increasingly growing amount of work and its capability to be enhanced to accommodate that level of growth. In the context of a log, any log created should be able to handle a large amount of future work or efforts along with a potential to develop room for including any quantum of future actions.

7. *Adaptable*: Adaptable is the ability to change or to be transformed to fit or work well in different contexts. Here, a log should be adaptable to work better in any type of contexts and situations to provide additional efficiency to introduce any plan of actions. It should also be open for any type of changes and transformations to enable it to become more efficient and productive.

30.2.4 Solution

Figure 30.1 shows the solution without a media for the AnyLog design pattern. The model has one EBT named Recording and BOs are AnyParty, AnyEntry, AnyLog, UnformattedEntry, and FormattedEntry.

Figure 30.1 shows the AnyLog pattern with the AnyMedia pattern included. This pattern can be used when the media plays a very important role and hence needs to be illustrated.

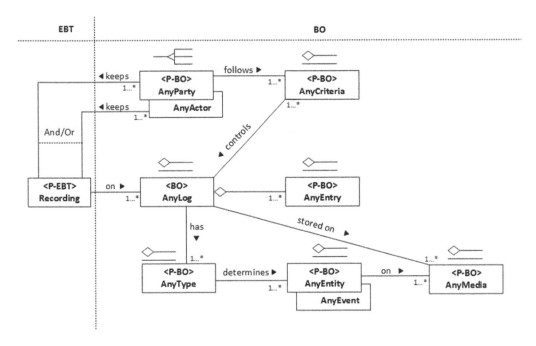

FIGURE 30.1   AnyLog design pattern.

30.3   SUMMARY

In computing, a log is a recorded file of events that occur in an operating system or other software processes or that emanate from sending messages from one user to the other. Logging is an intentional act of maintaining or keeping a log. A log file is almost similar to black box recordings of an airline or an aircraft. It can help unfold a sequence of events or process that might have occurred during the normal operation of a computer system. A log is essentially a record of events and it could be used as a guide to suggest future corrective actions that can enhance important performance parameters. AnyLog design pattern is important because it can capture real-time events and help a developer to develop numerous applications that are likely to utilize different features of logs.

30.4   REVIEW QUESTIONS

1. (T/F) AnyLog design pattern is domain specific.
2. (T/F) AnyLog can never be a person.
3. (T/F) AnyLog cannot be inanimate.
4. (T/F) The AnyLog pattern describes a system.
5. (T/F) The AnyLog design pattern can be applied and extended to any domain.
6. Define AnyLog design pattern.
7. List two major applications of the AnyLog design pattern.
8. Why does the AnyLog pattern have "Any" as a prefix?
9. List three challenges in formulating the AnyLog pattern.
10. List four constraints during the formulation of the AnyLog pattern.
11. Draw the design pattern solution for the AnyLog design pattern.
12. List the participants (EBTs and BOs) in the AnyLog design pattern.
13. Illustrate the detailed model with a diagram for the AnyLog design pattern.
14. Draw the CRC cards for two of the BOs in the AnyLog design pattern.
15. List four main benefits and usage notes for the AnyLog design pattern.

16. List the use cases in classifying patient data applying the AnyLog design pattern.
17. List the use cases in classifying laser printer in an inventory applying the AnyLog design pattern.
18. List some of the patterns related to the AnyLog design pattern.
19. List two design issues for the "Recording" EBT during the process of linking from the analysis phase to the design phase.
20. List two important implementation issues for the "AnyEntity" BO during the process of linking from the design phase to the implementation phase.
21. List two scenarios that would not fit within the context of the AnyLog design pattern.
22. List three business issues in the context of the AnyLog design pattern.
23. Briefly explain how software stability concepts have been incorporated in the AnyLog design pattern.
24. List a couple of research issues relevant to the AnyLog design pattern.
25. How will the AnyLog design pattern be stable over time?
26. Explain briefly how the AnyLog design pattern provides a high level of extensibility.
27. What does the term "log" mean? Can the term "log" be used in any other context than what has already been mentioned in this chapter?
28. List the synonyms of "log." Can these terms be used interchangeably with respect to the any pattern?
29. What are the requirements for a "log"? Describe each of them.
30. List differences between AnyLog pattern and those that are created by using traditional methods.
31. What lessons have you learnt by studying the AnyLog pattern?
32. List some domains that you believe the AnyLog pattern would not apply. Explain your reasons.
33. List three challenges in formulating the AnyLog design pattern.
34. Is the AnyLog pattern incomplete without the use of other patterns? Explain briefly.
35. Present the sequence diagram for applicability of the AnyLog stable analysis pattern in the e-commerce domain.
36. What do you think are the implementation issues for the AnyParty BO when used in the AnyLog design pattern?
37. What do you think are the implementation issues for the AnyEntry BO when used with the AnyLog design pattern?
38. List some of the testing patterns that can be applied for testing the AnyLog design pattern.
39. List three test cases to test the class members of the AnyLog pattern.
40. Name the different types of entities that the AnyLog pattern may represent.

30.5 EXERCISES

Explain with an example the applicability of the AnyLog design pattern in classifying students who are enrolled in a university and their class activities per semester. Draw the class diagram including all the EBTs and BOs.

1. Illustrate with a class diagram the applicability of the AnyLog design pattern to a major league sports team and its talent scout, who searches worldwide for the next star player.
2. Think of any scenarios that have not been described in this chapter. List them with a short description.
3. Try to create a use case and interaction diagram for each of the scenarios you thought of in the above question.
4. Log versus record stable patterns
   a. What is the difference between log stable pattern and record pattern?

b. Create the record stable pattern.
c. Name and describe three contexts for each of them.
d. Show how to apply the two patterns including one context of your choice and including a use case, the application of each pattern with your selected context.

5. Log versus storage stable patterns
a. What is the difference between log stable pattern and storage pattern?
b. Create the storage stable pattern.
c. Name and describe three contexts for each of them.
d. Show how to apply the two patterns including one context of your choice and including a use case, the application of each pattern with your selected context.

6. Create five applications of AnyLog patterns in a table as shown in Table 23.1 (Chapter 23).

7. Create two applications of two contexts of AnyLog stable design pattern and provide the following for each one of the applications:
a. Context.
b. Application diagram using the hype stable design patterns.
c. Detailed use case.
d. A sequence diagram for the application.

30.6 PROJECTS

Video Game Concept: A video game has a main character, who is a Sherlock Holmes type of detective. The detective character takes notes during investigations and players of the game solve the crime based on the compilation of notes that were gathered. Notes can be gathered from the environment (written by Nonplayable characters), that occur during the investigation, and from conversations that occur between different characters. Create a system that aggregates the wide array of possible notes that may occur in the game. The system will require interaction between the player and the notes in a visual tree diagram that allows the player to connect the dots. Some notes may be unrelated to the investigation, and there is a certain pattern (game rules) that allow notes to be connected. Assume the game rules or consider AnyCriteria has been implemented for you. Create a class diagram, CRC cards, and a sequence diagram to describe this application.

REFERENCES

1. M.E. Fayad and A. Altman. Introduction to software stability, *Communications of the ACM*, 44(9), 2001, 95–98.
2. M.E. Fayad. Accomplishing software stability, *Communications of the ACM*, 45(1), 2002, 111–115.
3. M.E. Fayad. How to deal with software stability. *Communications of the ACM*, 45(4), 2002, 109–112.

31 AnyEvent Stable Design Pattern

Change is not an event, it's a process.

Cheryl James [1]

This document proposes how to document the AnyEvent stable design pattern using a standard documentation template provided in Appendix C. AnyEvent stable design pattern provides instructive information. In stable design patterns, the core EBT and BO remain same for any domain facing the same recurring problem.

31.1 INTRODUCTION

Event is an important aspect of every important step taken by everyone in life. Event in event management indicates that something is bound to occur or change. Event indicates the project duration, situation where it can be applied. In software project, event should be creative, technical, and logically implemented.

31.2 PATTERN DOCUMENTATION

31.2.1 Pattern Name: AnyEvent Stable Design Pattern

Event stable design pattern can be found anywhere including personal, educations, software project, religious, political, sports, etc. Event name is a BO.

The main goal of event is occurrence. Event occurs due to some situations that may lead to minor or major consequences or impacts. This pattern reuses the same problem under any given context.

31.2.2 Known As

- *Happening*: Another misinterpretation of event is happening. Happening of any event is the part of that event. In any case if something happens, then we can say it as event.
- *Instant*: Event sometimes conflicts with instant. It may be the case that during one particular instant event may occur or may not occur. So in software project management incomplete instant create problem. Instant may not exist timely. In project coding instant of one object can call the methods or function which will make the coding and implementation essay. However, event is much better instead of instant. Instant cannot be used to represent event in some context.
- *Occurrence*: Event can be defined as the things to be happening or facts of things occurring. Occurrence is sometimes used in context of event. Although event occurs due to some reason, occurrence can be defined as something that takes place.
- *Competition*: In event, people take part due to competition. It is not the case that due to competition, event will occur. So we cannot say that competition leads to the events.

31.2.3 Context

Event occurs based on whether we consider event as an individual or as a group activity or whose identity depends on where it can be applied. Events can be used based on their purpose and objectives. It can be classified into categories such as personal event, organization event, cultural event, leisure event, and system event.

Event design pattern can be applied to numerous applications.

- Event occurs in software computing, UML diagram, sets of outcomes, testing module, and time as well.
- Event helps us to communicate with various industries, groups, and societies.
- Event can help us to identify the project boundary, start time, end time, its performance. Through event, we can identify market strategies.
- Companies establish event to communicate with clients.
- Cultural event includes various religious, arts, dance, etc. events.
- Leisure event includes various sports and music events.
- To live and die is the personal event.
- System events include how to start the system, the working of the system, how to close the system.
  - *Job Fair Event*: Consider a job fair event. In job fair model, we have hiring manager and applicant as IOs for AnyParty (BO). The applicants submit their resumes to the hiring manager, and the manager interviews them, and gives the result. AnyEvent type contains schedule as an IO to show the job fair event as a scheduled event. AnyContext contains the context of the job fair, which is to meet the requirement of the company. Here, the BOs are in yellow and IOs are in gray.
  - *Winning a lottery*: This represents an unplanned event. The user will purchase the ticket and will become the ticket holder. To select the lucky winner, there will be a lucky draw. From the draw, the winner will be selected and awarded. In this AnyEvent type, AnyParty and AnyContext are the BOs. AnyEvent type indicates the type of events; here, the event type is lottery result. AnyContext is the LotteryLuckyDraw situation. Ticket holder, winner, lottery associations are the IOs to AnyParty.

31.2.4   Problems

Event is used in different context or in different domains. So it is necessary to identify the area where and at what time the particular event is applied or taken place without affecting the functionality. The AnyEvent pattern has the following functional and nonfunctional requirements.

31.2.4.1   Functional Requirements

1. *AnyType*: Event has many types such as cultural event, personal event, system event, leisure event, and organization event.
2. *AnyEntity*: AnyEntity is determined based on the event type. AnyEntity performs the event.
3. *AnyParty*: AnyParty includes organization, country, political parties, or combination of some of them. AnyParty may or may not be part of an event at giving time.
4. *AnyActor*: AnyActor may or may not be part of an event at any given time.
5. *AnyMedia*: Medias include newspapers, TV, books, etc. An event taken place within an existing media.
6. *AnyDuration*: Event occurs in duration. Event may occur for short term or long term. Some event can be performed for nanoseconds, seconds, minutes, hours, days, months, or years.
7. *AnyConsequences*: Event produces major or minor consequences.

31.2.4.2   Nonfunctional Requirements

1. *Timly*: Event occurs in timely manner.
2. *Traceability*: Event can be traceable. AnyParty/AnyActor or AnyEntity knows that how the event is performed and the event flow.
3. *Effective*: Event is effective enough to cause consequences.

5. *Achievable*: Event should be achievable. Through events, person can achieve a particular goal or objective.

31.2.5 SOLUTION

The event stable design pattern is developed using software stability model.

Class Diagram Description as shown in Figure 31.1.

The class diagram provides visual illustration of all classes in the model along with their static relationships with other classes. Description of event design pattern class diagram is as below:

- Event is a BO and occurrence is the EBT which causes the event.
- Occurrences (EBT) observed by one or more AnyParty or AnyActor.
- AnyEvent(BO) has one or more AnyType(BO).
- AnyDuration(BO) produces one or more AnyConsequences(BO).
- AnyEntity(BO) acts on AnyMedia(BO).

31.3 SUMMARY

Building a stable design pattern of event, which can be reusable was a challenging task. It is possible to design a pattern with the help of software stability model with enduring concepts that are reusable, extendable, and highly adaptable. The good and bad consequences of this pattern have been discussed. Unlimited number of applications can be developed using this pattern.

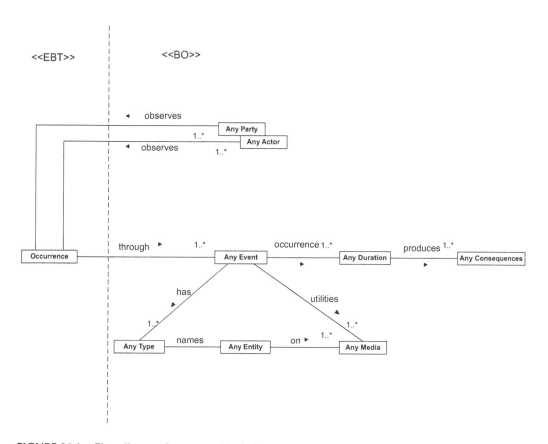

FIGURE 31.1 Class diagram for event stable design pattern.

31.4   EXERCISES

1. Event versus occasion stable patterns
   a. What is the difference between event stable pattern and occasion pattern?
   b. Create the occasion stable pattern.
   c. Name and describe three contexts for each of them.
   d. Show how to apply the two patterns including one context of your choice and including a use case, the application of each pattern with your selected context.
2. Event versus festival stable patterns
   a. What is the difference between event stable pattern and festival pattern?
   b. Create the festival stable pattern.
   c. Name and describe three contexts for each of them.
   d. Show how to apply the two patterns including one context of your choice and including a use case, the application of each pattern with your selected context.
3. Event versus competition stable patterns
   a. What is the difference between event stable pattern and competition pattern?
   b. Create the competition stable pattern.
   c. Name and describe three contexts for each of them.
   d. Show how to apply the two patterns including one context of your choice and including a use case, the application of each pattern with your selected context.
4. Event versus sport stable patterns
   a. What is the difference between event stable pattern and sport pattern?
   b. Create the sport stable pattern.
   c. Name and describe three contexts for each of them.
   d. Show how to apply the two patterns including one context of your choice and including a use case, the application of each pattern with your selected context.
5. Create five applications of AnyEvent Patterns in a table as shown in Table 23.1 (Chapter 23).
6. Create two applications of two contexts of AnyEvent stable design pattern and provide the following for each one of the applications:
   a. Context.
   b. Application diagram using the event stable design patterns.
   c. Detailed use case.
   d. A sequence diagram for the application.

31.5   PROJECTS

Use event in one of the following and create:

a. Event Management
b. Media Event
c. Disaster
Three contexts for each one of them
Stable architecture pattern for each of them
An application of each one of them

REFERENCE

1. Cheryl R. James (born March 28, 1966), last modified on June 22, 2017,https://en.wikipedia.org/wiki/Cheryl_James

32 Conclusions

The future belongs to those who prepare for it today.

Malcolm X [1]

This book presented a new and pragmatic approach for both requirements understanding and ultimate design of the problem and in utilizing stable design patterns for any field of knowledge and modeling the right and stable software systems, components, and frameworks.

This book provides three unique pattern documentation templates that are used for documenting and describing stable design patterns, which makes application of these patterns practically very easy and efficient.

Along with the core value of reusing the presented patterns, this book also helps readers attain the basic knowledge that is needed to analyze, design, and extract stable design patterns for their own domains of interests. Moreover, readers will also learn and master ways to document their own patterns in an effective, easy, and comprehensible manner.

In addition, this book also answered the following questions:

1. How one can achieve or reach the objective of software stability over a period and later build stable design patterns that can be effectively and unlimitedly reused?
2. In what ways the stable design pattern can capture and model the ultimate and optimized knowledge of the solution?
3. How can we achieve the necessary level of abstraction that makes the resulting design patterns effectively and widely reusable yet easy to comprehend and understand?
4. What are the necessary details that design patterns must provide in order to ensure a smooth transition from the design phase to the unlimited applications phase?
5. What is the most practical way to describe and narrate design patterns to make them easy to understand and reuse?

Throughout this book, we have furnished a number of answers to them and practical approaches to follow clear-cut processes that arise from these answers. The software stability concepts acted as the major backbone to all these questions [2–8].

By applying concepts of stability model to the assumptions of design patterns, we suggest the concept of stable design patterns. The main ideas behind using stable design patterns are to analyze and design the overall problem under question, in terms of its EBTs and the BOs, mainly with the goal of increased stability and broader reuse. By examining ensuing problem in terms of their EBTs and the BOs, the resulting pattern will form the core knowledge of the problem. The ultimate goal of this new concept is achieving ample stability. Accordingly, these stable patterns could be easily comprehended and reused to model the same underlying problem under any given situation.

32.1 SUMMARY

Software design patterns play a major and decisive role in reducing the overall cost and in condensing the time duration of software project life cycles. However, building reusable and stable design patterns is still a major challenge. Stable design patterns are the new and fresh tools for building stable and reusable design patterns based on the concept of software stability.

Software stability concepts have demonstrated great promise and immense hope in the area of software reuse and life cycle improvement. In practice, software stability models apply the concepts of "Enduring Business Themes" or "Goals" (EBTs) and "Business Objects" or "Capabilities to achieve the Goals" (BOs). These revolutionary concepts have been shown to produce and yield models that are both stable over time and stable across various paradigm shifts within a given domain or application context. By applying the enduring concepts of stability model to the notion of analysis patterns, this book proposes the concept of stable analysis patterns. Here, an attempt is made to analyze the problem under consideration, in terms of its EBTs and the BOs, with the ultimate goal of reaching increased stability and broader reuse. By analyzing the problem in terms of its EBTs and BOs, the resulting pattern models construct the core knowledge of the problem. The ultimate goal, therefore, is achieving *stability*. As a result, these stable patterns could be easily understood and reused to model the same problem in any context [2–6].

32.2 OUTSTANDING FEATURES

This book presents a pragmatic and a novel approach for understanding the problem domain and in proposing stable solutions called stable design patterns for engineering the right and stable software systems, components, and frameworks. Besides the value of reusing the presented patterns, this book also helps the developers to accomplish the knowledge needed to analyze and extract design patterns for their domain of interest.

Moreover, readers will learn how to

1. Achieve software stability over a period and also build stable design patterns that can be effectively reused.
2. Use stable design patterns to seek an accurate or precise solution to the underlying problem.
3. Achieve the necessary level of abstraction that makes the resultant design patterns effectively reusable yet easy to comprehend and understand.
4. Use the necessary information that design patterns provide in order to ensure and guarantee a smooth transition from the design phase to the implementation phase.
5. Transform stable pattern model to any other models such as formal model or implementation model.

32.3 REVIEW QUESTIONS

1. What are the advantages of using stable design patterns?
2. Present ways of enhancing stable design pattern's usage.

32.4 EXERCISE

1. Research ways of enhancing stable design pattern's usage.

REFERENCES

1. Malcolm X (May 19, 1925–February 21, 1965), last modified on June 23, 2017, https://en.wikipedia.org/wiki/Malcolm_X
2. M.E. Fayad and A. Altman. Introduction to software stability, *Communications of the ACM*, 44(9), 2001, 95–98.
3. M.E. Fayad. Accomplishing software stability, *Communications of the ACM*, 45(1), 2002, 111–115.
4. M.E. Fayad. How to deal with software stability, *Communications of the ACM*, 45(4), 2002, 109–112.
5. M.E. Fayad and S. Wu. Merging multiple conventional models into one stable model, *Communications of the ACM*, 45(9), 2002, 102–106.

6. H. Hamza and M.E. Fayad. A pattern language for building stable analysis patterns, in *Proceedings of 9th Conference on Pattern Language of Programs, (PLoP2002)*, Monticello, IL, September 2002.
7. D.C. Schmidt, M.E. Fayad, and R. Johnson. Software patterns, *Communications of the ACM*, 39(10), 1996, 37–39.
8. M.E. Fayad, H.A. Sanchez, S.G.K. Hegde, A. Basia, and A. Vakil. *Software Patterns, Knowledge Maps, and Domain Analysis*, Auerbach Publications, Boca Raton, FL, December 2014.

Appendix A: Detailed Pattern Documentation Template (Preferred)

- **Name:** Presents the name of the presented pattern.

- Provide short definition of the term (Name).
- Compare the name of the patterns with other selective name, and conclude with the right selection of the name.
- Why did you choose that specific name?
- Justify the name (such as why use "Any…" as a prefix for BO only).

- **Known As:** List all the terms that are similar to the name of the pattern. Two possible sources that one can use to fill this section are (1) similar patterns that are proposed in the literature and (2) other names that you may find relative to the developed pattern. In some cases, several names might make sense so you can keep a list of few of these names under this section.

Discuss the following cases briefly

1. Names match the pattern name: Just list similar names and why.
2. Names match with doubts: List them, describe, and indicate doubts, and why.
3. Names do not match, but people think they match the pattern name: list them, describe, and show why they do not match.

- **Context:** Gives possible scenarios for the situations in which the pattern may recur. It is important in this section that you motivate the problem you solve in an attractive way. For example, if I were writing a pattern about Trust, I would flush the trust in the context of e-commerce. Keep this section short yet exciting.

- Describe the boundaries.
- List basic scenario—context.
- Show by good examples where the pattern can be applied.
- For example "account" would have ownership and handler context, and can be applied to banking Internet providers, private clubs, etc.

- **Problem:** Presents the problem the pattern is concentrating on. This is one of the hardest parts in the pattern writing. Do not try to write it quite well in the first iteration, and most probably, you will not be able to. The problem should focus on the core purpose of the pattern and should be able to answer the question: In what situation I may benefit from your pattern? Try and keep this section as short as possible otherwise reader may get confused.

Length: 1/4 to 1/3 Pages

- Has to be about a specific problem and description = actual requirements of the pattern (functional and nonfunctional requirements of the pattern described in the template).
- It must be within the domain. There are two basic domains Analysis/Design and Own Fields of existence.
- Discuss all the elements of the goal of the pattern.
- You may create a list of the subgoals for requirements of the pattern.

- **Challenges and Constraints:** Illustrates the challenges and the constraints that the pattern needs to resolve. You may create two subsections: (1) Challenges and (2) Constraints. In particular, in this section, you try to say, this is not a trivial problem and that trivial solution may not work. Be clear and brief. One major mistake in writing this section is that you mix the problem statement with the forces themselves. After writing this section, try to read the problem statement again and make sure that they are not the same. It always happens.

- Describe some of the challenges that must be overcome by the pattern.
- Describe the constraints related to the pattern, such as multiplicities, limits, and range.
- Make sure to list the challenges and constraints as bullets.

- **Solution:**
  - **Pattern Structure and Participants:** Gives the class diagram of the pattern (EBT or BO). It also introduces briefly each class and its role. Associations, aggregations, dependencies, and specializations should be included in the class diagram. Association classes, constraints, interfaces, tagged values, and notes must be included in the class diagram. Include the hooks (show each of the BOs connections to IOs). A *full description of the class diagram should be included with the final submission.*
  - **CRC Cards:** Summarizes the responsibility and collaboration of each participant (class). Each participant should have only one well-defined responsibility in its CRC card. Participants with more than one responsibility should be presented with more than one CRC card when each CRC card will handle one of these responsibilities. Refer to Appendix B—CRC Card Layout.
  - **Behavior Model (whenever is possible):** If the abstraction of the pattern prevents you from writing an appropriate behavior model, then you can flush the dynamics of the pattern later on within the Example Section.

Description

- Describe the constraints related to the pattern such as multiplicities, limits, and range.
- Describe some of the challenges that must be overcome by the pattern.
- Note: Not all IOs and BOs may have inheritance.

Detail Models

- Describe the model, role story, such as scenarios, how they play together.

Participants

- Each name and its short description, and how it behaves within the model, such as classes and patterns in the patterns.

CRC Cards

- **Consequences:** How does the pattern (EBT or BO) supports its objectives or goals? What are the tradeoff and results of using the pattern? It is also important to highlight the things that the pattern does not cover and reason about why you choose to exclude them. Another point that I found useful in this section is to highlight other components that may arise from using the proposed patterns. For example, in AnyAccount pattern, we can say that using this pattern for banking systems will require the integration of entries and logs to keep track of the accounts. However, this does not mean that the pattern is incomplete, but this is the nature of patterns anyway, they need to be used with other components.

- List and briefly describe the good (the benefits) of this pattern.
- List and briefly describe the bad (side effects) of the pattern with suggested solutions.

- **Describe briefly and map five different applications using the pattern using the following table format:**

EBT	BOs	App-1 Name—IOs	App-2 Name—IOs	App-3 Name—IOs	App-4 Name—IOs	App-5 Name—IOs

- **Applicability with Illustrated Examples:** Provides clear and detailed two case studies for applying the pattern in different contexts. The following subelements represent the required details in one case.
  - *Case Studies*: Shows the scenario of two cases studies from different contexts.
  - *Class Diagram*: Presents the EBTs, BOs, and IOs.
  - *Use Case Template*: Gives detailed description for a complete use case. Include test cases for the EBT and all the BOs—abstraction of actors, roles, and classes, classes' type, such as EBT, BOs, IOs, attributes, and operations. Refer to Appendix B—use case template.
  - *Behavior Diagram*: Map the above use case into a sequence diagram.

- Show 2–3 distinct scenarios.
- Description of the problem statement of the particular problem.
- Describe the Model—class diagram.
- Use case description with test cases (do not need to do use case diagrams).
- Sequence diagram/use case.

- **Related Patterns and Measurability:** Shows other patterns that usually interact with the described pattern and those who are included within the described pattern. Related patterns can be classified as *related analysis or/and related design patterns*. Related patterns usually share common forces and rationale. In addition, it is possible that you might give some insights of other patterns that can or need to be used with the proposed patterns; for example, in the case of AnyAccount pattern, we might point out to the AnyEntry pattern as a complementary pattern. There are rooms for contrasting and comparing the existing patterns with the documented pattern. This section also provides a few metrics for measuring several things related to the pattern structure, such as complexity and size, cyclomatic complexity, lack of cohesion, coupling between object classes, etc.

This section is divided into two parts:

1. **Related Pattern**
   Two approaches:
   a. Search for an existing traditional pattern on the same topic. Compare with traditional existing pattern's model with reference with ours.
   b. If existing patterns do not exist, select a single definition of the name of our pattern, develop a traditional model class diagram, and describe it briefly.

2. **Measurability**
   Measurability compares our pattern to other models on the number of behaviors and number of classes. Justification of why the number of behaviors or classes are so high or low.

   You may compare and comment on other quality factors, such as reuse, extensibility, integration, scalability, applicability, etc.

   Two Approaches: Compare the traditional with stability models in two of the following approaches:
   a. **Quantitative Measurability such as**
      i. Number of behaviors or operations per class
      ii. Number of attributes per class
      iii. Number of associations
      iv. Number of inheritance
      v. Number of aggregations
      vi. Number of interactions per class
      vii. Number of EBTs versus number of requirements classes in TM
      viii. Number of classes
      ix. Documentation—number of pages
      x. Number of IOs
      xi. Number of applications
      xii. Estimation metrics

xiii. Measurement metrics
  xiv. Etc.
 b. **Qualitative Measurability**
   i. Scalability
   ii. Maintainability
   iii. Documentation
   iv. Expressiveness
   v. Adaptability
   vi. Configurability
   vii. Reuse
   viii. Extensibility
   ix. Arrangement and rearrangement
   x. Etc.

- **Modeling Issues, Criteria, and Constraints:** There are a number of modeling issues, criteria, and constraints that you need to address, in such a way as to explain them, and make sure that the model satisfies all the modeling criteria and constraints.

Modeling Issues are

- **Abstraction:** Describe the abstraction process of this pattern, list, and discuss briefly the abstractions within this pattern.
  - Show the abstractions that are required for the patterns (EBT, BOs, and IOs).
  - Elaborate on the abstraction of why EBTs and BOs are selected?
  - Show examples of unselected EBTs and why?
  - Show examples of unselected BOs and why?
- **Static Models:** Illustrate and describe one or two of the static models of this pattern, and list and discuss briefly the complete story of the pattern's model using actual objects.
  - Determine the sample model that you are planning to use: CRC cards, class diagram, component diagram, etc.—show the model.
  - Tell a complete story of the pattern's model using objects.
  - Repeat a complete story with other objects.
- **Dynamic Models:** Illustrate and describe one or two of the dynamic models of this pattern, and list and discuss briefly the behavior of the pattern through the selected dynamic models.
  - Determine the sample model that you are planning to use: interaction diagram or state transition diagram. Show the model.
- **Modeling Essentials:** Examine the pattern using the modeling essentials, and list and discuss briefly the outcome of this examination.
  - List or reference to the model essentials, and use them as criteria to examine the pattern.
  - Elaborate on how to examine the model of the pattern by using the model essential criteria.
  - Briefly describe the outcome.

- • **Concurrent Development:** Show the role of the concurrent development of developing this pattern. Describe.
  - • Describe and show with illustration of the concurrent development of this pattern.
- • **Modeling Heuristics:** Examine the pattern by using the modeling heuristics, and list and discuss briefly the outcome of this examination.
  - • List or reference to the modeling heuristic and use them as a criteria to examine the pattern.
  - • Elaborate on how to examine the model of the pattern using the modeling heuristics.
  - • Briefly describe the outcome.
  - • Modeling heuristics: such as
    - – No dangling
    - – No star
    - – No tree
    - – No sequence
    - – General enough to be reused in different applications.
    - – Etc.

- • **Design and Implementation Issues:** For each EBT, discuss and elaborate on the important issues required for linking the analysis phase to the design phase and for each BO, discuss the important issues required for linking the design phase to the implementation phase, for example, hooks. Describe the design issues (EBT), for example, hooking issues. Alternatively, discuss the implementation issues (BO), for example, why use relationship rather than inheritance, hooking, hot spots problems. Show segments of code here.

Here is a list of analysis issues

- • Divide and conquer
- • Understanding
- • Simplicity
- • One unique base that is suitable to many applications
- • Goals
- • Fitting with business modeling
- • Requirements specifications models
- • Packaging
- • Components
- • Type (TOP) (A)
- • Actors/roles
- • Responsibility and collaborations
- • Generic and reusable models
- • Etc.

Design Issues (EBT)

- • For example, hooking issues
- • Implementation issues (BO)

- For example, why using aggregation or delegation rather than inheritance
- For example, hooking, hot spots problems
- Can show code here

Here is a sample list of design and implementation issues

- Framework models (D)
- Classes (TOP) (D)
- Collaborations (D)
- Refinement (D)
- Generic and reusable designs (D)
- Precision (I)
- Hooks (I)
- Pluggable parts (I)
- Navigation (I)
- Object identity (I)
- Object state (I)
- Associations/aggregations (I)
- Collections (I)
- Static invariants (I)
- Boolean operators (I)
- Collection operators (I)
- Dictionary (D) (I)
- Behavior models (D) (I)
- Pre-Post-conditions specify actions (I)
- Joint actions (use cases) (D)
- Localized actions (I)
- Action parameters (I)
- Actions and effects (I)
- Concurrent actions (I)
- Collaborations (I)
- Interaction diagrams (D)
- Sequence diagrams with actions (D) (I)
- Pattern 1: Continuity
- Pattern 2: Performance
- Pattern 3: Reuse
- Pattern 4: Flexibility
- Pattern 5: Orthogonal Abstractions
- Pattern 6: Refinement
- Pattern 7: Deliverables
- Pattern 8: Recursive Refinement
- Package (D) (I)

Here is a list of Java Patterns

- Fundamental design patterns
    - Delegation (when not to use inheritance)
    - Proxy

- Creational patterns
  - Abstract Factory
  - Builder
  - Factory Method
  - Object Pool
  - Prototype
  - Singleton
- Partitioning patterns
  - Composite
  - Filter
  - Layered Initialization
- Structural patterns
  - Adaptor
  - Bridge
  - Cache Management
  - Decorator
  - Dynamic Linkage
  - Façade
  - Flyweight
  - Iterator
  - Virtual Proxy
- Behavioral patterns
  - Chain of Responsibility
  - Command
  - Little Language/Interpreter
  - Mediator
  - Null Object
  - Observer
  - Snapshot
  - State
  - Strategy
  - Template Method
  - Visitor

- **Testability:** Describes the test cases, test scenarios, testing patterns, etc. (This is a very important point, but sometimes it is very hard to write for an isolated pattern. I am not sure what is the best way to write this part.) You can use three ways to document testability: (1) test procedures and test cases within class members of the patterns; (2) propose testing patterns that are useful for this pattern and other existing patterns; and (3) check if the pattern fits with as many scenarios as possible without changing the core design.

- Mention to people to try to find scenarios within the context that cannot work with this pattern.
- Show how you can test the requirements and the design artifacts within use cases.
- Can also use exhaustive testing of behaviors (may require more pages) by using testing patterns.

- **Formalization Using Z++, Object Z, or Object-Constraints Language (OCL) (Optional):** Describes the pattern structure by using the formal language (Z++ or Object Z), BNF, EBNF, and/or XML.
- **Business Issues:** Cover one or more of the following issues.
  - **Business Rules:** Describe and document the business rules, and how you can extend them in the context and scenarios that are listed.
    - Define the business rules, business policies, business facts, in relation to the pattern.
    - Illustrate the business rules that are derived from the pattern.

Check the following links

- http://en.wikipedia.org/wiki/Business_rules
- http://www.businessrulesgroup.org/bra.shtml
- http://www.businessrulesgroup.org/first_paper/br01c0.htm—pdf format file.
- http://www.businessrulesgroup.org/brmanifesto.htm

Define the business rules in relation to the pattern.
Illustrate the business rules that are derived from the pattern.

- Business Models: Issues:
  - Business Model Design and Innovation
  - Business Model's Samples http://en.wikipedia.org/wiki/Business_model:
    - Subscription business model
    - Razor and blades business model (bait and hook)
    - Pyramid scheme business model
    - Multilevel marketing business model
    - Network effects business model
    - Monopolistic business model
    - Cutting out the middleman model
    - Auction business model
    - Online auction business model
    - Bricks and clicks business model
    - Loyalty business models
    - Collective business models
    - Industrialization of services business model
    - Servitization of products business model
    - Low-cost carrier business model
    - Online content business model
    - Premium business model
    - Direct sales model
    - Professional open-source model
    - Various distribution business models
  - Describe the pattern. If it is part of or it is a business model.
  - Describe the direct impacts of the pattern on the business model.
  - Describe the indirect impacts of the pattern on the business model.

Describe the same for the following business issues:

- **Business Standards**
  - Vertical standards versus horizontal standards

- **Business Integration**
  - Data integration
  - People integration
  - Tools integration
- **Business Processes or Workflow:** Here are some of the business processes issues:
  - Business process management (BPM) is a systematic approach to improving those processes.
  - Business process modeling and design.
  - Business process improvement.
  - Continuous business process improvement.
  - Business process categories: Management processes, operational processes, and supporting processes.
  - Business process ROI.
  - Business process rules.
  - Business process mapping.
- **E-Business**
  - **E-Commerce**
  - **E-Business Models** http://en.wikipedia.org/wiki/E-Business
    - E-shops
    - E-commerce
    - E-procurement
    - E-malls
    - E-auctions
    - Virtual Communities
    - Collaboration platforms
    - Third-party marketplaces
    - Value-chain integrators
    - Value-chain service providers
    - Information brokerage
    - Telecommunication
  - **E-Business Categories** http://en.wikipedia.org/wiki/E-Business
    - Business-to-business (B2B)
    - Business-to-consumer (B2C)
    - Business-to-employee (B2E)
    - Business-to-government (B2G)
    - Government-to-business (G2B)
    - Government-to-government (G2G)
    - Government-to-citizen (G2C)
    - Consumer-to-consumer (C2C)
    - Consumer-to-business (C2B)
  - **Web Applications**
- **Business Patterns**
  - Business modeling with UML
  - Business knowledge Map
- **Business Strategies**
  - Business strategy modeling
  - Business strategy frameworks
  - Strategic management
  - Strategic analysis
  - Strategy implementation
  - Strategy global business

- **Business Performance Management (BPM)**
  - Methodologies
  - BPM framework
  - BPM knowledge map
  - Assessment and indication
- **Business Transformation**
- **Enduring Business Themes**
- **Security and Privacy**

- **Known Usage:** Give examples of the use of the pattern within existing systems or examples of known applications that may benefit from the proposed pattern. Mention some projects that used it.
- **Tips and Heuristics:** List and briefly describe all the lessons learned, tips, and heuristics from the utilization of this pattern, if any.

- What did you discover?
- Why did you include or exclude different classes?
- Are there any tips on usage such as scaling, adaptability, flexibility?

Appendix B: Midsize Pattern Documentation Template

1. Determine which one is an EBT and which one is a BO.
2. Define and document the context section (Requirements) of each of the concepts—make sure to name and document two scenarios.
   a. **Context:** Gives possible scenarios for the situations in which the pattern may recur. It is important in this Section that you motivate the problem you solve in an attractive way. For example, if I am writing a pattern about Trust, I would flush the trust in the context of e-commerce, for example. Keep this section short yet exciting.

Length: 1/4 to 1/3 Pages

- Describe the boundaries.
- List basic scenario—context.
- Show by good examples where the pattern can be applied (three scenario limit).
- For example "account" would have ownership and handler context, and can be applied to banking Internet providers, private clubs, etc.
- Discuss briefly a few unique context of the pattern.

---

3. Define and document the problem section (Requirements) of each of the concepts—make sure to define the functional and nonfunctional requirements.
   a. **Problem:** The problem should focus on the core purpose of the pattern and should be able to answer the question: In what situation I may benefit from your pattern?

Summary

   Has to be about *a specific problems and descriptions = actual requirements of the pattern*

   Must be in a domain. There are two basic domains Analysis/Design & Own Fields of existence.

List the requirements of the pattern (concept) and describe them briefly.

   Identify and document the functional requirements.
   Identify and document the nonfunctional requirements that are related to quality factors and must be enduring as well.

---

4. *Solution*: **Pattern Structure and Participants.** Gives the class diagram of the pattern (EBT or BO). It also introduces briefly each class and its role. Associations, aggregations, dependencies, and specializations should be included in the class diagram. Association classes, constraints, interfaces, tagged values, and notes must be included in the class diagram. *A full description of the class diagram should be included with the final submission.*

Summary

- Generate and model the class diagram of the pattern.
- Describe the model, role story, such as scenarios, how they play together.

5. Describe briefly and map five different applications using the pattern using the following table format:

		App-1	App-2	App-3	App-4	App-5
EBT	BOs	Name—IOs	Name—IOs	Name—IOs	Name—IOs	Name—IOs

6. **Applicability With Illustrated Examples:** Provides clear, significant, and detailed *Two Case Studies* for applying the pattern in different contexts. The following subelements represent the required details in each case study.

   a. *Case Studies*: Show the scenario of one case study from different contexts

   b. *Class Diagram*: Pattern + IOs

   c. *Use Case*: Presents the different use cases and the actors for each case study, and shows the relation between the different use cases, and the relation between these use cases and the actors of the system. You need to insert test case for EBTs and BOs only.

   d. *Use Case Description*: Gives detailed description for each use case. (Sometimes each would be so long for a paper but at least a sample of these use cases! Just my opinion.)

Length: 2–4 pages total ~1 page each

Show 2–3 distinct scenarios.
Description of the problem statement of the particular problem.
Describe the model.
Use case description (do not need to do use case diagrams).
Sequence diagrams.

Appendix C: Short Pattern Documentation Template

LEADING PARAGRAPH

1. *Pattern Name*: Determine which one is an EBT and which one is a BO. Briefly describe. Please Note: AnyConcept will be written in the requirements as AnyConcept (*Concept*) and it will be written in the entire book as *Concept* for readability purpose.
2. *Context*: Define and document the context section (Requirements) of each of the concepts—make sure to name and document two scenarios.
   a. **Context:** Gives possible scenarios for the situations in which the pattern may recur. It is important in this section that you motivate the problem you solve in an attractive way. For example, if I am writing a pattern about Trust, I would flush the trust in the context of e-commerce. Keep this section short yet exciting. *Document Two Contexts with EBT and BOs in mind.*

Length: 1/4 to 1/3 Pages

- Describe the boundaries.
- List basic scenario—context.
- Show by good examples where the pattern can be applied (three scenario limit).
- For example "account" would have ownership and handler context, and can be applied to banking Internet providers, private clubs, etc.
- Discuss briefly a few unique context of the pattern.

---

3. Requirements:
   a. Functional Requirements
   b. Nonfunctional Requirements
   Define and document the problem section (Requirements) of each of the concepts—make sure to define the functional and nonfunctional requirements.
   - **Problem:** The problem should focus on the core purpose of the pattern and should be able to answer the question: In what situation I may benefit from your pattern?

Summary

Has to be about *a specific problem and description = actual requirements of the pattern.*
Must be in a domain. There are two basic domains Analysis/Design and Own Fields of existence.

List the requirements of the pattern (concept) and describe them briefly.

Identify and document the functional requirements.
Identify and document the nonfunctional requirements that are related to quality factors and must be enduring as well

---

4. *Solution*: **Pattern Structure and Participants.** Gives the class diagram of the pattern (EBT or BO). It also introduces briefly each class and its role. Associations, aggregations, dependencies, and specializations should be included in the class diagram. Association classes, constraints, interfaces, tagged values, and notes must be included in the class diagram. *A full description of the class diagram should be included with the final submission.*

Summary

- Generate and model the class diagram of the pattern.
- Describe the model, role story, such as scenarios, how they play together.

---

5. Describe briefly and map five different applications using the pattern using the following table format: (Optional)

EBT	BOs	App-1 Name—IOs	App-2 Name—IOs	App-3 Name—IOs	App-4 Name—IOs	App-5 Name—IOs

Index

Printed and bound by CPI Group (UK) Ltd, Croydon, CR0 4YY

23/10/2024

01777691-0010